Radar for Indoor Monitoring

Radar for Indoor Monitoring

Detection, Classification, and Assessment

Edited by
Moeness G. Amin

CRC Press is an imprint of the
Taylor & Francis Group, an **informa** business

CRC Press
Taylor & Francis Group
6000 Broken Sound Parkway NW, Suite 300
Boca Raton, FL 33487-2742

© 2018 by Taylor & Francis Group, LLC
CRC Press is an imprint of Taylor & Francis Group, an Informa business

No claim to original U.S. Government works

Printed on acid-free paper

International Standard Book Number-13: 978-1-1387-4609-1 (Paperback)
International Standard Book Number-13: 978-1-4987-8198-5 (Hardback)

Visit the Taylor & Francis Web site at
http://www.taylorandfrancis.com

and the CRC Press Web site at
http://www.crcpress.com

Printed and bound in the United States of America by Sheridan

Contents

Preface

Radar has emerged as a leading technology supporting large sectors of commerce, defense, and security. Over the last decade, we have witnessed the birth of cognitive radar, medical and biometric radar, passive radar, automotive radar, and urban radar. These new applications add to the more established areas of over-the-horizon radar, ground-penetrating radar, and synthetic aperture radar. Irrespective of the application, the three primary goals in using radar are target detection, localization, and classification. These goals, respectively, address the following questions: Is there something there? If yes, then where is it and what is it?

This edited book captures a new and an exciting radar application that touches on our daily living—radar for indoor monitoring. The safety, reliability, and affordability of radar devices have made them a prime candidate for use inside office buildings, homes, schools, and hospitals, with the main purpose of monitoring regular and abnormal motion activities. Although radar imaging of stationary targets and background scenes, using physical and synthesized apertures, is a well-known and mature radar application, indoor radar sensing has heavily relied on Doppler frequencies stemming from the targets in motions.

Monitoring of human daily activities can be achieved using different sensing modalities, including cameras, acoustics, and infrared. Because of its preservation of privacy, insensitivity to light and heat, and penetration of many visually opaque objects, including walls, radar has presented itself as a viable alternative to existing contactless indoor monitoring technologies. The choice of radar parameters and system specifications depends on the monitoring objectives and the sensing missions. Whereas the resolution of moving targets in range requires the use of range-Doppler radar, antenna arrays or distributed radar apertures provide angle information and thus aid in target localization.

The timing of this book is motivated by recent advances in machine learning, sensor fusion, bistatic radars, and multiple-input multiple-output system configurations. It builds on the successful demonstrations of *seeing* through-the-wall radar technology over the past decade and is propelled by the vital role radar can play in the detection of falls, recognition of barometers of progressive disorders, and identification of markers of abnormal gait and breathing patterns. These attributes translate into reducing injuries, saving lives, improving health-care delivery, and cutting the costs of hospitalizations and home care.

The book strives to capture recent and important contributions to the area of indoor monitoring using radar. It consists of 15 chapters that cover various aspects of motion detection, localization, and classification. The chapters are written by leaders in this field and provide detailed explanations and demonstrations of the offerings of the electromagnetic sensing-based technology in the identification of human motion articulations and gross motor activities. Both continuous-wave and range-Doppler radars, with single- and multiantenna configurations, are considered. The book includes two introductory chapters (Chapters 1 and 2) on the fundamentals of radar theory and radar hardware, which should be helpful to the readers who are less familiar with radar systems. Chapter 3 focuses on the kinematics modeling of human motions, and thus provides an understanding of the anticipated radar Doppler and micro-Doppler motion signatures discussed in the chapters that follow. Chapters 4 through 7 and 12 cover fall detection, gait analysis, and motion tracking that make up the core of this book. They exploit the exposition of velocity, acceleration, and higher order motions in the joint-variable time–frequency domain.

Chapters 8, 10, and 11 incorporate multiantenna monitoring systems through colocated and distributed apertures. Chapter 9 addresses the important problem of vital sign monitoring, whereas Chapter 13 broadens radar indoor monitoring to disease detection. Chapters 14 and 15 offer alternative indoor sensing technologies based on wireless devices and infrared sensors.

I hope you find this book informative and useful to your research and pedagogy. I thank all authors for their excellent contributions, which include important analysis, and extensive simulations and experimentations.

Editor

Moeness G. Amin earned his PhD degree in electrical engineering from the University of Colorado, Boulder, Colorado in 1984. Since 1985, he has been with the Faculty of the Department of Electrical and Computer Engineering, Villanova University, Villanova, Pennsylvania, where he became the director of the Center for Advanced Communications, College of Engineering, in 2002.

Dr. Amin is a fellow of the Institute of Electrical and Electronics Engineers (IEEE); fellow of the International Society of Optical Engineering (SPIE); fellow of the Institute of Engineering and Technology (IET); and fellow of the European Association for Signal Processing (EURASIP). He is the recipient of the 2017 Fulbright Distinguished Chair in Advanced Science and Technology; the 2016 Alexander von Humboldt Research Award; the 2016 IET Achievement Medal; the 2014 IEEE Signal Processing Society Technical Achievement Award; the 2009 European Association for Signal Processing Individual Technical Achievement Award; the 2015 IEEE Aerospace and Electronic Systems Society Warren D White Award for *Excellence in Radar Engineering*; the 2010 The North Atlantic Treaty Organization (NATO) Scientific Achievement Award; and the 2010 Chief of Naval Research Challenge Award. Dr. Amin is the recipient of the IEEE Third Millennium Medal. He was a distinguished lecturer of the IEEE Signal Processing Society, 2003–2004. Dr. Amin has more than 700 journal and conference publications in signal processing theory and applications, spanning the areas of wireless communications, radar, sonar, satellite navigations, ultrasound, health care, and Radio-Frequency Identification (RFID). He has coauthored 22 book chapters and is the editor of three books: (1) *Through-the-Wall Radar Imaging*, (2) *Compressive Sensing for Urban Radar*, and (3) *Radar for Indoor Monitoring*, published by CRC Press in 2011, 2014, and 2017, respectively.

Contributors

Fauzia Ahmad
Department of Electrical and Computer
    Engineering
Temple University
Philadelphia, Pennsylvania

Moeness G. Amin
Center for Advanced Communications
Villanova University
Villanova, Pennsylvania

Ilangko Balasingham
Department of Electronic Systems
Norwegian University of Science and
    Technology
Trondheim, Norway

Abdesselam Bouzerdoum
School of Electrical, Computer and
    Telecommunications Engineering
University of Wollongong
Wollongong, Australia

Zachary Cammenga
Department of Electrical and Computer
    Engineering
The Ohio State University
Columbus, Ohio

A. Enis Cetin
Department of Electrical and Computer
    Engineering
University of Illinois at Chicago
Chicago, Illinois

Rohit Chandra
Department of Electronic Systems
Norwegian University of Science and
    Technology
Trondheim, Norway

Victor C. Chen
Ancortek Inc.
McLean, Virginia

Traian Dogaru
RF and Electronics Division
U.S. Army Research Laboratory
Adelphi, Maryland

Fatih Erden
Department of Electrical and Electronics
    Engineering
Bilkent University
Ankara, Turkey

Francesco Fioranelli
School of Engineering
Glasgow University
Glasgow, United Kingdom

Gianluca Gennarelli
Istituto per il Rilevamento
    Elettromagnetico dell'Ambiente
Consiglio Nazionale delle Ricerche
Napoli, Italy

Hugh Griffiths
Department of Electrical Engineering
University College London
London, United Kingdom

Changzhan Gu
Google ATAP (Advanced Technology and
    Projects)
Mountain View, California

Sevgi Zubeyde Gurbuz
Department of Electrical and Computer
    Engineering
Utah State University
Logan, Utah

Aboulnasr Hassanien
Department of Electrical Engineering
Wright State University
Dayton, Ohio

Braham Himed
Multi-Spectral Sensing and Detection
    Division
Air Force Research Laboratory
Dayton, Ohio

Dominic K. C. Ho
Department of Electrical and Computer
    Engineering
University of Missouri
Columbia, Missouri

Tien-Yu Huang
Department of Electrical and Computer
    Engineering
University of Florida
Gainesville, Florida

Sanaz Kianoush
Institute of Electronics, Computer and
    Telecommunication Engineering
Consiglio Nazionale delle Ricerche
Milano, Italy

Jeffrey Krolik
Department of Electrical and Computer
    Engineering
Duke University
Durham, North Carolina

Changzhi Li
Department of Electrical Engineering
Texas Tech University
Lubbock, Texas

Jenshan Lin
Department of Electrical and Computer
    Engineering
University of Florida
Gainesville, Florida

Anthony Martone
RF and Electronics Division
U.S. Army Research Laboratory
Adelphi, Maryland

Bijan G. Mobasseri
Department of Electrical and Computer
    Engineering
Villanova University
Villanova, Pennsylvania

Ram M. Narayanan
Department of Electrical Engineering
Pennsylvania State University
University Park, Pennsylvania

Monica Nicoli
Dipartimento di Elettronica, Informazione
    e Bioingegneria
Politecnico di Milano
Milano, Italy

Tomoaki Ohtsuki
Department of Information and Computer
    Science
Keio University
Yokohama, Japan

Shobha Sundar Ram
Department of Electronics and
    Communication Engineering
Indraprastha Institute of Information
    Technology Delhi
New Delhi, India

Vittorio Rampa
Insitute of Electronics, Computer and
    Telecommunication Engineering
Consiglio Nazionale delle Ricerche
Milano, Italy

Matthew Ritchie
Department of Electrical Engineering
University College London
London, United Kingdom

Stefano Savazzi
Insitute of Electronics, Computer and
    Telecommunication Engineering
Consiglio Nazionale delle Ricerche
Milano, Italy

Ann-Kathrin Seifert
Signal Processing Group, Institute of
    Telecommunications
Technische Universitat Darmstadt
Darmstadt, Germany

Christopher Sentelle
L-3 Communications
Security and Detection Systems
Orlando, Florida

Stephan Sigg
School for Electrical Engineering
Department of Communications and
    Networking
Aalto University
Aalto, Finland

Graeme E. Smith
Department of Electrical and Computer
    Engineering
The Ohio State University
Columbus, Ohio

Nicholas C. Soldner
Infotainment Hardware
Lucid Motors
Menlo Park, California

Francesco Soldovieri
Istituto per il Rilevamento
    Elettromagnetico dell'Ambiente
Consiglio Nazionale delle Ricerche
Napoli, Italy

Umberto Spagnolini
Dipartimento di Elettronica, Informazione
    e Bioingegneria
Politecnico di Milano
Milano, Italy

David Tahmoush
RF and Electronics Division
U.S. Army Research Laboratory
Adelphi, Maryland

Fok Hing Chi Tivive
School of Electrical, Computer and
    Telecommunications Engineering
University of Wollongong
Wollongong, Australia

Çağatay Tokgöz
Phillip M. Drayer Department of Electrical
    Engineering
College of Engineering
Lamar University
Beaumont, Texas

Chi Xu
Department of Electrical and Computer
    Engineering
Duke University
Durham, North Carolina

Yimin D. Zhang
Department of Electrical and Computer
    Engineering
Temple University
Philadelphia, Pennsylvania

Huiyuan Zhou
Department of Electrical Engineering
Pennsylvania State University
University Park, Pennsylvania

Abdelhak M. Zoubir
Signal Processing Group
Institute of Telecommunications
Technische Universitat Darmstadt
Darmstadt, Germany

1

Fundamentals of Indoor Radar

Aboulnasr Hassanien and Braham Himed

CONTENTS

1.1 Introduction

Radar, an acronym for radio detection and ranging, is an apparatus that detects radio frequency (RF) signals scattered from distant objects. The basic idea of radar has its roots in the electromagnetic radiation experiments conducted by the German physicist Heinrich Hertz (1886–1888) (Galati, 2016). He developed a device generating radio waves and demonstrated that radio-wave reflection could happen when distant metallic objects are located in their path. In 1904, Christian Hulsmeyer patented a collision prevention device for ships (Galati, 2016; Sarkar et al., 2016). The device was given the name *Telemobiloskop*, which is considered the first and most primitive form of radar (Sarkar et al., 2016). Robert Watson-Watt is often given credit for inventing radar following his contribution to building the British Chain Home (CH) radar system in the late 1930s (Sarkar et al., 2016). The CH system consisted of a network of transmission towers used in World War II to provide early detection of aircraft fleets. In parallel to the British efforts, research and experimental activities aiming at developing radar took place in Germany (Pritchard,

1989), the United States (Galati, 2016), and other nations (Swords, 1986; Chernyak and Immoreev, 2009; Guarnieri, 2010). In particular, experiments with radio transmission and reception at the Naval Research Laboratory in Washington, DC, were carried out by Lawrence Hyland, Albert Taylor, and Leo Young in 1930 (Galati, 2016). The experiment involved placing communication stations on opposite sides of the Potomac River. Interference caused by ships passing between these two communication stations proved to be a reliable indicator of the presence of an object. However, the developed technique for detecting objects stayed short of determining location and velocity. Motivated by the desire to measure range, experiments using pulsing techniques were conducted in 1934 (Sarkar et al., 2016). Later, the use of continuous-waves (CWs) in tandem with encoding techniques enabled extracting information about the location and velocity of moving objects (Sarkar et al., 2016). It is worth noting that the term *radar* was first used by the U.S. Navy in 1940 (Nguyen and Park, 2016).

Although radar was originally motivated and fostered by defense applications, the radar technology has been continuously developing and found numerous applications in both military and civilian areas. The need for advanced radar technologies has been driven by the radar's ubiquitous applicability, ranging from microscale radars applied in biomedical engineering to macroscale radars used in radio astronomy. Classical military applications of radar include detection and search, aviation, air traffic control, maritime surveillance, target acquisition, target tracking, missile guidance, weather, and airborne synthetic aperture radar. Civilian applications of radar include automotive, collision and avoidance, medical imaging, detection of abnormal conditions of structures, nondestructive testing, and detection of subsurface objects. During the past two decades, radar has found its applications in new areas such as assisted living for elders, homeland security, through-the-wall detection, and indoor imaging. Sparsity-aware radar sensing has recently been used for utilizing the presence of only few targets in low-scattering scene backgrounds and for coping with reduced measurements due to sampling rate restrictions and data gathering logistics difficulties (Amin, 2015).

1.1.1 The Problem of Indoor Detection and Imaging

During the past two decades, the emerging concept of indoor radar is becoming increasingly appealing and has received a considerable amount of attention from researchers and practitioners alike. Indoor radars are compact noncontact sensing devices that can improve our quality of life and enhance our security. They can provide fingerprinting and environmental profiling, offering the means for determining animate and inanimate object locations inside buildings and enclosed structures, thus enabling indoor target localization for civilian, security, and defense applications (Munoz-Ferreras et al., 2015). Recently, the use of radar technology for health monitoring has received additional attention (Mercuri et al., 2013; Porter et al., 2016a, b). Radar-based human monitoring has been successfully applied for classification of human motion articulations toward wellness and aging-in-place applications, most notably fall detection. The worldwide elderly population is growing and is projected to increase to one billion in 2030 (Amin et al., 2016). Investing in technology-based elderly assisted living has a potential to increase the life expectancy while minimizing the cost of providing care for elders. Radar technology can also be used for noncontact vital sign detection (Gu et al., 2010). Indoor radar applications include elderly care, rehabilitation, intruder detection, breath and/or vital sign detection, real-time health care, fall detection, and sleeping infant detection, among others. Indoor radar-based localization and positioning technologies hold promise for many ambient intelligence and security applications. More indoor radar applications are expected to emerge as the respective technology is proving safe, reliable, and cost effective, and most importantly, its deployment is being accepted by the end user.

Research on indoor radar imaging began as early as 1989 when a three-dimensional (3D) imaging millimeter-wave (mm-wave) radar system was developed for sensor-based locomotion (Lange and Detlefsen, 1989). A low-power X-band indoor radar was developed in 1995 for indoor intrusion alarm applications (Battiboia et al., 1995). The introduction of through-the-wall radar (TWR) imaging started in the late 1990s and has become a viable technology for providing high-quality imagery of enclosed structures (Amin, 2011). In 2000, a method for detecting, locating, and identifying building occupants using robot-mounted TWR was developed (Falconer et al., 2000). Intensive research followed on TWR with diverse applications, including safety and home security. For example, imaging of building interiors enables an accurate description of a building to guide firefighters and law enforcement officers in hostage crisis situations. The use of indoor radar for human motion detection has been reported in a number of papers starting in the early 2000s. When individual humans are subjected to radar signals, the bounced off signal components contain biometric information, which can further be used to detect and identify contraction of hearts, heart beating, blood vessels, lungs, and breathing. High-precision real-time localization using indoor radar can also provide valuable information enabling human–machine interface, biomedical and health monitoring, indoor navigation, cognitive prosthetic, and situated-assisted technologies. In addition, the emerging technology of Internet of things requires an accurate knowledge of physical location of different machines for efficient and reliable machine-to-machine communications.

Based on recent advances in indoor radar technology, several start-ups and commercial products began to emerge during the past few years. For example, Novelda AS, Oslo, Norway developed the so-called XeThru Impulse Radar which can be used as a high-precision electromagnetic sensor for human vital sign monitoring, personal security, and environmental monitoring. The concept of indoor radar has also attracted the attention of governmental agencies. Motivated by the need to model and understand the internal structure of buildings prior to an indoor mission, the U.S. Defense Advanced Research Project Agency (DARPA) initiated a program called *VisiBuilding* (Baranoski, 2006). The goal of the program was to develop tools that are capable of determining building layouts and providing vision behind walls. DARPA also launched a follow-on program, *Harnessing Infrastructure for Building Reconnaissance*, whose aim was to translate VisiBuilding into practical terms. This led to the development of *wall-penetrating* radars and other hand-held devices. One such device is the Radar Scope, developed for DARPA (Baranoski, 2005), weighs less than 0.7 kg and is powered by AA batteries. The Radar Scope allows the detection of people hiding behind the walls made out of concrete. Other TWR sensing devices include those developed by the companies L3, New York, NY; IDS, Pisa, Italy; Cambridge Consultants, Cambridge, UK; Raytheon, Waltham, MA; and Camero, Israel.

The subsequent sections provide an overview of how radars operate, discuss various types of radar waveforms, and highlight several indoor radar applications. It is worth noting that Chapter 2 provides additional information on the basics of radar but with emphasis on hardware and system implementations.

1.2 Radar Fundamentals and Radar Range Equation

The essence of conventional radar operation is to emit short pulses with very high peak power and listen to the echoes reflected by targets of interest. Figure 1.1 shows a simplified block diagram of a radar system that uses a single antenna. Transmitting pulses with maximum peak power and very short pulse duration usually results in low average power, which then limits the maximum achievable detection range. The development of pulse compression techniques permits reducing the required pulse peak power while maintaining high average power. The radar antenna focuses the transmit energy into narrow beams to localize the target in angle, then intercepts the target returns for additional processing. The receiver enhances the received signal through amplification, down-conversion of the radar signal to a low intermediate frequency, applying matched filter to the radar returns, detection of the signal envelope, and signal digitization for further processing. The postprocessing stage is intended to further separate target returns from disturbance (clutter + interference) signals, and estimate target parameters, including range, angle, and velocity. If the object is moving, either closer or farther away from the radar, then there is a slight change in the frequency of the radio waves, and this effect is referred to as the *Doppler effect*. The tracker then processes these detections to provide target history over time and predict future positions. Most radar systems are equipped with displays that show the operator the location of all detectable objects within the operating range of the radar system.

1.2.1 Radar Range Equation

Similar to conventional radar, the radiation pattern for indoor radar can be classified into near field and far field. Typical operating ranges for indoor radar applications start from below 1 m and up to several tens of meters. The carrier frequency of indoor radar ranges from less than 1 GHz up to beyond 100 GHz. The corresponding wavelength is typically in the mm-wave range or the microwave range. Therefore, one can fairly assume that, in most indoor radar applications, the targets of interest are located in the

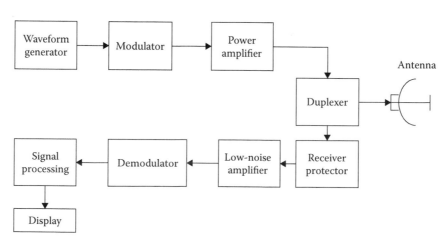

FIGURE 1.1
A basic radar block diagram.

far field. The radar range equation is used to size an indoor radar system to mission requirements. It allows us to calculate the maximum range beyond which a target cannot be detected. Assume that the transmit power, P_t, is radiated by a point source. Then, the power density uniformly distributed by the source over the surface of a sphere of radius R is

$$P_D = \frac{P_t}{4\pi R^2} \tag{1.1}$$

Assume a transmit antenna of gain G_T is used to radiate power toward a target cross section σ located in the peak of its main beam at range R. Then, the fraction of the radar power reflected by the target is given by

$$P_{\text{target}} = \frac{P_t G_t \sigma}{4\pi R^2} \tag{1.2}$$

The power density of the reflected signal at the radar receiver is, therefore,

$$P_\Delta = \frac{P_{\text{target}}}{4\pi R^2} = \frac{P_t G_t \sigma}{\left(4\pi R^2\right)^2} \tag{1.3}$$

The fraction of power intercepted by the radar receiver is given by

$$P_r = P_\Delta A_r \tag{1.4}$$

where:
  $A_r = \lambda^2 G_r / 4\pi$ is the radar receive aperture
  λ is the wavelength
  G_r is the receive antenna gain

Substituting Equation 1.3 in 1.4, the received radar power is expressed as

$$P_r = \frac{P_t G_t G_r \lambda^2 \sigma}{\left(4\pi\right)^3 R^4} \tag{1.5}$$

Equation 1.5 demonstrates that the radar signal attenuates by a factor of $1/R^4$ as it travels to and from a target at distance R. The radar received power, P_r, can also be written in terms of signal-to-noise ratio (SNR), S/N, as follows (Skolnik, 2002):

$$P_r = \left(\frac{S}{N}\right) K T_0 B N_F \tag{1.6}$$

where:
  $K T_0 B N_F$ is the thermal noise power
  K denotes Boltzmann's constant
  T_0 is the temperature of the radar
  B is the noise bandwidth of the radar receiver
  N_F is the noise figure of the radar receiver

In order to achieve detection performance commensurate with a given SNR, the equality condition between Equations 1.5 and 1.6 must be satisfied, that is,

$$\left(\frac{S}{N}\right)KT_0BN_F = \frac{P_tG_tG_r\lambda^2\sigma}{(4\pi)^3 R^4} \tag{1.7}$$

Therefore, the maximum range R_{max} upon which a target can be reliably detected can be computed using the formula

$$R_{\text{max}} = \sqrt[4]{\frac{P_tG_tG_r\lambda^2\sigma}{(4\pi)^3 KT_0BN_F\left(\dfrac{S}{N}\right)}} \tag{1.8}$$

It is worth noting that the radar Equation 1.8 depends on the average power of the transmit signal regardless of the waveform shape. This implies that the transmit waveform can be designed or selected to have other attributes that meet the demands on range and/or Doppler resolution or exploit prior information of the frequency response of the target of interest (Ahmad and Amin, 2010).

1.3 Pulse-Doppler Radar

A pulse-Doppler radar is a radar system that is designed to simultaneously determine the range and velocity of a target, that is, it combines the features and benefits of pulse and CW radars. The target range is determined using pulse-timing techniques via measuring the time required for the transmitted pulse(s) to hit the target and for the reflected wave to arrive at the radar receiver. At the same time, the Doppler effect of the returned signal is used to determine the velocity of the target. Another essential feature of pulse-Doppler radar is its ability to detect small targets observed in the background of strong clutter. Although pulse-Doppler radar was introduced during World War II, it was first widely used on fighter aircrafts starting in the 1960s. Since then, pulse-Doppler radar has been continuously developed, and sophisticated pulse-Doppler systems have been built for both military and civilian applications. In military applications, pulse-Doppler radars permit reducing the transmitted power while achieving acceptable performance. Among the civilian applications, pulse-Doppler radar systems have successfully been employed in meteorology, remote sensing and mapping, air traffic control, as well as in conventional surveillance applications. During the last decade, pulse-Doppler radar has received much attention in automotive and health-care applications. In particular, it has been successfully applied for elderly assisted living, fall risk assessment, and fall detection (Mercuri et al., 2013; Su et al., 2015; Jokanovic et al., 2015; Amin et al., 2016; Diraco et al., 2016; Erol and Amin, 2016b; also see Chapter 6).

In early radars, rough estimates of target speed and/or direction could be obtained through successive measurements of range over several scans. The introduction of Doppler radar enabled a direct and accurate way of range rate estimation via measuring the Doppler shift. Coherent detection methods are typically used for Doppler shift estimation. Recent advances in hardware, software, and computational engineering permitted Doppler sensing in near real time. A pulsed radar system transmits a series of modulated RF pulses where the number of pulses within 1 s refers to the pulse repetition frequency (PRF). The maximum unambiguous target range is directly related to the pulse repetition interval (PRI = 1/PRF). Therefore, lower PRF rates correspond to higher unambiguous range and vice versa. The minimum range difference between two targets such that their echos are received separately defines the range resolution of a pulsed radar. Traditionally, pulsed radar employs low PRF rates to maximize the radar detection range. Doppler sensing capabilities are achieved by collecting multiple pulses within a certain coherent processing interval (CPI) and sampling the collected data at a sampling frequency that is equal to the PRF. The maximum Doppler frequency that can be detected equals half the sampling frequency. Therefore, to increase the maximum target velocity that can be uniquely detected, the radar PRF should be increased. Hence, increasing the maximum detectable range comes at a price of reducing the maximum detectable velocity and vice versa.

1.3.1 Radar Data Cube

Radar processing can occur over N consecutive pulses, as long as they lie within the CPI, considered as *slow time*. The L radar samples collected during the PRI are binned, which corresponds to range. The PRI is referred to as *fast time*. Doppler processing occurs across the N samples in the L range bins.

Doppler processing operates on an array of data of dimension N by L to determine the target velocity at a given range. Space–time processing radar, on the contrary, operates on a cube of data as shown in Figure 1.2. The extra dimension, M, comes from the M distinct inputs from the array antennas (for both elevation and azimuth). This produces a radar data cube of dimension M (number of array antenna inputs) by L (number of range bins in fast time) by N (number of pulses in a CPI in slow time). In this case, target velocity can be measured for a given range and angle. In other words, the parameters of interest can be determined by 3D spectrum analysis of the data cube. Chapters 10 and 11 deal with multiantenna radar systems for indoor human monitoring.

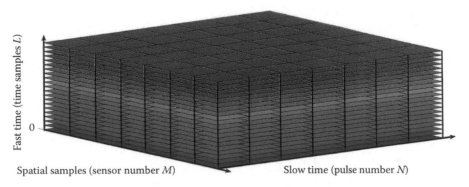

FIGURE 1.2
Coherent radar measurements from M antennas, N pulses, and L samples per pulse per channel arranged in a data cube.

1.4 Frequency-Modulated CW Radar

Ordinary pulsed radar systems usually have very short pulse duration, and, therefore, a high transmit signal peak is required to detect the range to a target. The duty cycle of the radar is the ratio of the transmitted pulse duration to the PRI. The shorter the duty cycle of the pulse, the higher the instantaneous power needed to achieve a certain average transmit power. This makes pulsed radar complex to manufacture and expensive to operate. However, CW radar uses a known stable frequency, making it inexpensive and energy efficient. It uses Doppler processing to detect targets and determine their speeds. In the CW radar, the maximum range limit is determined by the transmit power level, while there is no minimum range limit. A large number of samples can be used along with signal integration in order to extend the detection range without increasing the transmit power. The main shortcoming of the fixed frequency CW radar is its inability to determine the target range.

Frequency-modulated CW (FMCW) radar can achieve an average power higher than that of pulsed radar. It enables measuring the range and Doppler simultaneously while enjoying the advantages of CW radar, for example, smaller instantaneous transmit power and physical size. FMCW radar systems continuously emit periodic pulses whose frequency content varies with time. There are several types of FMCW radars such as linear FM sweep, sawtooth modulation, sinusoidal frequency modulation, or frequency hopping. In the FMCW radar, the range to the target is found by detecting change in the frequency spectrum, that is, the frequency difference between the received and emitted radar signals. The range to the target is proportional to this frequency difference or spectrum spreading. The change in frequency is sometimes referred to as the beat frequency and is seen to be proportional to distance. Therefore, the measurement of the range is basically done by beating the reflected wave with the generated one. FMCW radar has many short-range and indoor applications such as altimeters, proximity sensors, through-the-wall imaging, transportation and transit tracking, parking sensors, and anticollision in automotive radar.

In the FMCW radar, range measurement accuracy and range resolution depend on the frequency sweeping rate accuracy. Unfortunately, it is difficult in practice to achieve a fixed and accurate frequency sweeping rate over a wide extent. This results in the so-called range smearing. In addition, the noise figure of the receiver degrades due to the wide bandwidth used. As a result, the use of FMCW radar may not be suitable for applications that require very high range accuracy and wideband operation. Moreover, FMCW radar suffers in the presence of multiple targets and multipath reflections commonly encountered in indoor environments.

1.5 Stepped Frequency Radar

Stepped frequency radars are those radar systems that transmit stepped frequency CW (SFCW) signals. Similar to FMCW radar, SFCW radar acquires target range information using continuous waveforms. The main advantage of SFCW radar compared to FMCW radar is the enhanced range resolution. High range resolution is a desirable feature in many indoor radar applications. Apart from providing the ability to resolve closely spaced targets or events, it also enhances the range accuracy, enables clutter rejection, and provides the means to reduce multipath. The essence of SFCW radar is to transmit consecutive intervals of CW signals, where the carrier frequency at each interval is different. The frequency change from interval to interval follows a step function, and the separation between two adjacent frequencies is fixed. Figure 1.3a shows the stepped frequencies versus time for a typical SFCW radar waveform. The corresponding waveform in the time domain is illustrated in Figure 1.3b.

SFCW radar finds its applications in indoor areas such as localization and identification of concealed weapons, detection of hidden illegal drugs, detection and localization of personnel, through-the-wall imaging and surveillance, building surveillance, and monitoring. Other applications include detection, identification and assessment of abnormal conditions in bridges, buildings, and buried underground pipes. Moreover, SFCW radar can be utilized for collision and obstacle avoidance, autonomous driving, wayfinding of indoors medical imaging, detection of tumors, and vital sign detection.

Therefore, an SFCW radar has a large effective bandwidth spread over a train of pulses, yet it has a narrow instantaneous bandwidth corresponding to an individual stepped frequency for each pulse. The narrow-band instantaneous bandwidth leads to a reduction in the noise spectrum, and thus achieves a higher SNR. Also, it eases the requirement on the analog-to-digital converters and reduces the number of bits per sample without sacrificing the received signal quality. As a result, slow processors and inexpensive commercial off-the-shelf components can be used to build affordable SFCW radar systems suitable for indoor applications.

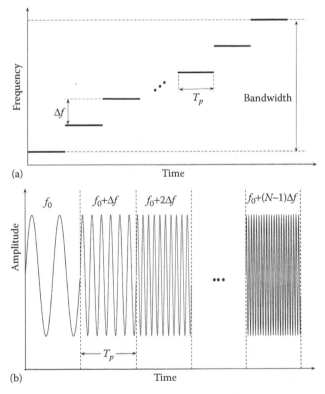

FIGURE 1.3
SFCW radar waveform: (a) frequency versus time and (b) amplitude versus time.

At each stepped frequency, the signals reflected from targets located within the operational range of the SFCW radar are collected at the receiver and demodulated into the baseband. The in-phase and quadrature components of the received baseband signals contain the amplitude and phase information of the signals reflected by the targets. The aggregate signals corresponding to all stepped frequencies are transformed into the time domain using the inverse Fourier transform yielding a synthetic radar pulse. The target(s) information is then extracted via postprocessing the synthetic pulse, similar to conventional radar.

Assume that the total bandwidth of the SFCW radar is B. The bandwidth can be partitioned into N equidistant stepped frequencies, that is,

$$f_n = f_0 + n\Delta f, \quad n = 0,\dots,N-1 \tag{1.9}$$

where:
f_0 is the initial frequency
$\Delta f \triangleq B/N$ is the stepped frequency separation

The signal transmitted during the nth pulse is given by

$$s_n(t) = A_n e^{j2\pi(f_0 + n\Delta f)t}, \quad nT_p \leq t \leq (n+1)T_p \tag{1.10}$$

where:
A_n is the amplitude of the signal during the nth pulse
T_p is the PRI

The signal bounced off a target located at range R impinges on the radar receiver after time delay $\tau = 2R/c$, where c is the speed of light. The received signal can be modeled as

$$r_n(t) = \tilde{A}_n e^{j2\pi(f_0 + n\Delta f)\left(t - \frac{2R}{c}\right)}, \quad n = 0,\dots,N-1 \tag{1.11}$$

Applying frequency down-conversion and demodulation yields the baseband discrete signals

$$y[n] = \tilde{A}_n e^{j2\pi\Delta f \frac{2R}{c}n}, \quad n = 0,\dots,N-1 \tag{1.12}$$

The signal model (Equation 1.12) shows that the phase of the received signal is a function of range R. It is worth noting that the instantaneous bandwidth B_{inst} of the transmitted stepped frequency signal is proportional to the inverse of the pulse duration. The range resolution of the radar system is proportional to the bandwidth, that is,

$$\Delta R = \frac{c}{2B_{inst}} = \frac{cT_p}{2} \tag{1.13}$$

If threshold detection is applied directly to the magnitudes of the signals in Equation 1.12, then the receiver will suffer from poor range resolution. However, the joint processing of the N discrete samples can lead to great improvement in range resolution. To show this, we consider the case when the target's range remains unchanged during the entire processing interval, for example, stationary target. Then, the received discrete signal can be rewritten as

$$y[n] = \tilde{A}_n e^{j2\pi F_s n T_p}, \quad n = 0,\dots,N-1 \tag{1.14}$$

where:

$$F_s \triangleq \frac{2R\Delta f}{cT_p} \tag{1.15}$$

is the baseband signal frequency. Therefore, the range of the target can be estimated via determining the frequency F_s. The simplest way to find F_s is to take the discrete Fourier transform (DFT) of Equation 1.14, that is,

$$Y[k] = DFT\{y[n]\}. \tag{1.16}$$

The discrete sequence $Y[k]$ can be thought of as a synthetic pulse. Range information can be obtained from $Y[k]$ with enhanced range resolution:

$$\Delta R_e = \frac{c}{2N\Delta f} \tag{1.17}$$

If $\Delta f T_p$ is chosen to be unity, then the enhanced range resolution can be rewritten as

$$\Delta R_e = \frac{cT_p}{2N} = \frac{\Delta R}{N} \tag{1.18}$$

Therefore, Equation1.18 shows that SFCW radar using N stepped frequencies can enhance the range resolution by a factor of N. SFCW radar offers a trade-off between the instantaneous bandwidth and the observation time, which enables producing the same resolution as a narrow pulsed system.

1.6 Signal Processing and Target Detection

Depending on the specific radar application, there exists various signal processing and/ or detection tools, which can be used to analyze the received data and extract information regarding events of targets involved. In this section, we focus on time–frequency analysis and detection of events in the time–frequency domain.

1.6.1 Signal Processing

Consider a CW radar operating at a carrier frequency f_c. The baseband radar return from a point target located in the far field can be expressed as

$$r(t) = \xi(t)e^{-j\varphi(t)} \tag{1.19}$$

where $\xi(t)$ and $\varphi(t)$ are the amplitude and phase of the received signal, respectively. Note that the amplitude is range dependent, that is, it changes from range bin to range bin. If the point target under consideration is in motion, the return signal will be Doppler shifted, and the corresponding Doppler frequency can be determined by the derivative of the phase, that is,

$$f_D(t) = -\frac{1}{2\pi}\frac{\partial\varphi(t)}{\partial t} \tag{1.20}$$

In practical indoor radar applications, the point target assumption is usually violated due to the extended size of certain targets, for example, human body. Such an extended target can be considered as a collection of point scatterers. Therefore, the baseband signal at the output of the radar receiver can be considered as a sum of returns from all point scatterers comprising the target extent (Wu et al., 2015), that is,

$$x_r(t) = \sum_i r_i(t) \qquad (1.21)$$

where $r_i(t)$ is the radar return due to the ith scatterer. In this case, the Doppler signature of the extended target will be the summation of the Doppler signals of all point scatterers. If the extended target is a moving human, then the Doppler frequencies will be time varying constituting what is referred to as micro-Doppler signature (Jokanovic et al., 2015). Therefore, time–frequency analysis is an essential tool for processing the radar returns from human targets and is used in Chapter 4 to analyze human fall and in Chapters 5 and 12 to analyze human gait.

Let $x_0(t)$ be the background signal, that is, the signal at the output of the radar receiver in the absence of target. Background subtraction can be applied to all data observations yielding $x(t)$. The simplest time–frequency analysis tool that can be used is the short-time Fourier transform (STFT). The spectrogram of the signal $x(t)$ can be obtained by performing the discrete-time STFT, that is,

$$X(n,k) = \sum_m x[m]w(n-m)\exp\left(-j2\pi m\frac{k}{K}\right), \quad k = 0,1...,K-1 \qquad (1.22)$$

where:
$x[n]$ is the discrete version of $x(t)$
k is the discrete frequency index
$w(n)$ is a window function

It is worth noting that the length and shape of the window can be selected according to the desired trade-off between time and frequency resolutions. The simplest window function that can be used is the rectangular window that translates into a sinc function in the frequency domain causing poor frequency resolution. Other window functions that can be used to enhance the time–frequency signal representation include Slepian sequences, Hermite functions, and various types of time–frequency kernels (Jokanovic et al., 2015).

1.6.2 Event Detection

In many indoor radar applications, the desired target is an event such as a fall of an elderly, a vital sign in a wounded person, a heartbeat of a person in a coma, or a breath of an intruder. These types of events can be analyzed in a time–frequency graph. Once the spectrogram of the data is obtained, a proper detection algorithm can be applied to detect the presence of an event and to determine exactly when important events have occurred. One simple, yet powerful tool for event detection is the so-called power burst curve (PBC) (Wu et al., 2015). The essence of the PBC detector is based on calculating the summation of signal power within a specific Doppler frequency band at time instant n, that is, between two discrete frequencies k_1 and k_2. As such, the PBC detector can be expressed as

$$\text{PBC}[n] = \sum_{k=k_1}^{k_2} |X[n,k]|^2 + |X[n,K-k]|^2 \qquad (1.23)$$

The detection of an event can be determined by comparing PBC[n] to an appropriately selected threshold. Another method for event detection is based on applying classical supervised detector, which can be implemented by using a support vector machine (Diraco et al., 2016). Such a classifier can be trained and accustomed for an individualized elderly.

After detection is performed, postprocessing can be applied for feature extraction and classification. Spectrogram segmentation and morphological processing techniques can be applied to extract information and/or classify events. Image segmentation using graph cuts is a commonly used segmentation technique (Tao et al., 2008). Micro-Doppler features and time extent of certain events are commonly used for accurate classification, for example, type of an elderly fall.

1.7 Image Formation

In Section 1.3, we showed that pulsed Doppler radar permits forming a 3D data cube at the receiver output. Range-Doppler images can be formed by applying two-dimensional (2D) spectrum analysis to the fast-time, slow-time slices of the data cube. Also, for every radar pulse, a range–cross-range image can be formed by applying 2D frequency analysis to the space–slow-time slices of the cube. This allows tracking the location of targets of interest in slow time.

For indoor applications where a single antenna radar is used, the data cube reduces to a 2D data array with one index corresponding to fast time and the other index corresponding to pulse number. For illustration, Figure 1.4 depicts a through-the-wall imaging scenario using a single antenna on a slowly moving platform. The platform velocity v is slow enough to allow the illumination of the cell $\Delta x \Delta y$ with a burst of pulses within a CPI. In this case, the collected data is 2D (in slow time, fast time). Therefore, for each range bin, a 2D time graph can be produced showing the time–frequency signatures of targets located

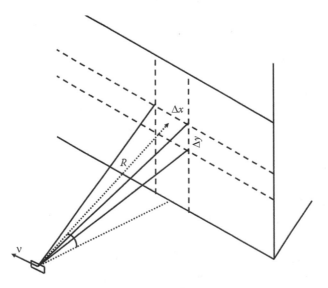

FIGURE 1.4
An illustrative diagram for indoor radar imaging.

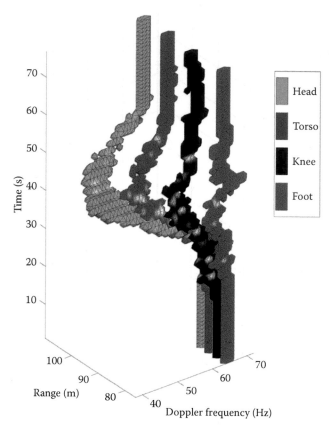

FIGURE 1.5
3D time–frequency–range analysis of human motion arranged in a data cube.

at that range. For example, a radar-based fall detection exploiting time–frequency features was reported by Rivera et al. (2014). Another radar-based fall detection utilizing Doppler time–frequency signatures with application to assisted living was developed by Wu et al. (2015). A multiwindow time–frequency signature reconstruction from undersampled CW radar measurements for fall detection was developed by Jokanovic et al. (2015). A wavelet transform-based Doppler radar technique for fall activity detection was reported by Su et al. (2015). A method for automatic human fall detection in fractional Fourier domain with application to assisted living was reported by Liu et al. (2016). Figure 1.5 shows an example of such time–frequency analysis where the time–frequency signature of a moving individual is depicted. The figure shows how the time–frequency signatures of different body parts of the individual under test change along the range index.

1.8 Indoor Radar Applications

Recent advances in indoor radar systems show that their use as noncontact sensory devices can improve our security, safety, and quality of life at home, offices, hospitals, recreation centers, or shopping malls. This section highlights some indoor radar applications and sheds light on recent advances in short-range radars and their civilian applications. In particular, we focus on security and health care as two areas of exploitation of indoor radar technology.

1.8.1 Through-the-Wall Radar

In urban environments, sometimes it would be advantageous to detect and locate all humans inside a building or an urban structure. TWR technology enables detecting, identifying, classifying, and tracking of moving objects behind the wall (Baranoski, 2006; Amin, 2011). During the past two decades, through-the-wall sensing has emerged as a viable area that addresses the desire to see inside buildings. It has been shown that this technology can be used for determining the room layouts, discerning the nature of activities inside buildings, detecting and localizing the occupants, and even identifying and classifying objects within the building. The development of a reliable sensing through-the-wall technology has been greatly motivated and strongly desired by law enforcement agencies, police, first responders, fire and rescue, emergency relief workers, and urban military operations, to name a few. In essence, vision into otherwise obscured areas can be made available using sensing through-the-wall-based technology.

The development of through-the-wall sensing technology has faced many challenges. The main challenge involves the penetration of the electromagnetic waves through a lossy medium such as solid concrete walls, which results in signal attenuation especially at high-frequency ranges. Inherently, there is a trade-off between good resolution and good penetration properties. The use of low frequency in through-the-wall sensing has the advantage of good penetration through building structures. Therefore, low-frequency wideband radars represent a good choice for through-the-wall sensing and imaging. Another difficulty in through-the-wall sensing is caused by the propagation distortions encountered by the radar signals when they pass through walls and objects. This type of signal distortion represents a serious challenge because the signals have to propagate through the wall twice before they finally reach the radar receiver. This causes degradation and can lead to ambiguities in target detection and localization. Several other factors affect the way signals propagate through walls such as shadowing, attenuation, reflection, refraction, diffraction, and dispersion (Amin and Ahmad, 2013). If these factors are not accounted for properly, they can severely impact the radar system performance.

A common problem in indoor radar in general and through-the-wall imaging in particular arises due to signal reflections and multipath propagation. An illustrative diagram depicting the phenomenon of reflection and multipath propagation is shown in Figure 1.6a. The figure shows that several first- and second-order signal reflections due to a single target can be received by the imaging system. Unfortunately, signal reflection and multipath propagation result in the occurrence of incorrect localization of multiple false targets and ghosts. Figure 1.6b shows a scenario where a single target is present inside a room, yet multiple ghost targets can be observed at the radar receiver. Therefore, proper compensation techniques for multiple reflections and multipath propagation are needed to eliminate or exploit ghosts and false targets from the image (Leigsnering et al., 2014). Chapter 7 discusses human tracking indoor in the presence of multipath.

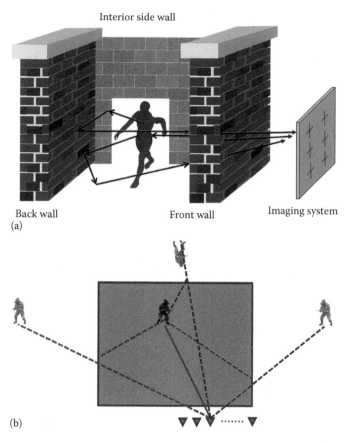

Interior side wall

Back wall
(a)

Front wall

Imaging system

(b)

FIGURE 1.6
Multipath in TWR: (a) 3D illustrative diagram and (b) direct path, first-order, and second-order multipath diagram.

1.8.2 Assisted Living Radar

In the area of elderly care, the negative consequences of human falls can be minimized via providing a timely detection of falls, thus immediately notifying caregivers (Amin et al., 2016; Su et al., 2015). A portable device for fall detection and indoor human tracking using coherent FMCW radar was developed by Peng et al. (2016). A radar-based technique for fall detection using Doppler time–frequency analysis with application to elderly care was reported by Wu et al. (2015). Although radar-based technology is shown to enable developing reliable noncontact devices for fall detection and health monitoring, a practical concern arises due to the problem of high false alarm. Recently, a method for reducing the false alarms in assisted-living radar using range information was reported by Erol et al. (2016), whereas the effects of range spread and aspect angle on the performance of radar fall detection and its practical implications were investigated by Erol and Amin (2016a).

Assisted living radar can also provide wayfinding assistance to humans in areas where global positioning system (GPS) navigation is not available, for example, shopping malls. For example, a radar-based technique has recently been developed for safe point-to-point navigation of impaired users by Mancini et al. (2015). A recent study on the use of mm-wave radar imaging for indoor navigation and mapping was reported by Moallem and Sarabandi (2014).

1.8.3 Concealed Weapon Detection

The reliable detection and accurate identification of weapons concealed underneath a person's clothing is necessary for the improvement of our security and the safety of public infrastructure such as airports, shopping malls, and governmental buildings. Indoor radar-based sensor can provide the means for law enforcement, military, and other security forces to instantly detect concealed weapons and contraband at a safe distance. An overview of different imaging techniques, including indoor radar imaging, for concealed weapon detection is given by Chen et al. (2005). Radar-based weapon detection is receiving additional attention due to the fact that radar waves, unlike X-rays, do not ionize the human tissue and, therefore, do not pose health risks (Sakamoto et al., 2016). In addition, radar-based concealed weapon detection provides better detection performance compared to other sensing modalities such as X-ray and infrared. A comparison between mm-wave radar-based and infrared-based concealed weapon detection is given in Figure 1.7. The original mm-wave

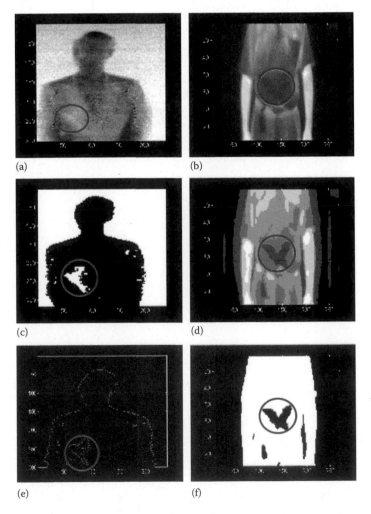

(a) (b)

(c) (d)

(e) (f)

FIGURE 1.7
Performance of mm-wave concealed weapon detection versus infrared-based detection: (a) original mm-wave image, (b) original infrared image, (c) processed mm-wave image, (d) processed infrared image, (e) thresholded mm-wave image, and (f) thresholded infrared image. (Courtesy of AFRL, Wright-Patterson Air Force Base, OH.)

and infrared images are shown in Figure 1.7a and b, respectively. The processed mm-wave and infrared images are shown in Figure 1.7c and d, respectively. Threshold-based edge detection is applied to the processed images yielding the final images shown in Figure 1.7d and f, respectively. It is clear from these figures that mm-wave radar technique has better detection performance than infrared-based techniques.

1.9 Conclusions

The focus of this chapter was to discuss the basics and fundamentals of the emerging concept of indoor radar and its applications. A brief historical note on radar was given followed by an overview of the problem of indoor detection and imaging. The importance and diverse applications of indoor radar were highlighted. In particular, the applicability of indoor radar in areas such as health care, indoor monitoring, and security was provided. The basics of radar and the radar range equation were reviewed in Section 1.2. The essence of pulse-Doppler radar was described in Section 1.3. The fundamentals of FMCW and SFCW radars and their use in indoor radar applications were concisely reviewed in Sections 1.4 and 1.5, respectively. Signal processing and target detection techniques were overviewed in Section 1.6, and their utilization in various indoor radar applications was also highlighted. The basics of image formation in indoor radar was described in Section 1.7. Finally, various civilian and military applications of indoor radar were discussed in Section 1.8 with a focus on TWR, assisted living radar, and concealed weapon detection.

References

Ahmad, F. and Amin, M. G. (2010). Matched-illumination waveform design for a multistatic through-the-wall radar system. *IEEE Journal of Selected Topics in Signal Processing*, 4(1):177–186.

Amin, M. G. and Ahmad, F. (2013). *Through-the-Wall Radar Imaging: Theory and Applications*. In S. Theodoridis and R. Chellappa (Eds.), Academic Press Library in Signal Processing, Academic Press, Cambridge, MA.

Amin, M. G., Ed. (2011). *Through-the-Wall Radar Imaging*. CRC press, Boca Raton, FL.

Amin, M. G., Ed. (2015). *Compressive Sensing for Urban Radar*. CRC press, Boca Raton, FL.

Amin, M. G., Zhang, Y. D., Ahmad, F., and Ho, K. C. (2016). Radar signal processing for elderly fall detection: The future for in-home monitoring. *IEEE Signal Processing Magazine*, 33(2):71–80.

Baranoski, E. J. (2005). Urban operations, the new frontier for radar. In *Systems and Technology Symposium*, Anaheim, CA, pp. 155–159.

Baranoski, E. J. (2006). VisiBuilding: Sensing through walls. In *Fourth IEEE Workshop on Sensor Array and Multichannel Processing, 2006*. Westin Hotel, Waltham, MA, pp. 1–22.

Battiboia, S., Caliumi, A., Catena, S., Marazzi, E., and Masini, L. (1995). Low-power X-band radar for indoor burglar alarms. *IEEE Transactions on Microwave Theory and Techniques*, 43(7):1710–1714.

Chen, H.-M., Lee, S., Rao, R. M., Slamani, M. A., and Varshney, P. K. (2005). Imaging for concealed weapon detection: A tutorial overview of development in imaging sensors and processing. *IEEE Signal Processing Magazine*, 22(2):52–61.

Chernyak, V. S. and Immoreev, I. Y. (2009). A brief history of radar in the Soviet Union and Russia. *IEEE Aerospace and Electronic Systems Magazine*, 24(9):B1–B32.

Diraco, G., Leone, A., and Siciliano, P. (2016). Radar sensing technology for fall detection under near real-life conditions. In *2nd IET International Conference on Technologies for Active and Assisted Living*, pp. 1–6.

Erol, B. and Amin, M. G. (2016a). Effects of range spread and aspect angle on radar fall detection. In *2016 IEEE Sensor Array and Multichannel Signal Processing Workshop*, pp. 1–5.

Erol, B. and Amin, M. G. (2016b). Fall motion detection using combined range and Doppler features. In *2016 24th European Signal Processing Conference*, Budapest, Hungary, pp. 2075–2080.

Erol, B., Amin, M. G., Zhou, Z., and Zhang, J. (2016). Range information for reducing fall false alarms in assisted living. In *2016 IEEE Radar Conference*, pp. 1–6.

Falconer, D. G., Ficklin, R. W., and Konolige, K. G. (2000). Robot-mounted through-wall radar for detecting, locating, and identifying building occupants. In *Proceedings 2000 ICRA. Millennium Conference. IEEE International Conference on Robotics and Automation. Symposia Proceedings (Cat. No.00CH37065)*, vol. 2, pp. 1868–1875.

Galati, G. (2016). *100 Years of Radar*. Springer, Cham, Switzerland.

Gu, C., Li, C., Lin, J., Long, J., Huangfu, J., and Ran, L. (2010). Instrument-based noncontact Doppler radar vital sign detection system using heterodyne digital quadrature demodulation architecture. *IEEE Transactions on Instrumentation and Measurement*, 59(6):1580–1588.

Guarnieri, M. (2010). The early history of radar. *IEEE Industrial Electronics Magazine*, 4(3):36–42.

Jokanovic, B., Amin, M. G., Zhang, Y. D., and Ahmad, F. (2015). Multi-window time-frequency signature reconstruction from undersampled continuous-wave radar measurements for fall detection. *IET Radar, Sonar Navigation*, 9(2):173–183.

Lange, M. and Detlefsen, J. (1989). 94 GHz 3D-imaging radar for sensor-based locomotion. In *IEEE MTT-S International Microwave Symposium Digest*, vol. 2, pp. 1091–1094.

Leigsnering, M., Amin, M. G., Ahmad, F., and Zoubir, A. M. (2014). Multipath exploitation and suppression for SAR imaging of building interiors: An overview of recent advances. *IEEE Signal Processing Magazine*, 31(4):110–119.

Liu, S., Zeng, Z., Zhang, Y. D., Fan, T., Shan, T., and Tao, R. (2016). Automatic human fall detection in fractional Fourier domain for assisted living. In *2016 IEEE International Conference on Acoustics, Speech and Signal Processing*, pp. 799–803.

Mancini, A., Frontoni, E., and Zingaretti, P. (2015). Embedded multisensor system for safe point-to-point navigation of impaired users. *IEEE Transactions on Intelligent Transportation Systems*, 16(6):3543–3555.

Mercuri, M., Soh, P. J., Pandey, G., Karsmakers, P., Vandenbosch, G. A. E., Leroux, P., and Schreurs, D. (2013). Analysis of an indoor biomedical radar-based system for health monitoring. *IEEE Transactions on Microwave Theory and Techniques*, 61(5):2061–2068.

Moallem, M. and Sarabandi, K. (2014). Polarimetric study of MMW imaging radars for indoor navigation and mapping. *IEEE Transactions on Antennas and Propagation*, 62(1):500–504.

Munoz-Ferreras, J. M., Peng, Z., Gomez-Garcia, R., Wang, G., Gu, C., and Li, C. (2015). Isolate the clutter: Pure and hybrid linear-frequency-modulated continuous-wave (LFMCW) radars for indoor applications. *IEEE Microwave Magazine*, 16(4):40–54.

Nguyen, C. and Park, J. (2016). *Stepped-Frequency Radar Sensors: Theory, Analysis, and Design*. Springer Briefs in Electrical and Computer Engineering. Springer, Cham, Switzerland.

Peng, Z., Munoz-Ferreras, J. M., Tang, Y., Gómez-García, R., and Li, C. (2016). Portable coherent frequency-modulated continuous-wave radar for indoor human tracking. In *2016 IEEE Topical Conference on Biomedical Wireless Technologies, Networks, and Sensing Systems*, pp. 36–38.

Porter, E., Bahrami, H., Santorelli, A., Gosselin, B., Rusch, L. A., and Popović, M. (2016a). A wearable microwave antenna array for time-domain breast tumor screening. *IEEE Transactions on Medical Imaging*, 35(6):1501–1509.

Porter, E., Coates, M., and Popović, M. (2016b). An early clinical study of time-domain microwave radar for breast health monitoring. *IEEE Transactions on Biomedical Engineering*, 63(3):530–539.

Pritchard, D. (1989). *The Radar War: Germany's Pioneering Achievement 1904–1945*. Patrick Stephens Ltd., Wellingborough, UK.

Rivera, L. R., Ulmer, E., Zhang, Y. D., Tao, W., and Amin, M. G. (2014). Radar-based fall detection exploiting time-frequency features. In *2014 IEEE China Summit International Conference on Signal and Information Processing*, pp. 713–717.

Sakamoto, T., Sato, T., Aubry, P., and Yarovoy, A. (2016). Fast imaging method for security systems using ultrawideband radar. *IEEE Transactions on Aerospace and Electronic Systems*, 52(2):658–670.

Sarkar, T. K., Palma, M. S., and Mokole, E. L. (2016). Echoing across the years: A history of early radar evolution. *IEEE Microwave Magazine*, 17(10):46–60.

Skolnik, M. (2002). *Introduction to Radar Systems*. 3rd ed. McGraw-Hill, New York, NY.

Su, B. Y., Ho, K. C., Rantz, M. J., and Skubic, M. (2015). Doppler radar fall activity detection using the wavelet transform. *IEEE Transactions on Biomedical Engineering*, 62(3):865–875.

Swords, S. S. (1986). *Technical History of the Beginnings of Radar*. Peter Peregrinus, London.

Tao, W., Jin, H., Zhang, Y., Liu, L., and Wang, D. (2008). Image thresholding using graph cuts. *IEEE Transactions on Systems, Man, and Cybernetics—Part A: Systems and Humans*, 38(5):1181–1195.

Wu, Q., Zhang, Y. D., Tao, W., and Amin, M. G. (2015). Radar-based fall detection based on Doppler time-frequency signatures for assisted living. *IET Radar, Sonar Navigation*, 9(2):164–172.

2

Radar Hardware for Indoor Monitoring

Çağatay Tokgöz and Nicholas C. Soldner

CONTENTS

2.1 Introduction

A radar is a device that can detect echo signals, and then processes these signals to measure the range and radial velocity of objects within a certain resolution and over a specific region of space. Christian Hülsmeyer has been credited with building the first practical radar used to prevent ship collisions in foggy conditions. Hülsmeyer was issued a German patent for his invention which used a spark-gap (broadband) generator as its signal source (Hülsmeyer 1904; Hülsmeyer 1906). Since then, military, automotive, safety, and weather applications have driven the development of radar hardware, with lower cost adaptations finding its extensive use in indoor security applications over the past three decades. Continual miniaturization, semiconductor process improvements, and integration have reduced the cost of radars leading to its increasing use in automotive and home applications. As of 2017, base vehicles are beginning to include side collision radars as standard with higher end vehicles using as many as six frequency-modulated continuous wave (FMCW) radars for autonomous driving. These radars have a range of 10–100 m, a sub-millimeter resolution, and a high-angular accuracy of <5°. In addition, bedside sleep monitoring radars are available for purchase at big box retail stores, and radar-based gesture control is being rapidly developed for smartphones. Fortunately, these developments also drive cost down and make radars more capable for indoor use.

It is technically challenging to design and develop radar signal transmit waveforms, radar hardware, and signal processing algorithms that produce reliable decisions on the motion of people, pets, or objects. In any indoor situation, the radar platform will need to be capable of identifying and removing undesired objects that would constitute false alarms. Understanding both the targets and noise sources is necessary to evaluate trade-offs inherent in radar hardware architecture such as maximum range, range resolution, radial resolution, pulse rate, tracking rate, antenna design, and output power. For instance, range resolution needs to be compromised in order to increase the total range of a radar and vice versa. In addition, target size, geometry, and material greatly affect the return signal direction and strength, which strongly influences the choice of frequency, transmit power, and receiver sensitivity. It is important to remember that a radar will measure the range, radial velocity, and echo signal power for all objects reflecting energy in the observation area, not just the desired target; this often drives the need for a costly directional antenna array and advanced target classification algorithms.

This chapter focuses on short-range radars (<10 m) which are well suited for indoor monitoring as a result of their lower implementation cost, safe emission levels, resolution, and adaptability to different form factors. The impact of radar hardware architecture on range accuracy, hardware cost constraints, minimum range, peak transmitted power, the ability to resolve multiple targets, the ability to reject clutter, and the transmitter (TX)–an RX isolation characteristics are considered. The following sections also discuss indoor monitoring radars capable of providing the range and/or radial velocity signals needed to monitor human activity indoors. This chapter can be viewed as a complement of Chapter 1 with emphasis on hardware aspects and challenges of indoor radar technology.

2.2 Continuous Wave Radar

The continuous wave (CW) radar represents one of the simplest hardware implementations in which a fixed frequency signal, typically sinusoidal in low-cost applications, is transmitted, and reflections of that signal from any object within the range are received and mixed with the transmitted carrier. A CW radar detects the radial velocity of a moving object, which alters the frequency of signals it reflects, known as the Doppler shift. Objects moving closer to or away from a CW radar cause the reflected signals to move up and down in frequency, respectively, compared to the signal transmitted by the radar. It is interesting to note that the Doppler component could be only a few hertz shift on top of a multi-gigahertz carrier signal. To separate the Doppler components from the carrier, a straightforward radiofrequency (RF) mixer can be employed that accepts a reference signal (usually the carrier) and the received signal as inputs then produces the Doppler frequency as an output including harmonics. The mixer can be implemented any number of ways with a low-cost implementation using a low barrier (<0.2V) RF diode. The desired output for human monitoring is typically in the range of 0.1–100 Hz after applying a low-pass filter. The frequency range of 0.1–100 Hz captures most of the energy reflected from a sitting (e.g., respiration/heartbeat), walking, or running human. In some cases, frequencies as low as 0.05 Hz are needed to characterize human vital signs, given that respiration rate can be as low as one cycle every 20 s. This low-frequency range poses a challenge for post-mixer filters, whether they are implemented as analog circuits or as an algorithm, the filter order is typically 4, but can be as high as 10. With orders this high, filters can take tens of seconds to settle and any large motion can perturb their stability and reset their settling time. One strategy to mitigate the issue of filter settling time is to break the filters into smaller frequency ranges and place them in parallel with the mixer output. For example, one filter can tackle the 0.05–2 Hz range and another the 10–100 Hz range. This strategy works for indoor human detection, because walking in a house can be easily detected with the fast settling 10–100 Hz filter. Once someone sits down, there is typically enough time to allow the 0.05–2 Hz filter to settle so that vital signs could be detected. It is recommended to implement these filters in the hardware domain as much as possible to ease the burden on a digital signal processor (DSP). For CW or pulsed CW radars, this strategy works well when the low-frequency noise is eliminated by analog filters allowing the DSP to run adaptable filtering such as creating subfrequency or wavelet transform routines that focus on tracking a signal of interest.

It may be surprising that an effective CW radar can be built with less than $5 in discrete components. In fact, most radar front ends used in consumer market motion detectors fall within this price class. The major components of a low-end CW or pulsed CW radar are a crystal oscillator, 10–20 digital logic gates, one or two low barrier mixer diodes, filter capacitors, and one or two stages of operational amplifier-implemented filters. The major downside to this low-cost implementation is limited receiver sensitivity that tends to be between −47 and −40 dBm due to the use of low-barrier diodes as detectors/mixers. Adding a low-noise amplifier (LNA) before the mixer typically increases cost by 10%–30% with the advantage of improving receiver sensitivity to −70 dBm or lower. For CW or pulsed CW radar applications in which detection of large human motion is desired, −47 dBm receiver sensitivity is adequate for up to 10 m range using simple dipole antennas at 5–10 GHz. For detection of vital signs at a range of 3 m or greater, an active receiver is required due to the small radar cross section associated with vital signs.

2.2.1 Homodyne versus Heterodyne Receivers

A reference signal is typically used in CW radars to down-convert the received signal. The easiest way to achieve this is to couple to the receiver a small portion of the local oscillator (LO) signal that was used to generate the signal at the TX and use it as a reference signal. Such a receiver that uses a reference signal at the same frequency as the carrier signal is called a homodyne receiver. Again, the reference signal is mixed with the received signal using a mixer for down-conversion. Then, the mixer output passes through a low-pass filter to significantly reduce high-frequency signals and harmonics resulting from the mixing operation to obtain the down-converted signal as a baseband signal. An LNA and an RF band-pass filter may be used to amplify and filter the received signal prior to the mixing operation to improve the signal-to-noise ratio (SNR). One challenge with this architecture is dealing with the DC component of the baseband signal that relates to reflections from stationary objects and energy from the transmitted signal itself. In some cases, the DC component can be used to identify the changes in the environment and is actually very powerful when trying to detect low-frequency information such as vital signs. The issue comes in correcting for DC drift due to thermal and other environmental changes. A simple way to deal with this is to selectively use capacitive coupling to eliminate the DC component. Because the DC component can be very high, typically several hundreds of millivolts at the mixer output, it can be difficult to separate the relatively small micro-volt Doppler signals. Selective capacitive coupling allows the radar to adapt to the sensing situation and environment. Considering the spectrum of a typical baseband signal, each peak in the spectrum ideally represents a moving object with radial velocity proportional to

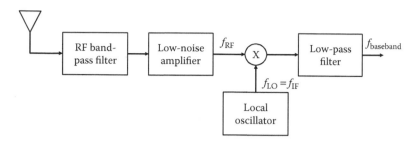

FIGURE 2.1
A basic homodyne receiver architecture.

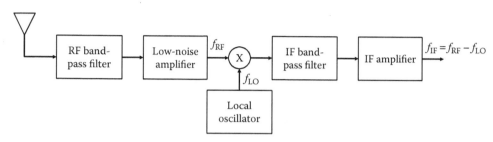

FIGURE 2.2
A basic heterodyne receiver architecture.

the frequency at the peak location. In this example, the presence of a high-DC component usually makes it challenging to amplify the baseband signal, and for this reason, some homodyne receivers amplify the baseband signal after AC coupling, eliminating the DC component. A basic architecture of a homodyne receiver is shown in Figure 2.1.

A heterodyne receiver differs from a homodyne receiver in that it uses an LO at a frequency different from the carrier signal as the reference signal for down-conversion. Then, the down-converted signal becomes a signal modulated on an intermediate frequency (IF) rather than being a baseband signal, after filtering is applied. As with the homodyne, an LNA may be used to amplify the received signal prior to the mixing operation. The presence of an IF stage allows for the use of high-quality components tuned to the carrier frequency, and band-pass filters at the RF and IF stages help eliminate unwanted signals and improve SNR. In addition, an IF amplifier may be used to amplify the IF signal prior to detection. The issue with the DC component pertaining to a homodyne receiver is eliminated in a heterodyne receiver; however, the architecture of a heterodyne receiver is more complicated compared to that of a homodyne receiver. A basic architecture of a heterodyne receiver is shown in Figure 2.2.

2.2.2 Single-Channel versus Dual-Channel Homodyne Receivers

Basic block diagrams for single-channel and dual-channel CW radar homodyne transceivers are shown in Figures 2.3 and 2.4, respectively. In a single-channel homodyne transceiver, the signal generated by the LO is transmitted, and the return signal reaching the receiver is directly down-converted to baseband by mixing it with the signal coupled from the same LO used at the TX. However, in a dual-channel or quadrature homodyne receiver, the reference signal is split into two branches: one with 0° phase shift and the other with

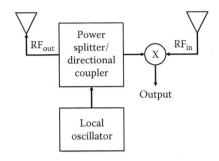

FIGURE 2.3
Block diagram for a single-channel CW radar homodyne transceiver.

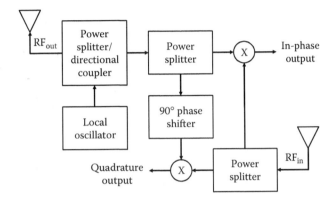

FIGURE 2.4
Block diagram for a dual-channel (quadrature) CW radar homodyne transceiver.

90° phase shift. The received signal is also split into two branches that are in phase and are mixed with 0° and 90° phase-shifted reference signals to generate in-phase and quadrature components, forming a dual-channel baseband signal.

The in-phase and quadrature components of baseband signals in dual-channel receivers will experience at least 3 dB loss compared to the baseband signals in single-channel receivers due to the splitting of reference and received signals. This loss will also result in reduced baseband signal SNR; however, having both in-phase and quadrature components of the baseband signal enables determination of whether an object is moving closer to or away from the radar, and also provides phase information. It is expected that the quadrature component will either lead or lag the in-phase component by 90° depending on whether the object is moving closer to or away from the radar. In cases of small magnitude reflections from slowly moving objects, in-phase and quadrature components may be observed to have 0° or 180° phase difference between them. This behavior can be verified using a small argument approximation for the in-phase and quadrature components. The small argument approximation is also useful to remove or detect phase ambiguity, which can result in incorrect object direction and classification. Phase information between the in-phase and quadrature components of the baseband signal is also very useful as an input to signal classification algorithms in addition to single-channel time–amplitude information.

Single-channel CW radars, generated by mixing the reference signal with the received signal, are common, inexpensive, and very useful. Although a single-channel CW radar will be unable to identify the direction of movement, it will still provide a sensitive Doppler component. Regardless of whether a single-channel or dual-channel receiver is employed, the major limitation of a CW radar is the fact that it does not have the capability to unambiguously detect the distance to a moving object.

There are also circuit effects in CW radars that must be accounted for. Even though the circuitry for a homodyne receiver is simple, there are challenges such as DC offset and low-frequency noise from components such as power supplies, LOs, mixers, and amplifiers (Park et al. 2007). CW radars have been used for vital sign monitoring indoors for many years, and the reader is referred to several publications on this topic (Droitcour et al. 2004; Xiao et al. 2006; Park et al. 2007).

2.2.3 Challenges with CW Radars

Besides the inability to detect distance, there are other limitations associated with CW radars. The signal radiated by the transmitting antenna of a CW radar may couple strongly to its received antenna, which will affect the DC component of the baseband signal at the receiver and may be interpreted as a reflection from a stationary object requiring the TX–RX isolation to be sufficiently high. Low-cost CW radars also suffer from multipath interference that can cause Doppler spread, essentially blurring the desired motion of the target object, and making it harder to pinpoint a frequency or to perform target classification.

Another important consideration is whether the CW radar is built as DC coupled or AC coupled. This can be implemented by simply using a shunt or a series capacitor to filter the baseband signal prior to signal processing. Even though arctangent demodulation in the absence of the DC offset was studied (Park et al. 2007), it usually requires the DC offset to be kept in the baseband signal. Thus, a DC-coupled CW radar is preferred to apply arctangent demodulation and to extract the phase information linearly proportional to the motion of an object. However, the DC offset may be much stronger in magnitude than the Doppler shifted signal of a moving object, especially when the movement is small in magnitude such as chest displacements of 0.2 mm–10 cm from heartbeat or respiration, and when there are also strong reflections from stationary objects. The presence of a strong DC offset makes it challenging to amplify the DC-coupled baseband signal to identify small movements. Automatic gain control may be used to adjust the level of amplification in accordance with the strength of the DC offset. This problem significantly limits the range of low-cost (~$30) DC-coupled CW radars to <10 m.

An AC-coupled CW radar makes it much easier to amplify the baseband signal, because the high-energy DC component is removed. However, arctangent demodulation in the absence of DC offset will be very challenging to achieve, if not impossible. Other techniques such as autocorrelation of the baseband signal and fast Fourier transform may also have issues in accurate prediction of the radial velocity of a moving object. Some AC-coupled CW radars have built-in baseband amplifiers following the AC coupling, which is recommended to significantly improve the range of the radar.

Some CW radars have pulsed operation capability allowing the distance to a moving object to be determined. The pulsed operation is usually defined by a pulse duration and a repetition frequency. The limitation of a CW radar to detect the distance can also be eliminated by modifying the radar via frequency modulation (FM) of the CW signal or transmitting CW signals sequentially at multiple frequencies. FMCW radars, discussed next, are useful for eliminating the range ambiguity at the expense of a more complicated transceiver architecture and need to be calibrated to remove nonlinear frequency effects of the transmit circuitry.

2.3 FMCW Radar

An FMCW radar can be seen as an extension to the CW radar and enables measurement of both the range and the Doppler shift. In the FMCW radar, the frequency of the transmitted signal is moved up and down continuously over a fixed time period by a modulating signal of triangular (linear FM), sawtooth, sinusoidal, or other waveforms. A common FMCW radar is the linear FMCW (LFMCW) radar, which linearly increases and decreases the frequency of the CW signal. Linear FM is achieved by periodically moving the frequency up and down following a triangular waveform as shown in Figure 2.5. The solid line represents the transmitted waveform, whereas the dashed line shows the waveform received from a stationary target at range R. An LFMCW radar determines the range to an object based on the frequency difference between the received and transmitted radar signals, which is called the beat frequency, f_b, shown in Figure 2.5 (Skolnik 2002). The range to an object is proportional to the beat frequency, which increases with time delay. In the case of a stationary object, a constant beat frequency is observed at the receiver. A portion of the signal from the TX is coupled to the receiver as a reference signal and mixed with the received signal to generate a signal with beat frequency. The propagation time, τ, of the return signal between the times it was transmitted and received can be expressed as

$$\tau = \frac{2R}{C}$$

where C is the speed of light. The frequency of the triangular waveform is the modulation frequency, which is given based on Figure 2.5 as

$$f_m = \frac{1}{2t_0}$$

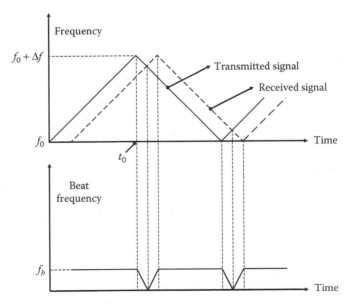

FIGURE 2.5
Modulated waveforms at the TX and RX, and beat frequency for a stationary target.

The chirp rate, r_c, which is the rate of change of the carrier frequency with time, is defined as

$$r_c = \frac{\partial f}{\partial t} = \frac{\Delta f}{t_0} = 2f_m \Delta f$$

where Δf is the maximum frequency deviation. The beat frequency can be found as

$$f_b = r_c \tau = \frac{4 R f_m \Delta f}{C}$$

for a stationary object. However, if an object is moving, its radial velocity with respect to a radar will cause a phase shift of

$$\phi = \frac{4\pi V t}{\lambda}$$

to the signals that are incident on the object. Hence, the Doppler shift in the frequency of the signals will be observed at the radar as

$$\frac{1}{2\pi} \frac{\partial \phi}{\partial t} = \frac{2V}{\lambda}$$

Therefore, when an object is not stationary, the beat frequency of an FMCW radar involves both range and radial velocity information. Hence, the beat frequency includes both a Doppler shift and a frequency shift due to the travel time of the echo signal. Each segment of the modulated signal with positive or negative slope is denoted as chirp. The Doppler shift will subtract from and add to the beat frequency during the segments of the modulated signal with positive and negative slopes, respectively, as shown in Figure 2.6. The beat frequency during positive and negative slopes, respectively, can be determined as

$$f_{bp} = \frac{4 R f_m \Delta f}{C} - \frac{2V}{\lambda}$$

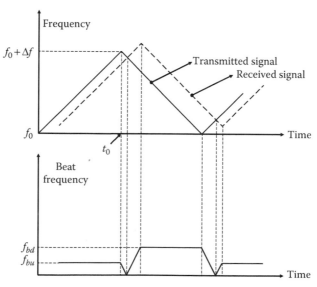

FIGURE 2.6
Modulated waveforms at the TX and RX, and beat frequency for a moving target.

$$f_{bn} = \frac{4Rf_m\Delta f}{C} + \frac{2V}{\lambda}$$

Then, the range and radial velocity of the moving object can be computed by adding and subtracting f_{bp} and f_{bn}, respectively, as

$$R = \frac{C}{8f_m\Delta f}\left(f_{bn} + f_{bp}\right)$$

$$V = \frac{\lambda}{4}\left(f_{bn} - f_{bp}\right)$$

Therefore, the LFMCW radar makes it possible to predict both the range and the radial velocity by conducting beat frequency measurements in two consecutive chirps.

The range of the FMCW will be a fraction of the wavelength of the frequency that is the difference between the lowest and highest transmitted frequencies. Hence, the range and range resolution of the FMCW radar can be increased by making the range of frequencies it sweeps (bandwidth) broader and narrower, respectively. A basic block diagram for an FMCW radar is shown in Figure 2.7.

The radar signal does not limit the minimum detection range of the FMCW radar. The minimum range is limited by the TX–RX isolation. Because the TX and RX are continuously on, the TX–RX isolation is a significant design challenge for FMCW radars (Skolnik 2002). FMCW radars use separate TX and RX antennas to accomplish the desired TX–RX isolation, and modulation techniques can be employed to enable a single-antenna FMCW radar (Saunders 1961) at the expense of more sophisticated signal processing algorithms and hardware implementation. FMCW radars do not have clutter rejection capabilities, because their receivers are always on.

FMCW radars have also been successfully used for the detection of motion and human life signs such as breathing and heartbeat (Anitori et al. 2009; Postolache et al. 2011; Adib et al. 2014; Adib et al. 2015). One limitation of FMCW in this application is that in the case of multiple objects, the range and radial velocity cannot be resolved unambiguously by measuring beat frequencies for two consecutive chirps. This is because ghost

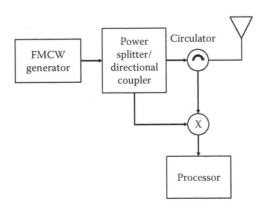

FIGURE 2.7
Block diagram for an FMCW radar.

objects are introduced, but this issue can be resolved by transmitting and measuring multiple chirps with different slopes using the LFMCW radar. In addition, the FMCW radar can use multiple receive antennas to allow for both range and angle discrimination so that targets can be spatially separated.

Besides FMCW radars, which sweep frequency continuously, there are multiple frequency radars that provide distance prediction by hopping through multiple discrete frequencies. They transmit two or more unmodulated signals that alternate in time and have different carrier frequencies. One type of the multiple frequency radar is the frequency shift keying (FSK) radar (Ahmad et al. 2009; Setlur et al. 2010), which is based on the CW radar. Determination of Doppler shift and range is possible using an FSK radar; however, the range resolution will be poor, because it uses discrete frequencies as opposed to the continuous frequency sweep of the FMCW radar. Depending on the frequency difference between the transmit signals, the unambiguous range can be very large and on the order of several meters.

It is desired to measure the range and radial velocity simultaneously in numerous radar applications. In the indoor location, radial velocity provides target motion information and range that allows the radar to focus only on the target and ignore objects at other distances. LFMCW and FSK radars satisfy these requirements; however, the LFMCW radar requires several measurement cycles and signal processing algorithms to resolve ambiguities, whereas the FSK radar lacks range resolution. Hence, the features of LFMCW and FSK radars were combined to introduce the multiple FSK (MFSK) radar, which shifts frequency in an interleaved manner. Each frequency sweep has a certain bandwidth and duration. Similar to the FSK radar, the MFSK radar measures the phase difference between the interleaved signals to determine the range and radial velocity. Ghost targets can be eliminated using the MFSK radar, because it combines the benefits of LFMCW and FSK radars. MFSK radars are typically more expensive than CW Doppler radars, but they may become an option for widespread indoor use as the high-volume automotive applications drive cost down. Stepped-frequency radar is another class of low-cost radar architecture that should be considered for its implementation simplicity. (See Chapter 1 for information on stepped-frequency radars.)

2.4 Pulsed Radar

A pulsed radar periodically transmits pulses in a certain waveform and receives reflected pulses. The pulse propagation time, τ, for a single pulse can be measured to determine the range, R, as

$$R = \frac{C\tau}{2}$$

The unambiguous range is defined as the maximum range that can be detected by a pulsed radar. Because the pulsed radar transmits pulses periodically, the return pulse needs to be received before the next pulse is transmitted. Otherwise, if the return pulse is received after the next pulse is transmitted, it cannot be assigned to the original pulse, which will cause range ambiguity. Therefore, the maximum unambiguous range of a pulsed radar system can be determined as

$$R_{\max} = \frac{CT_r}{2}$$

where:

$R_{\max}$ is the maximum unambiguous range

$T_r = 1/f_r$ is the pulse repetition period

Pulses transmitted and received by a pulsed radar as well as the radar parameters, T_r and τ, and the pulse width of the transmitted signal, T_p, are shown in Figure 2.8.

Besides the maximum unambiguous range, range resolution is also important for pulsed radars. Range resolution, ΔR, is described as the minimum increment in range that allows two targets to be separated by the radar. When two targets are closer to each other compared to the range resolution, they cannot be identified by the radar as separate targets, because their echo signals overlap in time. Figure 2.9 shows two overlapping return pulses after the first transmitted pulse that are identified as one target. It also shows two slightly separated return pulses after the second transmitted pulse that can be resolved as separate targets. Short pulses have large bandwidth and can yield high-range resolution, and the range resolution can be expressed based on the pulse width of the transmitted signal as

$$\Delta R = \frac{CT_p}{2}$$

The waveform of the signal transmitted by a pulsed radar can be defined by the carrier frequency, pulse shape, pulse width, modulation, and pulse repetition frequency (PRF).

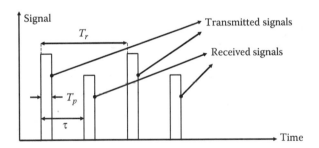

FIGURE 2.8

Pulses transmitted and received by a pulsed radar.

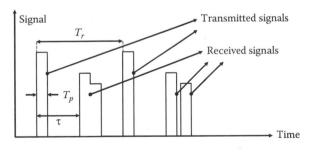

FIGURE 2.9

Examples of return pulses from two targets identified as a single target and two separate targets.

The carrier frequency can be selected based on design requirements. For instance, a lower carrier frequency yields better penetration, whereas a higher carrier frequency provides better range resolution. Various modulation techniques can be used to improve the performance and enhance the capabilities of the pulsed radar. If PRF is decreased, more accurate and longer range measurements can be accomplished, and ambiguities in range measurements can be mostly eliminated. However, there will be increased ambiguities associated with the radial velocity measurement based on the Doppler shift. If PRF is increased, the average transmitted power can be improved to provide better clutter rejection capabilities, and ambiguities in Doppler shift measurements can be mostly eliminated. At the same time, there will be more ambiguities in range measurements. Thus, PRF should be selected or adapted to avoid the ambiguities in range and radial velocity measurements as well as to maximize the average transmitted power.

Similar to the FMCW radar, if the pulsed transmitted signal is reflected by a moving object, then the reflected signal is Doppler shifted, which helps determine the radial velocity. The receiver can employ range gating at the RF front end, which makes it possible to select a single observable target range. The delay in the leading edge of the received signal compared to the transmitted signal is determined at the receiver yielding high accuracy.

Fine range accuracy requires a wide bandwidth, which can be achieved by employing a narrow pulse width. Hence, a pulsed radar needs to include a short pulse generator for short-range detection. The minimum detectable range of the pulsed radar is limited by factors such as the transmitted signal pulse width, poor isolation between the TX and RX, and clutter. However, the pulse width of the transmitted signal is the major limitation of the minimum detectable range among these factors. It is in fact easier to meet the isolation requirements between the TX and RX with a pulsed radar compared to CW radars.

In the pulsed radar, the TX and RX are not simultaneously on, unlike the FMCW radar. The pulsed radar does not turn its receiver on until the TX is done with transmission. If the pulsed radar has a rectangular pulse, then the average transmitted power is

$$P_{av} = \frac{T_p}{T_r} P_p$$

where:
P_p is the peak power of the transmitted signal
T_p/T_r is the duty cycle

Therefore, if the pulse width is made smaller compared to the pulse repetition period, then the peak power will have to be increased to achieve the same average power to compensate for the reduced duty cycle. Nevertheless, power requirements for short-range radars are relatively low, which may ease the concerns to increase the peak power. Reducing the pulse width of the transmitted signal provides range resolution, while range gating at the receiver improves clutter rejection.

Compared to the FMCW radar, the pulsed radar can detect the range to an object by transmitting a short pulse and measuring the time of flight of the echo signal, which requires the radar to have high-instantaneous transmit power and large physical size. Therefore, the FMCW radar can have the same performance as the pulsed radar with lower power and smaller physical size. FMCW radars generally use separate TX and RX antennas to accomplish the desired TX–RX isolation, whereas pulsed radars can use a single antenna for both transmission and reception.

2.5 Ultra-Wideband Radar

A signal can be classified as an ultra-wideband (UWB) signal if its fractional bandwidth is >0.2 and the bandwidth is ≥500 MHz. UWB signals yield high precision in ranging and are resistant to the effects of multipath propagation. This makes UWB radars a potential solution for indoor monitoring applications (Staderini 2002). UWB radars have been attracting significant attention, because the U.S. Federal Communication Commission (FCC) allowed for limited use of UWB signals by establishing the emission limits for safe operation of UWB devices within a specific frequency range (Immoreev and Tao 2008). The FCC restricted the unlicensed use of UWB systems to the frequency range between 3.1 and 10.6 GHz. In comparison with X-ray imaging, UWB radars use non-ionizing electromagnetic fields that do not harm human body, and as such are seen as an interesting alternative for medical surface and subsurface imaging.

Although time is needed to prove their usefulness in indoor environments, UWB radars are becoming simpler, less expensive, and lower power, and yielding higher data rates (Fontana 2004; Li et al. 2010). These types of radars are also well suited to biomedical applications (Fear et al. 2002; Staderini 2002; Lazaro et al. 2009; Leib et al. 2009). UWB radars have been employed to monitor breathing and heartbeat rates (Yarovoy et al. 2006; Immoreev and Tao 2008; Lazaro et al. 2010; Zito et al. 2011; Lazaro et al. 2014) as an alternative to CW radars. These studies demonstrated the feasibility of predicting breathing rates using a wireless, contactless, and noninvasive respiratory monitoring system with the UWB radar that can be used for indoor applications. Simultaneous detection of breathing and heartbeat rates using the UWB radar was also studied (Immoreev and Tao 2008). UWB radars can detect breathing rates with reasonable accuracy even through a wall (Yarovoy et al. 2006). The major disadvantage of UWB radars for indoor monitoring is the lack of low-cost integrated circuits (ICs) compared to low-cost CW radars. Nevertheless, complementary metal-oxide semiconductor (CMOS) ICs that can be used to design UWB radars for vital sign monitoring have been developed (Zito et al. 2011). Commercial development kits have also been introduced for UWB radars (Yu et al. 2012).

2.5.1 Features of UWB Radars

There are several attractive features of UWB radars for indoor monitoring applications, including low-power consumption, high data rates, coexistence with other equipment, low system complexity, low-cost, and resilience against interference and multipath (Taylor 1994; Pozar 2003; Immoreev and Tao 2008).

UWB systems do not require down-conversion or up-conversion via RF mixing, resulting in simpler transceivers. They can be implemented using CMOS technology, making them low-cost. Lower frequency components of UWB signals provide better penetration into materials. Broadband antennas such as Vivaldi and bowtie antennas are used to transmit and receive UWB signals.

2.5.2 Challenges with UWB Radars

Due to the frequency and multipath characteristics of the propagation channel, UWB signals experience significant pulse distortion. Antennas also contribute to these distortions. The pulse distortion causes the shape and frequency content of the received pulse to be significantly different compared to the transmitted pulse. Hence, matched filter correlation becomes challenging at the receiver. There is an ongoing research on the design of broadband antennas that maintain their input impedances and radiation patterns over a wide frequency range while maintaining low-cost, ease of manufacturing, and small form factor (Low et al. 2005). However, existing UWB antenna designs provide almost frequency-independent behavior only within a small portion of the UWB spectrum defined by the FCC. Due to the very large bandwidth of UWB signals, the received signal has to be sampled at the rate of several giga-samples per second to satisfy the Nyquist criterion. The requirement for such ultra-high speeds poses challenges for analog-to-digital conversion. These challenges are expected to be overcome by the development of ultra-high speed analog-to-digital converters with the rapid advances in very-large-scale integration (VLSI) technology.

Impulse radio UWB (IR-UWB) radar is a type of radar that transmits very short baseband pulses, typically on the order of a nanosecond, and has been used for ultra-fine motion detection (Win and Scholtz 1998; Staderini 2002; Lazaro et al. 2010; Lai et al. 2011; Schleicher et al. 2013).

2.5.3 Highly Integrated System-on-Chip UWB Radar

For consumer devices, UWB conforms to a −41.3 dBm/sqrt(Hz) power spectral density over a 500 MHz band. One advantage of using the UWB radar is that as long as the spectral power emission limits are adhered to, any center frequency can be chosen within the limitations of the CMOS technology. An UWB radar will diversify the multipath wave structure over at least 500 MHz of bandwidth, whereas a stationary narrowband radar will have a relatively fixed multipath structure in any given volume at any given frequency.

UWB radars often feature coherent integration which is advantageous when additional range, SNR, or resolution is the goal. For each doubling in integration time, an additional 3 dB of SNR is achieved. The disadvantage of using this approach is that at the extreme, integration time becomes so long that quickly moving objects are not sufficiently sampled. In indoor applications, this is typically not an issue, because large motions are associated with full body motion, which in turn has a larger radar cross section compared to vital sign-related returns. Targets with large radar cross section will return a larger signal, requiring less integration time and therefore allowing higher sample rates. When an indoor target is stationary, integration time can be increased adaptively to acquire smaller motions. The radar could, for example, set itself to a low integration time mode when a person is moving, then slowly increase the integration time as the amplitude or spectral power density decreases, as would be the case when someone sits down or goes to sleep.

UWB radars implemented as a system on chip (SoC) include the RF front end, pulse generator, filters, receive gain stage, LNA, digitization samplers, and universal serial bus (USB) or serial interface. This makes it very easy to interface the radar with modern computing platforms, and provides the flexibility to build signal processing, and classification algorithms in almost any computer environment. Once the algorithms are demonstrated externally, they can be compiled into more efficient routines that can run on a dedicated application coprocessor. The disadvantage of using this approach is the limited signal quality and TX power associated with regulated UWB radars. However, for indoor applications, the size, range, and simple data interfaces are attractive for localization, simple motion classification, and vital sign detection.

Hardware implementations of UWB radars have been simplified to a single 1 × 1 cm IC with only a handful of RF matching components. Solutions from XeThru have implemented 3.1–10.6 GHz UWB radars onto a single IC with the convenience of a direct USB connection (Hjortland 2006; Yu et al. 2012; Morawski et al. 2014). After several years of development, their current UWB radar has a range of ~10 m with standard Vivaldi antennas and provides a sub-millimeter-range resolution. Radars of this type are extremely versatile and, with the addition of a small 100 MHz–1 GHz multicore application co-processor, can be used to detect motion and vital signs. They can also be employed for localization in bistatic arrangements (Griffiths et al. 2017). (See Chapter 8 for information on bistatic configurations.)

One example of the modern UWB radar from XeThru provides native ranging with separate Vivaldi antennas for the TX and RX. This UWB radar is capable of generating Gaussian pulses within the 3.1–10.6 GHz frequency range and transmits them using the separate TX antenna. The RX antenna receives a pulse reflected from a target, samples it, and communicates the sampled data to a computer via a USB port for processing. The radar module uses adjustable gain comparators in series with a precisely controlled delay between each pair of consecutive comparators to sample a received pulse with 64, 128, 256, or 512 samples that are separated by 26, 52, or 280 ps corresponding to sampling frequencies of approximately 38.46, 19.23, and 3.57 GHz, respectively. The sampling frequencies of 38.46, 19.23, and 3.57 GHz correspond to range resolutions of 3.9, 7.8, and 42 mm, respectively. When the UWB radar uses 3.9 mm range resolution and 512 receive comparators, it could provide the maximum range of about 2 m. The frame stitching property of the UWB radar makes it possible to combine up to 15 consecutive frames to extend sampling duration and therefore the maximum range. The UWB radar could perform coherent integration of multiple received pulses in generating each frame consisting of one of the above-mentioned number of samples to improve SNR. This is an important consideration, because coherent integration could take significant time, and if the target is moving too quickly, insufficient samples will be available to classify the motion. Large integration times are typically needed when the goal is to extract very small motions associated with respiration or heartbeat near the chest cavity. Short integration times yielding sample rates of 100 Hz or more are typically sufficient to track whole-body or torso motions. The UWB radar can only determine the radial distance to an object with a single TX antenna and a single RX antenna. In order to perform two-dimensional or three-dimensional localization, a multistatic radar arrangement using two or more radars is required. Another approach would require multiple RX antennas to determine the phase of return signals and enable two-dimensional localization.

Power consumption from an UWB SoC radar is very low. It is one of the most efficient radars available for the provided range (<150 mW). The majority of power consumption comes from signal processing and target classification algorithms running on a local processor. Even with the radar transmitting continually and a typical 1 GHz four-core 64 bit processor running at full capacity, the entire radar can be powered by a USB 2.0 port consuming <1.5 W.

2.6 Challenges with Indoor Radar Hardware

Building cost-effective and reliable radar hardware for indoor use is challenging. When developing a new indoor radar platform, the application space and desired output from the radar are the determining factors of the hardware architecture. This point can be exemplified by considering two example applications: motion detection and multiple target classification.

For motion detection, the key technical requirements are likely to be maximum range, range adjustment, and minimum detectable radial velocity with an indication of motion or no motion as an output. Simple applications of human motion detection can be implemented using the low-cost CW radar. A radar of this type can be built using commercial off-the-shelf discrete components for less than $20 total cost, including radar front end, application processor, and packaging. From an architecture perspective, this radar has an inexpensive time base, and uses one or two RF detection diodes to down-convert target reflections directly to baseband. One of the most common examples is the Interlogix range-controlled radar (RCR) line of security sensors, which also include a classic passive infrared motion detector. When the goal is to classify multiple targets, several additional considerations are required, remembering that the CW radar can only detect motion in a room. In order to separate multiple targets, radar hardware that supports multiple receive antennas or multi-static configurations, or that can focus on fixed regions of space is typically employed. Most FMCW and UWB radars have their native ability to distinguish the regions of space and can focus on both a certain region and angle providing two-dimensional localization capabilities, with the addition of multiple receive antennas. Signals from individual targets can be isolated for further signal processing to extract gestures or vital signs with this capability.

There are challenges inherent in all indoor radar hardware architectures such as multipath fading, range, range resolution, obscuration, classification, and the ability to cover all angles. Some of these challenges are minimized with the use of radar technologies that are more immune to multipath and technologies such as the proven FMCW radar for high-resolution range detection.

Additional manufacturing and deployment challenges present themselves in radar hardware selection. Some of these challenges include maintaining performance across temperature, the ability to use common materials and manufacturing processes such as FR4 printed circuit board (PCB) material and CMOS semiconductors, power conditioning, antennas, signal processing, DSP coprocessor selection, installation and commissioning, and communications bandwidth. These points will be addressed in order next.

Temperature fluctuations have minimal impact on some radar platforms such as CW radar, and can have large impact on low-cost UWB and FMCW radars. Manufacturing a radar at frequencies <10 GHz enables the use of inexpensive and commonly available PCB materials such as FR4. Above 10 GHz, more expensive, temperature-stable, and lower loss tangent substrates are required to support RF front ends, mixers, and antenna systems. Although most radars for indoor applications can use CMOS processes up to 24 GHz, some ≥77 GHz radars would require much more expensive processes to incorporate, for example, gallium arsenide amplification and receiving systems. Most indoor radars will be line powered so that there is less concern over battery replacement. However, power conditioning is still a concern as it can directly affect the receive noise floor of the radar. In some cases, less efficient power conversion can actually result in lower noise floor, which is typically favored over reducing power consumption. A typical solution is to use linear power conditioning hardware as opposed to switched-mode power supplies.

The radar hardware should incorporate the correct LNAs, analog filters, and PCB design to achieve the best SNR possible. Although signal processing can implement adaptive noise filters, low-pass and band-pass filters, and complex filter windows, there is a huge advantage in low-cost radars to build as much analog filtering and gain into the front end. Signal processor selection is mostly dictated by the end application. When it comes to detecting whether there is motion or not, very little signal processing is required. This can be performed on processors with sampling frequency of <100 MHz and even 10 MHz in some cases. However, classifying motion or running complex algorithms such as neural networks, Gaussian mixture models, or supervised/machine learning algorithms can require 1 GHz class processors with significant memory to support timely classification. Installation and commissioning is a significant challenge given the variation in building layout, materials, and aspect ratios. Intrusion detection hardware is typically mounted up high and at the corner of a room to provide adequate antenna coverage. Commissioning means more than achieving good antenna coverage. It also includes any background subtraction, materials adjustment, or application-specific programming. An approach in which the radar firmware can be updated and specific performance attributes can be remotely monitored is recommended for advanced applications such as health, fall, and vital sign detection. Communications bandwidth varies considerably. In the case of motion detection, an indication of whether there is motion or not may suffice. When it comes to classifying motion, sampling rates on the order of 100 Hz with 16–24 bit resolution may be required for multiple radar channels. In advanced applications, ≥1 Mbps is easily achieved depending on how much signal processing is performed at the radar or remotely.

References

Adib, F., Z. Kabelac, and D. Katabi. 2014. *Multi-Person Motion Tracking via RF Body Reflections.* Massachusetts Institute of Technology Computer Science and Artificial Intelligence Laboratory Technical Report, MIT-CSAIL-TR-2014-008. Cambridge, MA.

Adib, F., H. Mao, Z. Kabelac, D. Katabi, and R. C. Miller. 2015. *Smart Homes That Monitor Breathing and Heart Rate.* ACM Conference on Human Factors in Computing Systems, Seoul, Korea, pp. 837–846.

Ahmad, F., M. G. Amin, and P. D. Zemany. 2009. Dual-frequency radars for target localization in Urban sensing. *IEEE Transactions on Aerospace and Electronic Systems* 45 (4): 1598–1609.

Anitori, L., A. de Jong, and F. Nennie. 2009. FMCW radar for life-sign detection. *IEEE Radar Conference.* Pasadena, CA.

Droitcour, A. D., O. Boric-Lubecke, V. M. Lubecke, J. Lin, and G. T. A. Kovac. 2004. Range correlation and I/Q performance benefits in single-chip silicon doppler radars for noncontact cardiopulmonary monitoring. *IEEE Transactions on Microwave Theory and Techniques* 52 (3): 838–848.

Fear, E. C., X. Li, S. C. Hagness, and M. A. Stuchly. 2002. Confocal microwave imaging for breast cancer detection: Localization of tumors in three dimensions. *IEEE Transactions on Biomedical Engineering* 49: 812–822.

Fontana, R. J. 2004. Recent system applications of short-pulse ultra-wideband (UWB) technology. *IEEE Transactions on Microwave Theory and Techniques* 52: 2087–2104.

Griffiths, H., M. Ritchie, and F. Fioranelli. 2017. Bistatic radar configuration for human motion detection and classification, Chapter 8, *Radar for Indoor Monitoring*, M. G. Amin (Ed.), CRC Press: Boca Raton, FL.

Hjortland, H. A. 2006. UWB impulse radar in 90 nm CMOS. Master's Thesis, University of Oslo Department of Informatics, Oslo, Norway.

Hülsmeyer, C. 1904. *Telemobiloscope.* Patent Publication DE 165546.

Hülsmeyer, C. 1906. *Wireless Transmitting and Receiving Mechanism for Electric Waves.* Patent Publication US 810150A.

Immoreev, I., and T.-H. Tao. 2008. UWB radar for patient monitoring. *IEEE Aerospace and Electronic Systems Magazine* 23 (11): 11–18.

Lai, J. C. Y., Y. Xu, E. Gunawan, E. C.-P. Chua, A. Maskooki, Y. L. Guan, K.-S. Low, C. B. Soh, and C.-L. Poh. 2011. Wireless sensing of human respiratory parameters by low-power ultrawideband impulse radio radar. *IEEE Transactions on Instrumentation and Measurement* 60 (3): 928–938.

Lazaro, A., D. Girbau, and R. Villarino. 2009. Wavelet-based breast tumor localization technique using a UWB radar. *Progress in Electromagnetic Research* 98: 75–95.

Lazaro, A., D. Girbau, and R. Villarino. 2010. Analysis of vital signs monitoring using an IR-UWB radar. *Progress in Electromagnetic Research* 100: 265–284.

Lazaro, A., D. Girbau, and R. Villarino. 2014. Techniques for clutter suppression in the presence of body movements during the detection of respiratory activity through UWB radars. *Sensors* 14: 2595–2618.

Leib, M., E. Schmitt, A. Gronau, J. Dederer, B. Schleicher, H. Schumacher, and W. Menzel. 2009. A compact ultra-wideband radar for medical applications. *Frequenz* 63 (1–2): 2–8.

Li, B., Z. Zhou, W. Zou, D. Li, and C. Zhao. 2010. Optimal waveforms design for ultra-wideband impulse radio sensors. *Sensors* 10: 11038–11106.

Li, C., J. Cummings, J. Lam, E. Graves, and W. Wu. 2009. Radar remote monitoring of vital signs. *IEEE Microwave Magazine* 10 (1): 47–56. doi:10.1109/MMM.2008.930675.

Low, Z. N., J. H. Cheong, and C. L. Law. 2005. Low-cost PCB antenna for UWB applications. *IEEE Antennas and Wireless Propagation Letters* 4: 237–239.

Morawski, R. Z., Y. Yashchyshyn, R. Brzyski, F. Jacobsen, and W. Winiecki. 2014. On applicability of impulse-radar sensors for monitoring of human movements. *20th IMEKO TC4 International Symposium*, pp. 754–759. Benevento, Italy.

Park, B., O. Boric-Lubecke, and V. Lubecke. 2007. Arctangent demodulation with DC offset compensation in quadrature doppler radar receiver systems. *IEEE Transactions on Microwave Theory and Techniques* 55 (5): 1073–1079.

Postolache, O. A., P. S. Girão, J. M. Dias Pereira, and G. Postolache. 2011. FM-CW radar sensors for vital signs and motor activity monitoring. *ICST Transactions on Ambient Systems* 11 (10–12): 1–10.

Pozar, D. M. 2003. Waveform optimization for ultrawideband radio systems. *IEEE Transactions on Antennas and Propagation* 31 (9): 2335–2345.

Saunders, W. K. 1961. Post-war developments in continuous-wave and frequency-modulated radar. *IRE Transactions on Aerospace and Navigational Electronics* 8 (1): 7–19.

Schleicher, B., I. Nasr, A. Trasser, and H. Schumacher. 2013. IR-UWB radar demonstrator for ultra-fine movement detection and vital-sign monitoring. *IEEE Transactions on Microwave Theory and Techniques* 61 (5): 2076–2085.

Setlur, B., M. Amin, and F. Ahmad. 2010. Dual-frequency doppler radars for indoor range estimation: Cramer-rao bound analysis. *IET Signal Processing* 4 (3): 256–271.

Skolnik, M. 2002. *Introduction to Radar Systems*. 3rd ed. McGraw-Hill, New York, NY.

Staderini, E. 2002. UWB radars in medicine. *IEEE Aerospace and Electronic Systems Magazine* 17 (1): 13–18.

Taylor, J. D., ed. 1994. *Introduction to Ultra-Wideband Radar Systems*. CRC Press, Boston, FL.

Win, M. Z. and R. A. Scholtz. 1998. Impulse radio: How it works. *IEEE Communications Letters* 2 (2): 36–38.

Xiao, Y., J. Lin, O. Boric-Lubecke, and V. M. Lubecke. 2006. Frequency tuning technique for remote detection of heartbeat and respiration using low-power double-sideband transmission in Ka-band. *IEEE Transactions on Microwave Theory and Techniques* 54 (5): 2023–2032.

Yarovoy, A. G., L. P. Ligthart, J. Matuzas, and B. Levitas. 2006. UWB radar for human being detection. *IEEE Aerospace and Electronic Systems Magazine* 21: 10–14.

Yu, Y., J. Yang, T. McKelvey, and B. Stoewe. 2012. A compact UWB indoor and through-wall radar with precise ranging and tracking. *International Journal of Antennas and Propagation*. Article ID 678590. doi:10.1155/2012/678590.

Zito, D., D. Pepe, M. Mincica, F. Zito, A. Tognetti, A. Lanata, and D. de-Rossi. 2011. SoC CMOS UWB Pulse radar sensor for contactless respiratory rate monitoring. *IEEE Transactions on Biomedical Circuits and Systems* 5 (6): 503–510.

3

Modeling and Simulation of Human Motions for Micro-Doppler Signatures

Shobha Sundar Ram, Sevgi Zubeyde Gurbuz, and Victor C. Chen

CONTENTS

3.1 Introduction

The indoor monitoring of human beings with radar, such as the detection and localization of a person in a room without radio frequency identification (RFID) tagging, detection of fall events, remote monitoring of human health, and detection and recognition of vital signs, has become an increasingly important research topic. The radar signal reflected from a surface of an object that has kinematic movements (such as vibration or rotation) contains time-varying phase shifts induced by the displacement of the surface. From these changes in phase, it is thus possible to measure the kinematic parameters of the object.

The same phenomenon occurs when radar signals impinge upon the body of a moving person. The mechanical movement of a human body part causes changes in the received radar signal phase or frequency and time delay. Based on these changes, it is possible to measure motion velocities and trajectories of human body parts. Kinematic parameters of human movements include linear position (or displacement), velocity, acceleration, and angular position (or orientation). To completely describe any human motion activity in a three-dimensional (3D) Cartesian coordinate system, the linear kinematic parameters of position, velocity, and acceleration define the manner in which the position of any point in the human body changes over time. Angular orientation of a body segment, called joint angle, is a very important kinematic parameter. Together with angular velocity and angular acceleration, the three angular kinematic parameters can describe the angular motion of human body parts.

To simulate human movements, we need two models: human body model and movement model. The body model is a skeletal model that describes the body segments by parts. The movement model describes the motion kinematics of each body segment. In biomechanical engineering, the skeletal model of human body parts can simplify the human body by including just the body segments needed as controlled by joint moments. The kinematic model can be an empirical mathematical model, which is formulated as a set of empirical human motion equations and, thus, constructs a computer model of the human movement. The mathematical model can then be used to study a single motion parameter isolated from other parameters.

To estimate the kinematic parameters of human body segments, Boulic, Thalmann, and Thalmann [1] proposed a global human walking model, which is based on an empirical mathematical parameterization using biomechanical experimental data. The global walking model is derived based on a large number of experimental data and is intended to provide 3D spatial positions and orientations of any segment of a walking human body as a function of time. Although the Boulic–Thalmann model is only applied to human walking, in principle the kinematic model is also suitable for other human movements.

As an alternative to empirical mathematical modeling, the time-varying positions of joints positions may be recorded and tracked in 3D space by active sensors, such as accelerometers, gyroscopes, magnetometers, acoustic sensors, and radar. Often, such systems employ the markers to indicate the points on the human body that are to be tracked. For instance, the Carnegie Mellon University (CMU) Graphics Laboratory, Pittsburgh, PA, has used multiple infrared (IR) cameras to capture motions by placing many retroreflective markers on a body suit worn by the subject [2]. The markers' positions and orientations in 3D space are tracked and stored in a motion capture (MOCAP) database.

Markerless MOCAP systems are also available. Recent advancement in the Microsoft's Kinect sensor can provide the depth and surface information of objects in the scene and have been used for low-cost markerless skeleton tracking [3].

From the MOCAP database, based on 3D motion positions of human body segments and joints, the reflected signals from the human body may be simulated as if an active radar has been used [4]. As is well known, radar is capable of detecting and tracking objects with high accuracy at a remote distance, day and night, as well as in all weather conditions. Because of these reasons, radar possesses unique advantages for indoor monitoring of human activities.

Human activities may be recognized with radar by identifying their time-varying Doppler patterns, known as the micro-Doppler signatures and caused by the micro-Doppler effect [5]. The micro-Doppler effect is induced by micro motions, such as vibrations or rotations, that generate frequency modulations in addition to the Doppler shift induced by the translational motion. During the last decade, applications of micro-Doppler effect in radar have been widely investigated. Micro-Doppler signature can be represented in the joint time–-frequency domain that provides time-varying properties associated with locomotion of a human body. The micro-Doppler signature reflects the kinematics of a human and provides a unique identification of the human movement. The signature possesses certain traits that can reveal the characteristics of the motion and disposition of the human. By carefully analyzing various patterns in the signature, features unique to different activities may be identified and served as a basis for discrimination and characterization of human movement. The important advantages of radar are that it can operate at far distances and is not sensitive to lighting conditions or background complexity—limitations that inhibit sensing with visual images. Thus, radar micro-Doppler signature has been successfully used as a basis to distinguish among various human motion activities and even human gestures [6–9]. Although the micro-Doppler signature is sensitive to the relative angle between the target's direction of motion and the radar line of sight, the micro-Doppler effect in interferometric radar has been investigated as a method to mitigate the angular effects and effectively capture the off-line-of-sight micro-Doppler effect [10].

Indoor human motion includes regular periodic and nonperiodic movements. Regular motion of a human body, such as human walking or running, is an articulated locomotion. Motion of the limbs in the human body can be characterized by repeated periodic movements. Different human movements have different body movement patterns and, thus, different micro-Doppler signatures. An interesting built-in feature of human gait pattern is its personalized characteristics. Although the general manner of walking is similar for everyone, people usually can recognize a friend at a distance just from his or her walking style. Moreover, some emotional aspects of human gait can often be observed. For example, the gait of a cheerful person is quite different from that of a depressed person. Therefore, recognizing emotionally affected gaiting can help in detecting anomalous behavior of a person. Micro-Doppler signature, then, has the potential of recognizing not just different activities but also anomalous behaviors and emotion-dependent gaits.

Besides regular repeated periodic movements, nonperiodic human motion represents another important class of regular human movement, such as standing up, sitting down, kneeling, and falling (see Chapter 4 for fall detection). Such aperiodic events can be an important indication of health—for example, a chronic limp, concussion, dizziness, or even critical event such as heart attack. The same principles of modeling, simulation, and feature extraction that were applied to analyze the periodic motion can also be applied to recognize important aperiodic events.

In this chapter, we provide an overview on the modeling and simulation of human body motions to observe their micro-Doppler effect in radar and extract micro-Doppler signatures. First, we discuss the kinematic modeling of human motion and introduce a commonly used radar cross section (RCS) prediction method. Then, we discuss the modeling and simulation of through-the-wall radar signatures. Finally, we present a number of applications of human motion signatures for the recognition of indoor human movements.

3.2 Kinematic Modeling of Human Motion

Different modeling techniques have been proposed and researched to simulate radar data of human motion. The process can be divided into two steps. The first step involves the generation of detailed kinematic descriptions of the motions of different body parts.

This could be accomplished through various techniques, three of which are discussed in Sections 3.2.1.1–3.2.1.3.

The second step involves converting the time-varying 3D trajectories of the different body parts to radar scattered electromagnetic (EM) data that can be used to generate radar signatures of human motion, including micro-Doppler signatures. This can be carried out with very simplistic methods such as model-based primitive modeling or more complex techniques using full-wave EM prediction. The latter technique captures the entire physics of the radar backscatter accurately. However, it does not lend itself to the modeling of scattering from dynamic motions because the modeling has to be repeated for each pose of the human. This can be significantly time consuming and complex because humans are spatially large, nonrigid, 3D objects. However, primitive-based modeling has proven to provide reasonably accurate radar backscatter at a very low computation cost. However, it does not capture the physics of the shadowing and multiple interactions between the different body parts. A detailed description of primitive-based modeling is given Section 3.2.1.

3.2.1 Modeling of Human Motion

3.2.1.1 Sinusoidal Model

In most Doppler radar configurations, for human walking, the highest Doppler returns arise from the swinging motion of the human legs. Therefore, some researchers have simulated micro-Doppler spectrograms of human walking by modeling the motion as a simple inverted pendulum, which is modeled by a small bob attached to one end of a string and the other end of the string is fixed to a fulcrum. The leg is represented by the string, and the fulcrum is at the hip. Though simple to execute, this model is restrictive and insufficient in the following reasons: first, it does not capture the scattered returns from the other parts of the human body such as the arms, torso, and head. The recognition of human activities often relies on identifying the differences in micro-Doppler signatures due to limb motion. Second, the pendulum model does not accurately map the human walking motion especially during those instances in the stride when both the feet are on the ground. Third, the model does not allow for the incorporation of any unique personification to the human walking motion (as a function of the height or weight of the human). Finally, the model cannot capture the physics of more complex motions, such as jumping and crawling, and is limited to the walking motion.

3.2.1.2 Boulic–Thalmann Model

The Boulic–Thalmann model [1] is an empirical model of the human walking motion with a relative walking speed (normalized by the height of leg) from 0 to 3 m/s. The model was derived from inverse kinematic studies from biomechanical experimental measurements of a large number of human subjects. The model defines the analytic expressions that describe the motion trajectories of 17 different body parts such as the head, shoulders, arms, and legs as function of two parameters—the height of the human and the relative velocity of the motion. Additional parameters are provided in the model that can be adjusted to incorporate unique personifications to the motion. Therefore, this model is more realistic in capturing the physics of human walking motion compared to the sinusoidal model.

However, the model possesses several disadvantages. First, the model only simulates walking and does not extend to other activities such as running, jumping, or skipping.

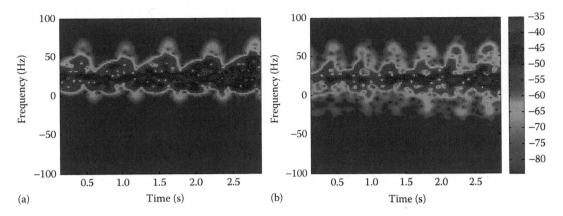

FIGURE 3.1
Micro-Doppler signature for a human walking (a) simulated by using the Boulic–Thalmann model and (b) measured with a 2.4 GHz CW radar.

Second, the model primarily depends on the walking speed and height of the person. Individual nuances in human gait cannot be represented in the kinematic model. Finally, the signatures generated by the Boulic–Thalmann model are quite *clean* with clear contributions from each body part being modeled. In measured signatures, individual components are not nearly so evident. For instance, in Figure 3.1a, the clean signature for a walking person simulated by the Boulic–Thalmann model is shown, whereas a measured result using a 2.4 GHz continuous-wave (CW) radar is shown in Figure 3.1b [11]. The Boulic–Thalmann model successfully captures the micro-Doppler signatures of the key components, such as the torso, legs, and arms. However, amplitudes and frequency distribution of the Doppler shifts are much more consistent than those in measured signatures.

A more practical method for capturing motion activities is to use computer-animated data gathered from motion data captured by active sensors to track animations of a skeletal representation of human body positions in a 3D space. It has been successfully used in industries that use computer vision, and especially, in film making.

3.2.1.3 Kinematic Modeling Using Marker Systems

Films exploit video-based facial gesture recognition and MOCAP systems comprising sensor suites to translate the actions and emotions of actors onto the computer-generated character to be animated. Gross, macroscopic motions are incorporated into computer animations by outfitting the person with garments bearing small, light, highly reflective IR markers or IR light sources placed at various points on the human body. Typically, these markers are affixed to a garment that tightly conforms to the body. Multiple IR cameras observe the motions of the subject, and then extract and track the location of each marker through triangulation. Recordings of marker position versus time are then used to supply the kinematic information required to animate the corresponding skeletal avatar in software.

The primary advantage of using MOCAP is that it enables virtually any activity to be simulated in such a way that individual nuances and statistical variation between trials are also incorporated. Each individual walks in his or her own unique way. With MOCAP, the signature generated from individuals is also unique. Moreover, even multiple recordings of the same activity enacted by the same individual differ. This provides an invaluable source of independent recordings that can be used to test algorithms for realistic simulations.

There are a few sources of MOCAP database for free, which can be exploited for simulation of human motions, including the Advanced Computing Center for Arts and Design at the Ohio State University, Columbus, Ohio [12] and the much larger MOCAP Library of the CMU Graphics Laboratory [2]. Commercial entities also sell their MOCAP database, such as the Japanese company Eyes, which provides both free and priced MOCAP database on the website http://www.mocapdata.com. These databases contain a truly wide range of human motion recordings, including but not limited to walking, running, jumping, climbing on a playground, dancing, walking on uneven terrain, slow walking, brisk walking, playing basketball, playing badminton, playing soccer or tennis, gymnastics, karate, and boxing, to list just a few. Both the CMU and Eyes databases include a large number of recordings of these activities, for example, 70 recordings of soccer, which is necessary to be able to train and test activity recognition algorithms.

3.2.1.4 *Capturing of Human Motion Activities Using Markerless Kinect Sensor*

The unveiling of Microsoft's Kinect sensor in November 2010 [13], which can provide the depth and surface information of objects in the scene, is an inexpensive and markerless MOCAP sensor. Originally designed for the purpose of enabling interactive gaming, the sensor uses gesture recognition to control electronic devices without requiring the hands to touch anything. Kinect is actually equipped with two sensors: a red–green–blue video camera as well as an IR camera. The IR camera operates in conjunction with an IR projector, which casts a grid of IR dots over objects in the camera's field of view. The camera is outfitted with an IR filter enabling it to sense IR returns. Thus, although the human eye cannot see the projected dots, the Kinect system can use these returns to calculate the depth and surface information of objects in the scene.

The technical specifications of the Kinect sensor are dependent in part upon image resolution, the software used to control the sensor, and the distance of the object. The camera has a horizontal field of view of 57° and a vertical field of view of 43° [14], while also equipped with a tilting capability of ±27°. If MATLAB® is used, data may be recorded at a rate of up to 20 frames per second (fps), while Processing™ supplied by Open National Instruments, Austin, TX, is limited to 29 fps. Kinect's color camera has a relatively low resolution of 640 × 408 pixels with 8-bit VGA, but when using a lower frame rate and alternative color formats such as UYVY, the resolution may be improved to as much as 1280 × 1024 pixels.

Several experimental studies have been conducted to independently evaluate the sensing limitations of the Kinect sensors. In [14], it is reported that at a distance of 2.2 m, a spatial precision of 4 pixels (15 mm) was observed in both the x and y directions. The depth resolution was observed to degrade with distance. In particular, measurements made in [15] showed that the random error of depth measurements increased quadratically with increasing distance, reaching as much as 4 cm at a distance of 5 m. Reported in [14], it showed slightly poorer results, with a depth precision of 4 cm at just 5 m distance. Both studies reported that the spacing between the projected IR dots also increases with distance. Although at 1 m the point spacing in depth is just 2 mm, at 3 m this increases to 2.5 cm and at 5 m it is as much as 7 cm [15]. These results are comparable to that found in [14], which reported a resolution of 12 mm at 2 m distance. It was thus recommended that the Kinect sensor be operated at a distance of 1–3 m from the objects of interest, although depth measurements were possible up to a distance of 9.7 m.

In addition to these technical specifications, it is also important to note that the Kinect sensor also suffers from the effects of shadowing—that is, the failure of IR energy to reach the areas blocked by objects—resulting in no depth information being measured from objects residing in the shadowed regions [16]. This becomes an important limitation in applying Kinect to human micro-Doppler simulation because it is possible for the motion of part of the body to shield, or hide, the motion of other body parts (self-shadowing), during activities such as crawling, thereby affecting skeleton tracking results.

Despite these drawbacks, a treadmill test setup was used with software modifications by Erol [3] to successfully simulate human micro-Doppler signatures in 2015. To ensure the operation within the sensing limits of the Kinect, a treadmill was placed between 2 and 3 m from the Kinect sensor, which was elevated to 92 cm from the ground so that the full height of the test subject would be captured, as shown in Figure 3.2a. In this way, the treadmill ensured that longer dwell times could be recorded for activities such as walking and running, without the subject moving out of the field of view. For in-place activities, such as leaping, boxing, and jumping, a treadmill was not required, and instead the subjects were located at a fixed distance of nominally 2 m from the Kinect, as shown in Figure 3.2b. Due to the decrease in elevation of the test subjects, the elevation of the Kinect sensor was likewise reduced to 80 cm from the ground.

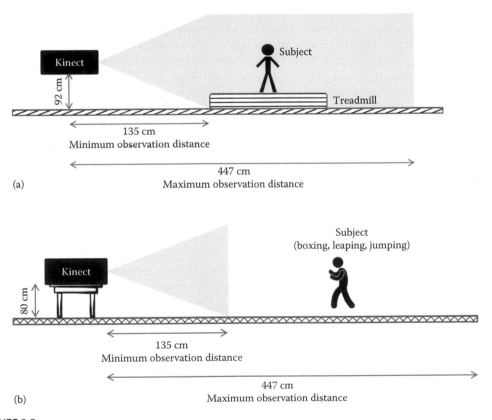

FIGURE 3.2
Experimental test setup for measurement of MOCAP data with Kinect. (a) Configuration for locomotion activities and (b) configuration for in-place activities.

FIGURE 3.3
Initialization of Kinect-based skeleton tracking software with PSI pose.

A key requirement for markerless MOCAP technology is the estimation of human pose and tracking skeleton. Because there are no markers for defining the skeleton, the Kinect-based MOCAP system first defines and initializes a skeleton based on an initial pose given by the tested subject, such as a person standing with legs slightly apart and hands raised upward bent at the elbows like the figure of ψ (psi). This position known as the PSI pose, and its resulting initial skeleton is shown in Figure 3.3.

Several software packages are available to enable skeleton tracking with the Kinect. For example, the Image Acquisition Toolbox Support Package for Kinect, a specific package for MATLAB software that allows users to import Kinect data into MATLAB, access the acquired depth map, and perform functions such as gesture recognition, and face and skeletal tracking. Generating accurate simulations of micro-Doppler signatures using Kinect, however, requires accurate tracking of extremities, such as the feet. The special package is able to supply adequate skeleton tracking but does not have an option for tracking the feet. Thus, in [3], modifications to the MATLAB toolbox were made to enable feet tracking and compensate for the tracking instabilities that occurred due to self-shadowing. The MOCAP data derived from the Kinect-based skeleton tracking is then utilized in generating the time-varying range profiles of each body part, which are required to compute the primitive-based descriptor modeling the expected radar return, as presented in Section 3.2.3. Figure 3.4 shows the micro-Doppler signatures obtained for a 15 GHz CW radar directly observing a person moving toward the radar.

3.2.2 Full-Wave EM Model of RCS of Humans

In addition to kinematics, one of the most important aspects of micro-Doppler signature modeling is computation of human RCS. Human RCS is difficult to ascertain as a person's posture, relative orientation, wave incidence angle, and distance to the radar are constantly changing in the course of walking (or any other type of motion). Moreover, human

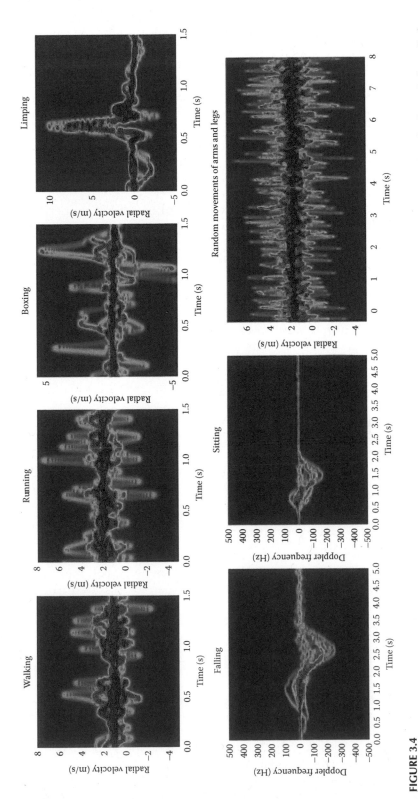

FIGURE 3.4
Example spectrograms generated from Kinect-based micro-Doppler simulator. (Courtesy of TOBB University of Economics and Technology, Ankara, Turkey; Villanova University, Philadelphia, Pennsylvania.)

sizes can significantly vary, from that of a small child to that of a tall man. Even the type of material comprising the clothes a person is wearing can affect the reflectivity properties and observed RCS [17].

In 2007, researchers from the Sensors and Electron Devices Directorate of the U.S. Army Research Laboratory, Adelphi, MD, used Xpatch [18] software developed by Science Applications International Corporation, McClean, VA, to generate computational EM models of the human body and conduct numerical computer simulations of the human body radar signature, analyzing the human RCS in different configurations as functions of aspect angle, frequency, and polarization [19,20]. Xpatch uses high-frequency approximations to model field propagation and scattering, as well as ray tracing and physical optics to compute the radar return from any target mesh the user may draw via a graphical user interface.

A number of important qualitative conclusions regarding the EM properties of human RCS were reported based on the following simulations:

- The RCS of the body as seen from the front varies between −10 and 0 dBsm.
- The kneeling position produces more RCS variations with azimuth angle than that of a person standing, as there are a larger number of scattering centers in the former case.
- There is a strong return from the back, as the body torso is more or less flat from that angle, creating a larger RCS compared to the more curved surfaces characteristic to other parts of the body.
- Angular RCS variation becomes more rapid as the transmitted radar frequency increases.
- The average RCS remains in a tight range of −4 to 0 dBsm for almost all frequencies and all body positions, possibly because the main contribution to the radar return is from the torso.
- Body shape does not have a major influence on average RCS, although there seems to be a dependence on the overall body size.
- Elevation angle can result in increased RCS, especially in the case of V–V polarization, as the body increasingly forms a *corner* resulting in higher reflections.

Additionally, some statistical studies [19] were conducted using these computer models to derive the probability distribution of the RCS for a single human measured from the ground plane, that is, zero elevation angle. In particular, for low frequencies, the RCS was found to follow a Swerling case 3 distribution, whereas for high frequencies, the distribution resembled a Swerling case 1 distribution. The transition between these two cases occurred between 1 and 2 GHz, although body position affected the exact transition frequency. For example, the transition for a kneeling man occurred at a lower frequency than that of a standing man, who continued to exhibit Swerling case 3 properties at 1.8 GHz.

Although computational EM modeling using applications, such as Xpatch, have contributed greatly to the understanding of the human RCS and modeling, it is computationally costly in terms of run-time and memory requirements. A less accurate, but simpler model, which has been shown to be effective especially in indoor applications, is the primitive-based predictor, discussed in Section 3.2.3.

3.2.3 Primitive-Based Predictor

MOCAP databases provide a recording of the position of each marker as a function of time. Together with the knowledge of the temporal characteristics of the recording, the MOCAP data may be used to simulate the micro-Doppler signature that would have been obtained, had a radar been observing the same activity. The earlier work on using MOCAP data to simulate human micro-Doppler signatures includes the work done by Blasch in 2006 [21] and Ram in 2008 [4]. More studies can be found in [22–24]. This is accomplished by first dividing the entire human body into a limited number of parts, *K*, which typically correspond to the total number of markers used by the MOCAP system to make recordings. For a pulsed Doppler radar with linear frequency modulation, each body part or marker is represented as a point scatterer (Figure 3.5) so that the entire human return, s_h, is written as the sum of returns from each individual point scatterer [25]:

$$s_h(n,t) = \sum_{i=1}^{K} a_{t,i} \mathrm{rect}\left(\frac{\hat{t} - t_{d,i}}{\tau}\right) \exp\{j[-2\pi f_c t_{d,i} + \pi\gamma(\hat{t} - t_{d,i})^2]\} \qquad (3.1)$$

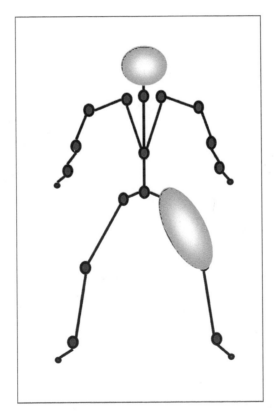

FIGURE 3.5
Primitive-based modeling of a human body as point targets whose RCS is modeled by spheres or ellipsoids.

where:
   τ is the pulse width
   γ is the chirp slope rate
   f_c is the transmitted carrier frequency

The total time elapsed, t, can be written in terms of the pulse repetition interval (PRI), T; the fast time, $\hat{t}$, denotes the time as measured from the start of each PRI; and the number of transmitted pulses n or the slow time are determined by $t = T(n+1)+\hat{t}$. The time delay, $t_{d,i}$ is measured in terms of the total elapsed time t and is defined as the round-trip travel time between the antenna and the ith body part. Thus, the time delay is related to the target range, R, as $t_d = 2R/c$, where c is the speed of light. In this way, the received return may be viewed as a two-dimensional (2D) signal that is a function of the fast time and the slow time. The amplitude, $a_{t,i}$ is defined for each point scatterer by the range equation as

$$a_{t,i} = \frac{G\lambda\sqrt{P_t\sigma_i}}{(4\pi)^{\frac{3}{2}} R_i^2 \sqrt{L_s}\sqrt{L_a}} \qquad (3.2)$$

where:
   G is the antenna gain
   λ is the wavelength
   P_t is the transmitted power
   σ_i is the RCS of the ith body part
   L_s represents system losses
   L_a is the propagation loss

Although this expression includes range- or geometry-dependent factors, such as the antenna gain and atmospheric losses, this work assumes that such factors are constant for each body part. The RCS is modeled according to the approximate shape of the body parts, that is, a sphere for the head, and cylinders or ellipsoids for the remaining parts. Thus, by grouping all factors except for the range, R_i, and the RCS of the ith body part, σ_i, into a constant, A, the amplitude may be expressed as $a_{t,i} = A\sigma_i/R_i^2$.

Once the received human return, as computed in Equation 3.1, is obtained, the result is stored as a matrix of the slow time and fast time. Through pulse compression, the fast-time samples are converted into range and, thus, the matrix becomes range profiles. Radar returns corresponding to the human subject are usually concentrated around a range bin. When the human moves through range cells, known as the range migration, the trajectory of the human motion is spread over range bins. This effect may be mitigated using motion compensation [26] by shifting so as to align the peaks of each pulse return into a single range bin.

A slice across the slow time at the range bin of the peak output is then taken to find the final form of the human return:

$$x_p[n] = \sum_{i=1}^{K} \frac{A}{R_{n,i}} \exp\left\{-j\frac{4\pi f_c}{c} R_{n,i}\right\} \qquad (3.3)$$

The MOCAP data are incorporated within this framework through the range variable, $R_{n,i}$. Note that the phase term is dependent upon the range, whose value changes with each pulse, and therefore results in Doppler and micro-Doppler shifts directly related to the average of the human moving velocity and periodic trajectories of human body parts, respectively. Moreover, depending on the source of MOCAP data, the temporal character-istics of the data may vary. For example, the CMU Motion Library data have a frame rate of 120 Hz. This rate is quite slow for micro-Doppler simulations; thus, to increase the time between each range sample interpolation methods, such as cubic spline, may be used. In [27], additional 10 samples were inserted between the original data points to yield a frame rate of 1200 Hz.

The micro-Doppler signature is a time–frequency representation of how the Doppler modulations vary with time. There are many possible time–frequency distri-butions [28] that could be used to represent the micro-Doppler signature, including the Gabor transform, Wigner–Ville distribution, Gabor–Wigner transform [29,30], and S-transform [31], among others. The Hilbert–Huang transform and empirical mode decomposition [32] in particular have been shown to yield sufficiently high resolution representations to reflect even minute movements, such as hand gestures or breath-ing. However, for most gross human motor movements, the spectrogram is the most ubiquitously used representation, which is defined as the modulus of the short-time Fourier transform (STFT):

$$\text{Spectrogram} \equiv \left| \sum_{-\infty}^{\infty} x_p[n+m]w[m]e^{-j\omega m} \right| \qquad (3.4)$$

where:
$x_p[n]$ is the human return as given in Equation 3.3
$w[m]$ is the windowing function

3.2.4 Joint Simulation and Measurement of Human Motion

In order to validate the simulation methodology, animation data and radar backscat-ter data of a moving human subject were simultaneously collected using an IR-based MOCAP system and a 2.4 GHz CW Doppler radar [4]. The MOCAP system consisted of 16 IR cameras that are used to locate the 3D positions of 48 sensors attached at dif-ferent joints on the human subject as shown in Figure 3.6a. The time-domain anima-tion data are converted to EM radar backscatter using the primitive-based predictor discussed earlier. Both the measured data and the animation data are processed by the STFT to generate the corresponding Doppler spectrograms of the human motions shown in Figure 3.6b and c. Over the 18 s collection interval, the subject undergoes a variety of movements. Doppler of the human is positive when the subject moves toward the radar and negative when the subject moves away from the radar. The motions of the arms and legs modulate the received signal and result in the micro-Doppler features that are observed in the Doppler spectrogram. The spectrograms from both the animation data and the measurement data show broad qualitative agreement. This is particularly discernible in the fine Doppler features arising from the motions of the human subject's legs (e.g., the feature within 6–10 s in the Doppler spectrograms). Due to spectral noise in the measurement data, a few differences in the spectrograms arise. Also, because IR sensors were not placed on the feet of the human

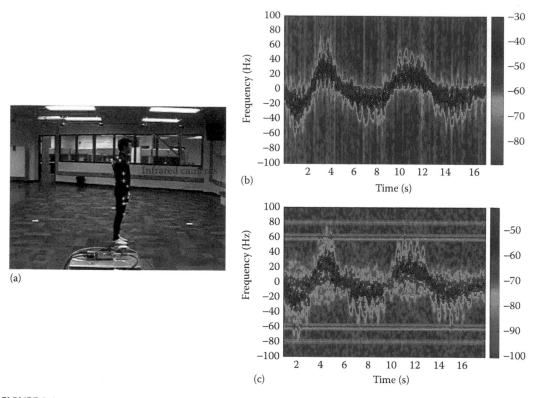

(a)

(b)

(c)

FIGURE 3.6

(a) Generation of IR MOCAP data and Doppler radar data of a moving human subject. Doppler spectrogram of human motions at 2.4 GHz generated from (b) MOCAP data and (c) measurement data.

subject during MOCAP, some of the highest Doppler shift that arise from the feet in Figure 3.6c are not observed in Figure 3.6b.

Using the techniques discussed previously, the micro-Doppler signatures of some common human motions, such as running and crawling, are presented here [4]. The animation data were obtained from Sony Computer Entertainment America, Los Angeles, California, The motions were subsequently replicated under laboratory conditions, and the micro-Doppler data were measured at 2.4 GHz. It is important to note that in these cases, the motions used for obtaining measurement data are not exactly identical to those used during simulation.

3.2.4.1 Running

In this case, the human subject runs around a circular path as shown in Figure 3.7a. The simulated and measured Doppler spectrograms of this motion are shown in Figure 3.7b and c, respectively. It is observed that the Doppler return of the torso is much higher than the torso Doppler while walking due to the increased speed of the body motion. The micro-Doppler spread is also much higher arising from the motions of the different limbs.

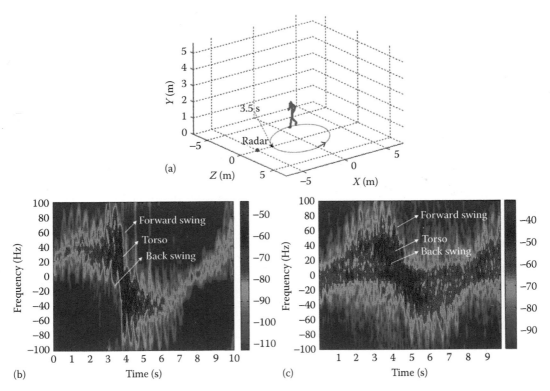

FIGURE 3.7
(a) Animation model of a human running motion from Sony Computer Entertainment America and (b,c) the corresponding Doppler spectrogram at 2.4 GHz.

In particular, it is now possible to observe both the front and the back swing of the lower legs and feet. At the 3.5 s time instant, the Doppler track shows a steep change from positive to negative Doppler shifts. This corresponds to the position of the closest approach to the radar, as indicated in Figure 3.7a.

3.2.4.2 Crawling

The radar is assumed to be at the coordinate position (5, 1, 0) m, and the human crawls toward the radar as shown in Figure 3.8a. Both the simulated and measured Doppler spectrograms of the motion (shown in Figure 3.8b and c) show considerable deviation from the spectrograms obtained from the walking and running motions. First, the torso Doppler is much lower (nearly zero) in this case. Also, the micro-Doppler arising from the legs is lower and is now comparable with the micro-Doppler from the arms. From a detailed analysis of the simulated spectrogram, it is possible to infer that the micro-Doppler of the left/right arm is slightly ahead of the micro-Doppler of the right/left leg.

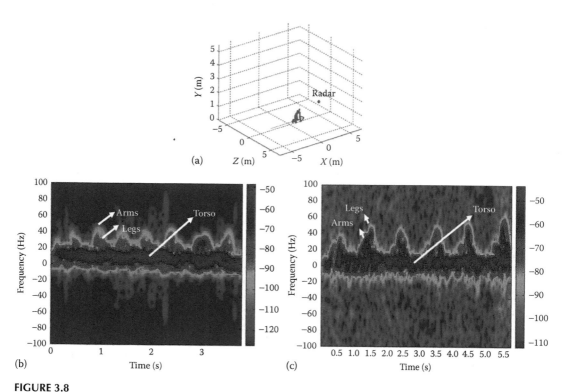

FIGURE 3.8
(a) Animation model of a human crawling motion from CMU Graphics Laboratory MOCAP database and the corresponding, (b) simulated Doppler spectrogram, and (c) the measured spectrogram at 2.4 GHz.

3.3 Modeling and Simulation of Through-the-Wall Radar Signatures

So far, the modeling and simulation of radar signatures has been under free-space conditions. However, in indoor environments, the propagation environment between the human and the radar may be quite complex and introduce significant distortions to the radar signatures. Walls, in particular, are very complex media for wave propagation that introduce attenuation, delay, and multipath to the radar returns. Here we discuss a simulation model that combines human motion scattering mechanisms with through-wall wave propagation physics [33].

Simple ray optic models are suitable for modeling homogeneous walls made of wood or adobe. However, if the walls are inhomogeneous due to the presence of metal reinforcements or air gaps, then a full-wave EM solver is required to model the wall propagation effects. In [33–35], finite-difference time-domain (FDTD) technique was used to model the propagation through three types of walls. They are as follows:

1. A homogeneous concrete wall with a dielectric constant of 7 and a conductivity of 0.0498 S/m

2. A reinforced concrete wall, with a dielectric constant of 7 and a conductivity of 0.0498 S/m, reinforced by square metal conductors that are 2.25 cm thick and 19.75 cm apart

3. An inhomogeneous cinderblock wall with air holes as shown in Figure 3.9

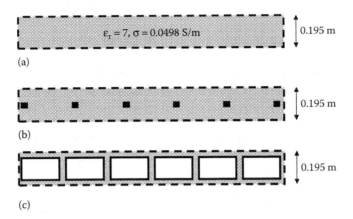

FIGURE 3.9
Models of three different types of walls for FDTD simulations: (a) homogeneous wall, (b) reinforced concrete wall, and (c) cinderblock wall.

The magnitude responses for the two walls at 2.4 GHz are shown in Figure 3.10a–d. Figure 3.10a shows the magnitude for a free-space case (i.e., without the wall) with a carrier frequency of 2.4 GHz. The magnitude response shows the electric field strength decaying as the distance from the source increases. Figure 3.10b shows the magnitude for the simulation space with the presence of a homogeneous concrete wall case. The strength of the electric field again decays as the distance from the pulse source increases. The wall, however, introduces an attenuation of ∼15 dB compared to free space. Also, the magnitude response shows some directionality due to the angle-dependent transmission. Figure 3.10c shows the magnitude response for the reinforced concrete wall case; here the wall introduces an attenuation of ∼10 dB. Also, the multiple scattering introduced by the metal reinforcements inside the wall interferes destructively in some regions. Figure 3.10d shows the magnitude response for the cinderblock wall. Again, the wall introduces significant attenuation on the transmitted signal. Moreover, the interior wall inhomogeneity introduces multipath components that interfere severely in certain regions (for instance, at azimuth angles of 55° and 83°).

Figure 3.11a–d shows the phase responses for free space and the three walls at 2.4 GHz. In Figure 3.11a, for the free space case, the phase response shows a regular circular spread. In Figure 3.11b, it is observed that beyond the homogeneous wall (Y: 0.5–1.5 m), the transmitted wavefront remains a well-behaved circular wavefront throughout the simulation space. This is very similar to the phase response that appears for a wave propagating in free space. However, when the reinforced concrete wall and the cinderblock walls are considered in Figure 3.11c and d, respectively, the complex wavefronts from the multiple reverberations within the wall give rise to significant phase distortions. These are especially severe near the wall.

The authors, in [33], proposed the following method to model the through-wall propagation effects by hybridizing the frequency responses shown in the plots with the EM models of the radar backscatter from the human. This can be done by substituting the two-way free-space propagation factor between the radar and the phase center of each human body primitive, $[\exp(-j2\pi f_c/c\,2r_b)]/r_b^2$, by the square of the complex wall transfer function, $[H_{wall}(f_c, r_b)]^2$, where r_b is the 3D phase center. If the wall transfer functions are derived from 2D EM simulations, the functions must be suitably rescaled to the desired

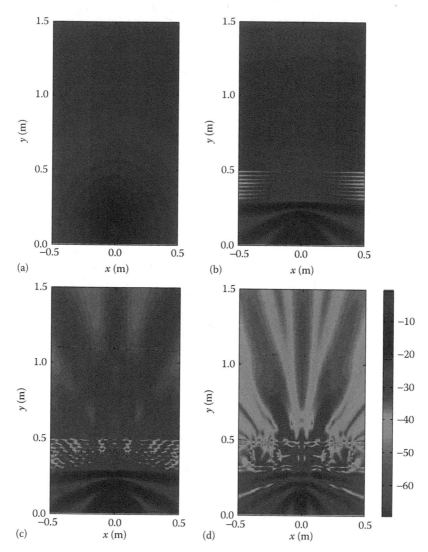

FIGURE 3.10
Results from FDTD simulations: Magnitude response at 2.4 GHz for (a) free space, (b) homogeneous concrete wall, (c) reinforced concrete wall, and (d) cinderblock wall.

3D modeling. First, it is noted that the 2D incident field due to a line source in free space is given by

$$E_i(\rho) = \frac{A'}{\sqrt{\rho}} \exp\left(-j\frac{2\pi f_c}{c}\rho\right)$$

(3.5)

where:
  A' is a constant proportional to the current excitation used to drive the FDTD
  ρ is the 2D position

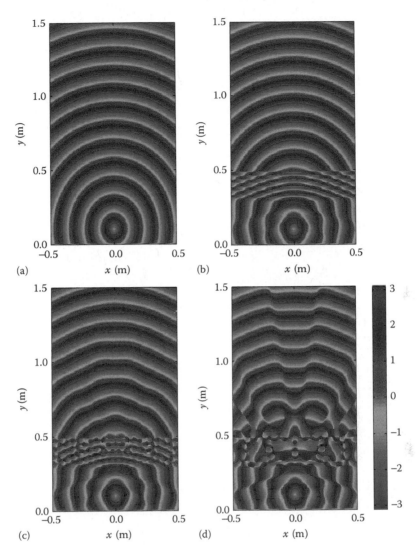

FIGURE 3.11
Results from FDTD simulations: Phase response at 2.4 GHz for (a) free space, (b) homogeneous concrete wall, (c) reinforced concrete wall, and (d) cinderblock wall.

Therefore, to translate the 2D FDTD modeling into three dimensions, the wall transfer function, $H_{wall}(f_c, \rho)$, from the FDTD is rescaled by the factor

$$C_{2D \to 3D} = \frac{\sqrt{\rho}}{r} \frac{1}{A'} \exp\left[-j\frac{2\pi f_c}{c}(r-\rho)\right] \qquad (3.6)$$

The rescaling is derived from free-space considerations and is carried out for the wall transfer function at every point ρ. Note that the FDTD simulation generates the wall transfer coefficients only at the FDTD grid positions. Hence, a bilinear interpolation is carried out to compute, $H_{wall}(f_c, \rho_b)$, more accurately, where $\rho_b(t)$ is the time-varying position coordinate of the bone primitive, $r_b(t)$, projected onto the 2D FDTD simulation space.

Finally, the scattered returns of the human behind the wall are generated by

$$E_s(t) = \frac{A}{\sqrt{4\pi}} \sum_{b=1}^{N} \sqrt{\sigma_b(t)} \quad \{C_{2D \to 3D} \, H_{wall}[f_c, \rho_b(t)]\}^2 \tag{3.7}$$

Here, we present the micro-Doppler signatures of a human walking behind the three types of walls in order to study the wall propagation effects on Doppler. However, in order to confine the motion of the human within the FDTD simulation space, the translation motion component is suppressed. As a result, the Doppler spectrograms only consist of the micro-Doppler components for the free-space case (Figure 3.12a) and the three through-wall cases (Figure 3.12b–d). The spectrograms indicate that the walls attenuate the radar signals to varying degrees. However, the Doppler signals remain almost unchanged with respect to the free space case. This is especially true in the homogeneous wall case. In the two inhomogeneous wall cases, there are however some instances, where the Doppler shifts are severely attenuated. This happens when the corresponding point scatterers on the human fall in those regions where the signal strength is very low due to destructive interference between the multipath components.

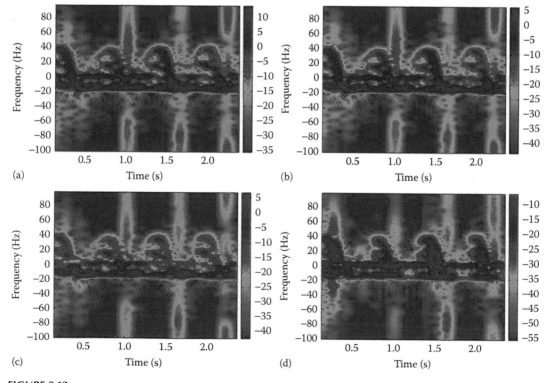

FIGURE 3.12
Micro-Doppler signatures at 2.4 GHz when a human is moving in (a) free space and behind, (b) homogeneous concrete wall, (c) reinforced concrete wall, and (d) cinderblock wall.

3.4 Application of Signatures to Human Indoor Activities

Simulated micro-Doppler signatures have been used in the place of measured data in many applications relating to human detection, classification, imaging, and tracking with radar. In [36,37], the Boulic–Thalmann kinematic model was used both to generate simulated data to test the performance of algorithms and to design a radar detector better matched to human returns than Doppler processing. More recently, micro-Doppler analysis has been heavily employed in especially indoor environments for activity recognition. As radars are often costly in comparison with alternative short-range sensors, it may be difficult and expensive to collect a large amount of data typically required to train and test classifiers for human activity recognition. Alternatively, MOCAP-based simulated micro-Doppler signatures can be easily generated for any desired radar configuration, transmit frequency, bandwidth, observation angle, target motion trajectory, signal-to-noise ratio (SNR), or observation duration (dwell time). In this way, algorithms can be tested for much more scenarios than that could be feasibly obtained through direct measurements. An example is the study in [38] that used Kinect-based MOCAP simulations to investigate the dependent of classification performance on a variety of operational parameters and selection of features.

However, a natural question that may then arise is, how well the simulated signatures actually emulate real data? There are many sources for errors to arise in the MOCAP process, including, but not limited to, the discrepancies that arise due to the human being modeled as a sum of point scatterers, errors due to the accuracy of depth and distance measurements by the MOCAP software, and errors due to self-shadowing or crisscross skeleton tracking, to name a few. Some variations of the basic methodology described here may be used to minimize these sources of error, such as the use of Xpatch [18] to model the EM properties of the human body, as was done in one of the early works on micro-Doppler simulation that utilized the Poser™ system for animation [21]. The answer to the real issue—what is *good enough*—is highly dependent upon the application and objective of the simulations.

Comparisons of algorithm performance may be easily performed on simulated data sets, although there may be a slight difference in absolute performance relative to that obtained on trials with measured data. When applied on the same data set, relative algorithm performance may still be evaluated. Studies involving both Kinect data [38,39] and CMU MOCAP-derived signatures [40,41] have shown that quite high rates of recognition are possible based upon simulated data.

3.4.1 Recognition of Human Motion Activities

To discriminate four classes of human motions—walking, running, creeping, and crawling—a 77 GHz linear frequency-modulated CW radar is used to collect a total of 56 measurements: 18 walking, 15 running, 12 creeping, and 11 crawling. One of the important problems of conducting classification studies using measured data is that typically it is very difficult, expensive, and time intensive to collect large amounts of data. The first step of the classification process usually involves the implementation of a training stage, in which the classifier learns how to categorize the data. A certain number of features are extracted from the training data, and then the distribution of these features are examined to determine the optimal boundaries of classes in the feature space. Some algorithms, such as the support vector machines, apply a linear boundary between classes, whereas classifiers such as *k*-nearest neighbors (kNNs) makes class decisions based upon the major class label possessed by the *k* neighbors nearest in distance to the new data point to be classified. Once the classifier is trained, the remaining data—known as the test data set—are then classified, and the number of correct or incorrect results is recorded as a confusion matrix. In contrast to methods that extract predefined features, deep learning structures, such as convolutional neural networks and auto-encoders, use backpropagation across multiple neural layers to learn features from the training data itself. Deep learning methods have been shown to be highly effective in classifying human activities [42–44], due to the ability of unsupervised feature learning to permit the distinguishing of nuances in the micro-Doppler signature not visually noticeable or differentiable with predefined features. Typically, large data sets are required to accurately gauge classification performance as roughly 60%–70% is separated for training and the remaining 30%–40% is used for testing. Deep learning methods require training data sets magnitudes larger than that required by supervised classifiers using predefined features, and the acquisition of sufficient training data remains the chief obstacle in the practical application of deep learning methods for micro-Doppler classification.

For the four-class problem being considered, a total of three features are extracted from the spectrograms (variance of upper envelope, variance of lower envelope, and mean torso velocity) and supplied to one-nearest neighbor classifier. Using 40 measured signatures for training and the remaining 16 signatures for testing, the resulting confusion matrix is shown in Table 3.1. Not surprisingly, the only confusion was observed between creeping and crawling, two very similar activities; however, overall a correct classification rate of 94% was achieved.

The same four-class problem can also be considered using MOCAP data. Using 378 signatures extracted from the CMU MOCAP database, three features are extracted and supplied to a kNN (with $k = 6$) classifier to achieve the results tabulated in the confusion matrix of Table 3.2. An overall correct classification performance of 92.9% is obtained, quite comparable with the result achieved from the measured signatures.

Another way of comparing the utility of simulated signatures with that of measured signatures is to compare the distribution of the feature extracted for use in the classification process. In this example, consider the distribution of values taken on by the three features extracted. In Figures 3.13 and 3.14, these values are plotted against each other in 3D space and labeled according to the class for CMU MOCAP-derived simulated data and measured data, respectively. Looking at the scatterplot for the simulated data, it may be observed that in the simulated data, walking and running are distinctly differentiable with these three features, appearing in entirely separate groupings. Walking is also mostly distinct from crawling, with just a few outliers from crawling residing in the same feature

TABLE 3.1

Confusion Matrix for Classifying Human Signatures Measured with 77 GHz Radar

	Walking	Running	Crawling	Creeping
Walking	5	0	0	0
Running	0	4	0	0
Crawling	0	0	3	0
Creeping	0	0	1	3

Source: Karabacak, C. et al., *IEEE Geosci. Remote Sens. Lett.*, 12, 2125–2129, 2015.

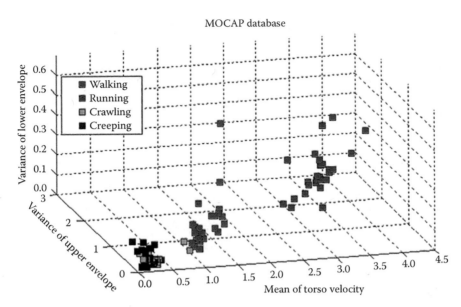

FIGURE 3.13
Scatterplot of feature values extracted from MOCAP-derived simulated signatures. (From Karabacak, C. et al., *IEEE Geosci. Remote Sens. Lett.*, 12, 2125–2129, 2015. With Permission.)

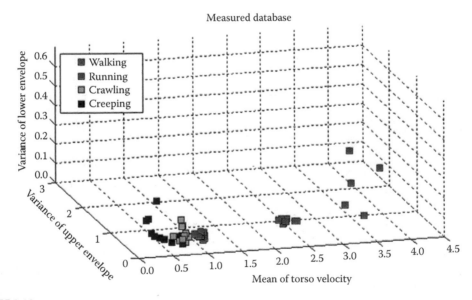

FIGURE 3.14
Scatterplot of feature values extracted from measured signatures. (From Karabacak, C. et al., *IEEE Geosci. Remote Sens. Lett.*, 12, 2125–2129, 2015. With Permission.)

space as walking. Significantly, and not surprisingly, creeping and crawling appear closely intertwined. These observations correspond to the classification results obtained from the simulated data (Table 3.2), which reveals that about 27% of the creeping and crawling signatures are confused with each other.

In contrast, the feature distribution from real, measured signatures reveals that only running separates as a clearly distinct grouping. Creeping, crawling, and walking are

TABLE 3.2

Confusion Matrix for Classifying Human Signatures Simulated from CMU MOCAP Data

	Walking	Running	Crawling	Creeping
Walking	98.72%	0	0	0
Running	3.57%	96.43%	0	0
Crawling	6.90%	0	72.41%	20.69%
Creeping	0	0	6.67%	93.33%

Source: Tekeli, B. et al., *IEEE Trans. Geosci. Remote Sens.*, 54, 2749–2762, 2016.

much more closely grouped. Crawling and walking although separable are quite close in distance, and several creeping and crawling signatures appear intertwined in the feature space, although not nearly as much as in the simulated feature space. Therefore, only 1 out of 16 signatures was confused in the real data.

A similar type of comparison may be made between different sources of MOCAP data as well. For example, consider a comparison of the Kinect-derived MOCAP signatures with the CMU database-derived MOCAP signatures. Table 3.3 shows a comparison of the mean and variance of features averaged over walking and running signatures. The means of features extracted from both the sources of MOCAP are comparable in most cases, with the exception of those features involving the lower envelope. Because MOCAP experiments using Kinect involve the use of a treadmill, as opposed to advancing along a fixed surface, these differences are expected. Interestingly, the variance of most features is higher for CMU database-derived signatures, with the notable

TABLE 3.3

Statistical Properties of Features Extracted from Various Sources of Simulated Spectrograms

	Feature	Kinect-Based Mean	Kinect-Based Variance	CMU-Based Mean	CMU-Based Variance
Walking					
1	Average torso radial velocity	1.0946	8.6997×10^{-5}	1.275	0.0614
2	Bandwidth of torso oscillation	0.5909	0.0041	0.3571	0.0248
3	Maximum of the upper envelope	3.7199	0.0813	4.5325	0.3472
4	Minimum of the lower envelope	0.0116	1.4282×10^{-5}	−0.7614	0.1679
5	Total Doppler bandwidth	3.7315	0.0831	5.2939	0.6812
6	Average of the upper envelope	2.6620	0.0459	3.4781	0.2399
7	Average of the lower envelope	−0.4420	0.0446	−0.4334	0.0371
Running					
1	Average torso radial velocity	1.9366	3.1715×10^{-5}	3.3313	0.3855
2	Bandwidth of torso oscillation	0.6100	0.0021	0.8444	0.3456
3	Maximum of the upper envelope	4.5660	0.1817	8.888	1.5294
4	Minimum of the lower envelope	−0.0044	1.3449×10^{-5}	−1.0233	2.9407
5	Total Doppler bandwidth	4.5704	0.1847	9.9113	6.5362
6	Average of the upper envelope	3.3132	0.0638	7.2686	1.2158
7	Average of the lower envelope	0.4783	0.0638	7/2686	1.2158

Source: Erol, B. et al., *IEEE Aero. El. Sys. Mag.*, 30, 6–17, 2015.

exception of the average torso radial velocity, which is miniscule for the Kinect-derived simulated data—again, an artifact of using a treadmill.

As could be seen from the previous example, however, a critical constraint in the classification of measured spectrograms is the small data size. In this regard, simulated signatures can have a significant contribution by serving as a source of training data. In the previous 77 GHz radar four-class recognition problem, just a total of 56 signatures were collected. Instead of using 40 of those signatures for training, MOCAP-derived simulated signatures were used. In particular, as a training set let us use instead a total of 112 CMU MOCAP-derived signatures with 28 signatures in each class. Then, as a test set, the entire set of 56 measured signatures is used. When the classifier trained on simulated data was tested on all 56 samples of training data, an overall correct classification rate of 93%, just 1% less than when only measured data were used for training (see Table 3.1). As before, the only confusions occurred between creeping and crawling, with four misclassifications as opposed to just one. But, in return, the entire set of 56 measurements was classified, not just a limited set of 16. Thus, in cases where there is a small amount of measurements, simulated signatures may be used to augment the data set and train the classifiers.

The fact that simulated data could be used to train classifiers of real data also serves as a testament to the quality of simulations being generated with MOCAP data. A similar methodology may be utilized to compare the quality of Kinect- and CMU-derived simulated signatures. Consider the three-class problem of walking, running, and leaping discrimination. Normally, this is not a very challenging problem and the data can be easily classified. To evaluate the quality of Kinect signatures, however, a total of 288 Kinect signatures, 96 signatures in each class, were used to train a kNN = 5 classifier. As a test data set, a total of 265 CMU MOCAP-derived signatures were utilized (93 walking, 32 running, and 140 leaping). In this case, despite the distinctness of the classes, confusion between walking and running is significant (25%), and there is even confusion between walking and leaping as well as running and leaping. The overall classification performance of 88% is easily 7%–8% lower than that obtained with comparative studies involving just CMU MOCAP-derived signatures (Table 3.2). This shows that the accuracy of Kinect signatures is less than that of the CMU signatures; nevertheless, the Kinect signatures are sufficiently representative of the underlying motion that high classification results can still be achieved. The trade-off between cost, flexibility, and performance should be considered when selecting sources of MOCAP data for simulating human micro-Doppler signatures.

3.4.2 Doppler-Enhanced Frontal Imaging of Human Indoor Activities

Most of the research and development works, over the last decade, relate to generating top-view (range and cross-range) through-wall images of humans using wideband waveforms and large radar apertures. However, a top-view image does not convey much information regarding human activity. A human is usually narrower along the down-range dimension and larger along the cross-range dimensions. Therefore, the more natural and informative viewing perspective of human activities is the frontal view (which is most similar to our human view). A frontal image of a human may be of great value to law enforcement personnel and for search and rescue missions especially in non-line-of-sight conditions.

We discuss three possible architectures to realize the frontal images of humans as shown in Figure 3.15. These architectures trade-off between cost and complexity and image quality.

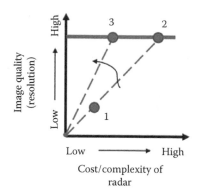

FIGURE 3.15
Radar architectures for frontal imaging of spatially large targets.

3.4.2.1 Case 1: Low-Resolution Imaging with Low-Complexity Radar

Lin and Ling in [46] explored the possibility of using a low-complexity CW Doppler radar with just three antenna elements to generate a frontal image of a human. First, they resolved multiple body parts based on their distinct micro-Doppler. Then, they estimated the azimuth and elevation of each distinct point scatterer, corresponding to a body part, using two-element interferometry. However, the resulting image quality is poor due to its low-resolution characteristics, especially when the micro-Dopplers of multiple body parts overlap.

3.4.2.2 Case 2: High-Resolution Imaging with High-Complexity Radar

High-resolution images of humans can be generated with 2D Fourier processing of spatial measurements gathered with a planar aperture with large number of sensors:

$$y = Fx + \eta \tag{3.8}$$

where:
    x is the image of a human that can be retrieved by Fourier processing of measurements at y elements along a planar aperture
    η is the noise in the measurements

3.4.2.2.1 Real Aperture

Based on Fourier principles, the spatial resolution is improved by a factor of N along both the cross-range dimensions (azimuth and elevation) when the number of elements in each dimension of the planar aperture increases by a factor of N. This is illustrated with an example [47]. Figure 3.16 shows some preliminary results of radar images generated from the simulation software described in Section no. 3.2. Figure 3.16a shows the frontal radar image of a 1.5×1.5 m large human at a stand-off distance of 10 m from a radar. The radar consists of a planar array (real aperture) operating at 30 GHz. The elements in the array are spaced half-wavelength apart to allow for beam scanning. The high-resolution image clearly shows the human torso, and the two arms and legs. The radar aperture is [$40\lambda \times 40\lambda$] resulting in 6400 antenna elements. Therefore, the cost and complexity of this radar is

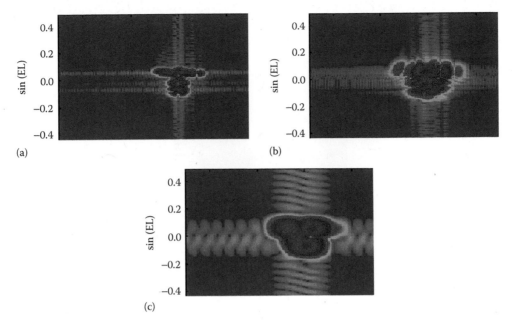

FIGURE 3.16
Frontal radar image of a human captured by array processing using (a) 6400 [80 × 80] elements operating at 30 GHz, (b) 6400 [80 × 80] elements operating at 7.5 GHz, and (c) 400 [20 × 20] elements operating at 30 GHz.

impractically high. Figure 3.16b and c shows the radar images generated with a [40λ × 40λ] aperture at 7.5 GHz and a [10λ × 10λ] aperture at 30 GHz. The reduction of either the carrier frequency (Figure 3.16b) or the aperture size (Figure 3.16c) by a factor of 4 results in significant deterioration of the image quality due to near-field effect and low spatial resolutions.

3.4.2.2.2 Synthetic Aperture

Alternately, synthetic aperture radars can be implemented with a single mobile element and a broadband signal. The image quality is governed by the number of sampling positions of the synthetic aperture radar (SAR) across the aperture and the carrier frequency. However, SAR images of moving targets are considerably distorted if the data acquisition duration, of a large number of sampling positions across a single aperture, is long. Therefore, this radar architecture is impractical for realizing high-quality radar images of moving humans.

3.4.2.3 Case 3: Desired High-Resolution Imaging with Medium- to Low-Complexity Radar

The desired goal is to generate high-resolution radar images that retain the image quality of Case 2 with significant reduction in the complexity of the radar architecture as shown in Figure 3.15. One technique to achieve this is by augmenting the 2D spatial Fourier processing in Case 2 with Doppler processing from Case 1. Here, the micro-Doppler data, arising from the dynamic movements of the arms and legs, are used to resolve some of the point scatterers on the human body. The additional Doppler dimension enables the relaxation of the resolution criteria across the cross-range dimensions. In [47], a CW Doppler radar, operating at 7.5 GHz with a uniform planar radar aperture with 20 × 20 elements where the elements are spaced half-wavelength apart, is considered. The radar data is Fourier

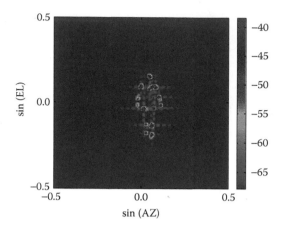

FIGURE 3.17
Radar image of a walking human generated by joint Doppler processing and 2D beam forming with a [20 × 20] array.

transformed across three dimensions—Doppler, azimuth, and elevation. Then the image is generated by the complex sum of the 2D cross-range point spread responses of each distinct scatterer in the 3D Fourier space. Figure 3.17a shows the Doppler-enhanced frontal radar image of a walking human generated from processing narrowband data over a duration of 0.5 s. Though the number of antenna elements has reduced from 6400 to 400, we are able to still distinguish the arms and legs of the human.

The main advantage of Doppler-enhanced frontal imaging is that radar operators may be able to directly infer a wide variety of human activities without the assistance of complicated machine learning algorithms [48]. For instance, Figure 3.18a shows the Doppler-enhanced image of a human crawling on the floor. The ground truth image is provided in the inset of the figure. It becomes apparent that the radar image is able to capture the head, one arm, the torso, and one leg. Figure 3.18b shows the radar image of a human skipping along with the ground truth image in the inset. This figure shows both arms, the torso, and one leg as well as the head. The most interesting feature of these images is that they capture the bobbing motion of the head during skipping.

The biggest limitation of Doppler-enhanced imaging is that it cannot be used to image stationary humans (with zero Doppler shifts). The performance also deteriorates when

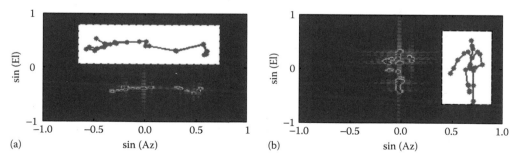

FIGURE 3.18
Doppler-enhanced images along with ground truth insets of (a) crawling and (b) skipping motions.

the human moves tangentially with respect to the radar. Further, when this technique is applied to through-wall imaging, the problem is complicated by external factors pertaining to the complex wall propagation medium [49]. Walls may introduce the attenuation of the radar signal resulting in low SNR levels, refraction of the radar signal resulting in error in the azimuth and elevation estimation, interference from multipath components resulting in clutter especially at short ranges, and nonlinear behavior because building materials are mostly dispersive.

References

1. R. Boulic, N. M. Thalmann, and D. Thalmann, A global human walking model with real-time kinematic personification, *The Visual Computer*, 6: 344–358, 1990.
2. Website of the Motion Research Laboratory, Carnegie Mellon University: http://mocap.cs.cmu.edu (accessed on January 3, 2016).
3. B. Erol, C. Karabacak, and S. Z. Gürbüz, A kinect-based human micro-doppler simulator, *IEEE Aerospace and Electronic Systems Magazine*, 30(5): 6–17, 2015.
4. S. S. Ram and H. Ling, Simulation of human microDopplers using computer animation data, *IEEE Radar Conference*, Rome, Italy, 2008, pp. 1–6.
5. V. C. Chen, *The Micro-Doppler Effect in Radar*. Artech House, London, 2011.
6. Q. Wan, Y. Li, C. Li, and R. Pal, Gesture recognition for smart home applications using portable radar sensors, *36th Annual International Conference of IEEE Engineering in Medicine and Biology Society*, 2014, pp. 6414–6417.
7. J. Lien, N. Gillian, M. Karagozler, P. Amihood, C. Schwesig, E. Olson, H. Raja, and I. Poupyrev, Soli: Ubiquitous gesture sensing with millimeter wave radar, *ACM Transaction on Graphics*, 35(4): 142, 2016.
8. P. Molchanov, S. Gupta, and K. Pulli, Short-range FMCW monopulse radar for hand-gesture sensing, *Proceedings of the IEEE Radar Conference*, 2015, pp. 1491–1496.
9. J. A. Nanzer and K. S. Zilevu, Dual interferometric-Doppler measurements of the radial and angular velocity of humans, *IEEE Transactions on Antennas and Propagation*, 62(3): 1513–1517, 2014.
10. J. A. Nanzer, Micro-motion signatures in radar angular velocity measurements, *2016 IEEE Radar Conference*, IEEE Aerospace and Electronic Systems Society, Philadelphia, PA, 2016, pp. 1–4.
11. S. S. Ram and H. Ling, Analysis of micro-Dopplers from human gait using reassigned joint time-frequency transform, *IET Electronics Letters*, 43(23): 1309–1311, 2007.
12. Website of the ACCAD, Ohio State University: https://accad.osu.edu/researchmain/gallery/project_gallery.html?tags=Motion+Capture (accessed on December 24, 2005).
13. Microsoft News Center, The future of entertainment starts today as kinect for Xbox 360 leaps and lands at retailers nationwide, November 4, 2010. https://news.microsoft.com.
14. M. R. Andersen, T. Jensen, P. Lisouski, A. K. Mortensen, M. K. Hansen, T. Gregersen, *Kinect Depth Sensor Evaluation for Computer Vision Applications*, Technical Report ECE-TR-6, Aarhus University, Department of Engineering, Aarhus, Denmark, February 2012.
15. K. Khoshelham and S. O. Elberink, Accuracy and resolution of Kinect depth data for indoor mapping applications, *IEEE Sensors*, 12: 1437–1454, 2012.
16. G. Borenstein, *Making Things See: 3D Vision with Kinect, Processing, Aduino and MakerBot*, Maker Media, Sebastopol, CA, 2012.
17. N. Yamada, Y. Tanaka, and K. Nishikawa, Radar cross section for a pedestrian in 76 GHz band, *2005 European Microwave Conference*, IEEE Paris, France, 2, October 4–6, 2005.

18. T. Dogaru and C. Le, Validation of Xpatch computer models for human body radar signature, *U.S. Army Research Laboratory Report ARL-TR-4403*, U.S. Army Research Lab, Adelphi, MD, March 2008.

19. T. Dogaru, L. Nguyen, and C. Le, Computer models of the human body signature for sensing through the wall radar applications, *U.S. Army Research Laboratory Technical Report ARL-TR-4290*, U.S. Army Research Lab, Adelphi, MD, September 2007.

20. C. Le and T. Dogaru, Numerical modeling of the airborne radar signature of dismount personnel in the UHF-, L-, Ku-, and Ka-bands, *U.S. Army Research Laboratory Technical Report ARL-TR-4336*, U.S. Army Research Lab, Adelphi, MD, December 2007.

21. E. Blasch, U. Majumder, and M. Minardi, Radar signals dismount tracking for urban operations, *Proceedings of the International Society of Optics and Photonics (SPIE)*, SPIE; Orlando, Florida, 2006, pp. 623504–623504.

22. L. Fei, H. Binke, Z. Hang, and D. Hao, Human gait recognition using micro-Doppler features, *5th Global Symposium on Millimeter Waves*, IEEE, Harbin, China, 2012, pp. 326–329.

23. Y. Yang and C. Lu, Human identifications using micro-Doppler signatures, *Proceedings of the Fifth IASTED International Conference on Antennas, Radar and Wave Propagation*, ACTA Press, Anaheim, CA, 2008, pp. 69–73.

24. J. Park, Multi-frequency radar signatures of human motion: Measurements and models, PhD Dissertation, Department of Electrical and Computer Engineering, Ohio State University, Columbus, OH, 2012.

25. P. van Dorp and F. C. A. Groen, Human walking estimation with radar, *IET Radar, Sonar and Navigation*, 150(5): 356–365, 2003.

26. W. G. Carrara, R. S. Goodman, and R. M. Majewski, *Spotlight Synthetic Aperture Radar: Signal Processing Algorithms*, Artech House, Norwood, MA, 1995.

27. C. Karabacak, S. Z. Gürbüz, and A. C. Gürbüz, Radar simulation of human micro-Doppler signature from video motion capture data, *Proceedings IEEE Signal Processing and Communications Applications Conference*, 2013.

28. L. Stankovic, M. Dakovic, T. Thayaparan, Time-Frequency Signal Analysis with Applications, Artech House, Boston, MA, 2013.

29. P. Suresh, T. Thayaparan, S. SivaSankaraSai, K. S. Sridharan, and K. Venkataramaniah, Gabor-Wigner transform for micro-Doppler analysis, *Proceedings of the 9th International Radar Symposium*, Bangalore, India, December 2013, pp. 10–14.

30. P. Suresh, K. Venkataramaniah, and T. Thayaparan, Analysis of micro-Doppler radar signatures of rotating targets using Gabor-Wigner transform, *International Journal of Innovative Research in Science, Engineering and Technology*, 3(1): 1295–1299, 2014.

31. Z. Sun, J. Wang, C. Yuan, Y. Bi, and H. Xiang, Parameter estimation of walking human based on micro-Doppler, *Proceedings of the 12th International Conference on Signal Processing (ICSP)*, IEEE, Hangzhou, China, October 2014.

32. D. P. Fairchild and R. M. Narayanan, Classification of human motions using empirical mode decomposition of human micro-Doppler signatures, *IET Radar, Sonar, Navigation*, 8(5): 425–434, 2014.

33. S. S. Ram, C. Christianson, and H. Ling, Simulation of high resolution range profiles of humans behind walls, *Proceedings of the IEEE Radar Conference*, IEEE, Pasadena, CA, 2009.

34. S. S. Ram, Radar simulator for monitoring human activities in non-line-of-sight environments, PhD Dissertation, Department of Electrical and Computer Engineering, University of Texas at Austin, Austin, TX, 2009.

35. S. S. Ram, C. Christianson, Y. Kim, and H. Ling, Simulation and analysis of human micro-Dopplers in through-wall environments, *IEEE Transactions on Geoscience and Remote Sensing*, 48(4): 2015–2023, 2010.

36. S. Z. Gürbüz, W. L. Melvin, and D. B. Williams, A non-linear phase model-based human detector for radar, *IEEE Transactions on Aerospace and Electronic Systems*, 47(4): 2502–2513, 2011.

37. S. Z. Gürbüz, W. L. Melvin, and D. B. Williams, Kinematic model-based human detectors for multi-channel radar, *IEEE Transactions on Aerospace and Electronic Systems*, 48(2): 1306–1318, 2012.

38. S. Z. Gurbuz, B. Erol, B. Cagliyan, and B. Tekeli, Operational assessment and adaptive selection of micro-Doppler features, *IET Radar, Sonar, and Navigation,* 9(9): 1196–1204, 2015.
39. B. Erol, M. Amin, Z. Zhou, and J. Zhang, Range information for reducing fall false alarms in assisted living, *Proceedings of the IEEE Radar Conference,* IEEE, Philadelphia, PA, 2016.
40. B. Tekeli, S. Z. Gürbüz, and M. Yüksel, Information theoretic feature selection for human micro-Doppler signature classification, *IEEE Transactions on Geoscience and Remote Sensing,* 54(5): 2749–2762, 2016.
41. B. Erol and S. Z. Gürbüz, Hyperbolically-warped cepstral coefficients for improved micro-Doppler classification, *Proceedings of the IEEE Radar Conference,* IEEE, Philadelphia, PA, 2016.
42. Y. Kim and T. Moon, Human detection and classification based on micro-Doppler signatures using deep convolutional networks, *IEEE Geoscience and Remote Sensing Letters,* 13(1): 8–12, 2016.
43. B. Jokanovic, M. Amin, and F. Ahmad, Radar fall motion detection using deep learning, *Proceedings of the IEEE Radar Conference,* 2016.
44. M. S. Seyfioglu, S. Z. Gurbuz, A. M. Ozbayoglu, and M. Yuksel, Deep learning of micro-Doppler features for aided and unaided gait recognition, *Proceedings of the IEEE Radar Conference,* 2017.
45. C. Karabacak, S. Z. Gürbüz, A. C. Gürbüz, M. B. Guldogan, G. Hendeby, and F. Gustafsson, Knowledge exploitation for human micro-Doppler classification, *IEEE Geoscience and Remote Sensing Letters,* 12(10): 2125–2129, 2015.
46. A. Lin and H. Ling, Doppler and direction-of-arrival (DDOA) radar for multiple-mover sensing, *IEEE Transactions on Aerospace and Electronic Systems,* 43(4): 1496–1509, 2007.
47. S. S. Ram and A. Majumdar, High-resolution radar imaging of moving humans using Doppler processing and compressed sensing, *IEEE Transactions on Aerospace and Electronic Systems,* 51(2): 1279–1287, 2015.
48. S. S. Ram, Doppler enhanced frontal radar images of multiple human activities, *Proceedings of the IEEE Radar Conference,* 2015.
49. S. S. Ram and A. Majumdar, Through-wall propagation effects on Doppler-enhanced frontal radar images of humans, *Proceedings of the IEEE Radar Conference,* 2016.

4

Continuous-Wave Doppler Radar for Fall Detection

Yimin D. Zhang and Dominic K. C. Ho

CONTENTS

4.1 Introduction

Falls are the leading cause of injuries for the elderly. Prompt fall detection saves lives, leads to timely interventions and most effective treatments, and reduces medical expenses (Sadigh et al. 2004, Igual et al. 2013). Driven by a pressing need to detect and attend to a fall, elderly fall detection has become an active area of research and development. Different sensing modalities, including inertial measurement unit-based wearable devices, video camera, and radar, have been proposed for this purpose (Bagala et al. 2012, Liu et al. 2012, Wu et al. 2013, 2015, Amin et al. 2016, Bennett et al. 2016). Among them, wearable devices have shortcomings that they are intrusive, are easily broken, must be carried, and are ineffective if forgotten to be worn. However, the benefits of vision-based methods are mitigated by privacy issues. As a result, radar has emerged as an important technology for health monitoring and fall detection in elderly assisted living. In particular, the attractive attributes of radar, related to its proven technology, nonobstructive illumination, nonintrusive sensing, insensitivity to lighting conditions, privacy preservation, and safety, have brought electromagnetic waves to the forefront of indoor monitoring modalities in competition with cameras and wearable devices.

A radar system, in general, performs target detection, localization, and classification by transmitting sensing signals and analyzing the received signals reflected or backscattered from targets and surroundings. The most important information measured by a radar system includes target range and Doppler frequency (Skolnik 2002). In the underlying application of fall detection, the Doppler frequency describes the change of frequency between the transmitted and received signals due to the relative motion between the human subject and the radar system (Chen and Ling 2002). In addition, the phenomenon of higher Doppler frequencies generated by the fast moving parts of the human body, such as the arms and legs, is referred to as a micro-Doppler effect (Chen 2011). Because such time-varying Doppler and micro-Doppler frequencies are directly associated with the motion and maneuvering behaviors of the human body and different parts, we can utilize these Doppler and micro-Doppler signatures to characterize and classify such motions and maneuvering patterns, and perform fall detection.

Depending on the signal bandwidth and the type of information collected and utilized, radar systems used for human motion classification and fall detection can be classified into two main categories: continuous-wave (CW) or narrowband radar that is primarily used to reveal the Doppler and micro-Doppler signatures (Liu et al. 2012, Wu et al. 2013, 2015) and (ultra-)wideband radar that can provide high-resolution range information (Shingu et al. 2008). This chapter focuses on the former, that is, narrowband radar, which is often also referred to as Doppler radar.

Radar backscatter from humans in motion generates changes in the radar frequencies caused by the Doppler effect. A Doppler radar obtains target Doppler information by observing the phase variation of the return signal from the targets corresponding to repetitively transmitted signals. As such, Doppler radars have the capability to separate moving targets from stationary background, whose signals are collectively referred to as clutter.

In Doppler radar, the radar signal returns of humans differ in their Doppler characteristics depending on the nature of the gross human motor activities. These signals are nonstationary in nature, inviting time–frequency analysis in its both linear and bilinear for processing. Their time–frequency representations play a fundamental role in motion identification, including fall motion detection and classification. A natural or unintentional human fall is a downward motion typically not exceeding a speed of about 5 m/s. The motion dynamics of a human fall creates a distinctive Doppler radar response that can be exploited for fall detection. The Doppler signatures determine the prominent features that underlie different human motions and activities.

In the following, we first provide in Section 4.2 the radar signal model and introduce several time–frequency analysis techniques that can be used for human fall detection. The radar Doppler signatures from various kinds of human fall and non-fall motions are then presented in Section 4.3. Section 4.4 describes a machine learning approach for fall detection that involves prescreening, feature extraction, and fall versus non-fall classification. Experimental results are then shown in Section 4.5. Section 4.6 elaborates on the benefit of fusing radar and infrared sensor measurements to improve fall detection, and a conclusion summarizes this chapter is provided in Section 4.7.

4.2 Signal Model and Signal Analysis Domains

4.2.1 Signal Model

For simplicity, we consider a monostatic CW radar that transmits a sinusoidal signal $s(t) = \exp(j2\pi f_c t)$ with a carrier frequency f_c over the sensing period. For a point target that is initially located at a distance of R_0 from the radar at time $t = 0$ and moves with a velocity $v(t)$ in a direction forming an angle $\theta(t)$ with the radar line of sight, the distance between the radar and the target at time instant t is given by

$$R(t) = R_0 + \int_0^t v(u)\cos(\theta(u))\mathrm{d}u \tag{4.1}$$

The radar return scattered from the target can be expressed as

$$x_a(t) = \rho \exp\left[j2\pi f_c\left(t - \frac{2R(t)}{c} \right) \right] \tag{4.2}$$

where:
  ρ is the target reflection coefficient
  c is the velocity of the electromagnetic wave propagation in free space

The Doppler frequency corresponding to $x_a(t)$ is given by $f_D(t) = 2v(t)\cos(\theta(t))/\lambda_c$, where $\lambda_c = c/f_c$ is the wavelength. A spatially extended target, such as a human, can be approximated as a collection of point scatterers. Therefore, the corresponding radar return is the integration over the target region Ω and is expressed as

$$x(t) = \int_\Omega x_a(t)\mathrm{d}a \tag{4.3}$$

In this case, the Doppler signature is the superposition of all component Doppler frequencies. Torso and limb motions generally generate time-varying Doppler frequencies, and the nature of this variation defines the Doppler signature associated with each human activity, including a fall.

The exact Doppler signature depends on the target shape and motion patterns. For instance, if the radar is mounted in the ceiling and the fall is directly underneath, at a carrier frequency of 6 GHz, the maximum Doppler frequency corresponding to a 5 m/s motion is about 200 Hz. If the radar is mounted on a wall at a height of 1 m and the fall is straight down 3 m away from the radar, the maximum Doppler shift reduces to 63 Hz. Indeed, the location of the radar affects the fall signature and the fall detection performance (Liu et al. 2012).

4.2.2 Time–Frequency Analyses

Time–frequency analysis is a natural tool that reveals the time-varying Doppler and micro-Doppler signatures with enhanced signal energy concentration. A number of methods are available to perform time–frequency analysis of the Doppler and micro-Doppler signatures. These methods can be generally divided into two classes: linear time–frequency analysis and quadratic time–frequency analysis. The short-time Fourier transform (STFT) is a commonly used technique to perform linear time–frequency analysis (Boashash 2003). Its squared magnitude, referred to as the spectrogram, is related to a large class of quadratic time–frequency representations (Boashash et al. 2015).

The STFT of the data $x(t)$ with a window $h(t)$ is expressed as

$$S(t,f) = \sum_{m=-\infty}^{\infty} h(m)x(t-m)\exp(-j2\pi fm) \tag{4.4}$$

where f is the frequency index. The spectrogram is obtained by computing the squared magnitude of STFT, expressed as follows:

$$D(t,f) = \left|S(t,f)\right|^2 = \left|\sum_{m=-\infty}^{\infty} h(m)\,x(t-m)\exp(-j2\pi fm)\right|^2 \tag{4.5}$$

However, the quadratic class of time–frequency distributions (TFDs) of a signal $x(t)$ is defined as the 2D Fourier transform of its kernelled ambiguity function, expressed as follows:

$$D(t,f) = \sum_{\theta=-\infty}^{\infty}\sum_{\tau=-\infty}^{\infty} \phi(\theta,\tau)A(\theta,\tau)\exp(j4\pi f\tau - j2\pi\theta t) \tag{4.6}$$

where:

$$A(\theta,\tau) = \sum_{u=-\infty}^{\infty} x(u+\tau)x^*(u-\tau)\exp(-j2\pi\theta u) \tag{4.7}$$

is the ambiguity function, and $\phi(\theta,\tau)$ is the time–frequency kernel. Here, θ and τ, respectively, denote the frequency shift (also referred to as Doppler frequency) and the time lag. The properties of a quadratic TFD are heavily affected by the applied kernel.

The Wigner–Ville distribution (WVD) is often regarded as the basic or prototype quadratic TFD, because the other classes of quadratic TFDs can be described as filtered versions of the WVD, whereas the effective kernel function of the WVD can be considered as unity across the entire ambiguity function. WVD is known to provide the best time–frequency resolution for single-component linear frequency-modulated signals, but it yields undesirable cross-terms when the frequency law is nonlinear or when a multi-component signal is considered.

Various reduced interference distribution (RID) kernels have been developed to reduce the cross-term interference. A RID kernel $\phi(\theta,\tau)$ acts as a filter and places different weightings on the ambiguity function. The majority of signals have auto-terms located near the origin in the ambiguity domain, whereas the signal cross-terms are distant from the time lag and frequency shift axes. As such, RID kernels exhibit low-pass filter characteristics to suppress cross-terms and preserve auto-terms. One of the kernels that are found useful in the analysis of fall detection is the extended modified B distribution (Amin et al. 2015, Boashash et al. 2015).

4.2.3 Wavelet Transform

Another important technique for nonstationary spectrum estimation is the time–scale analysis implemented using the wavelet transform (WT). Like the STFT, the WT uses the inner products to measure the similarity between a signal and an analyzing function. Compared to the STFT, which uses a fixed window function to capture the local frequency components, the WT uses multiresolution windows that change the position and scaling of the mother wavelet and therefore captures the short-duration, high-frequency components, and long-duration, low-frequency components (Mallat 2009). According to the uncertainty principle (Cohen 1995), the product of the time resolution and the frequency resolution is lower bounded, that is, we cannot achieve a high resolution in both the time domain and the frequency domain simultaneously. By changing the position and scaling of the mother wavelet function, the WT implements the multiresolution concept capturing both the short-duration, high-frequency components, and long-duration, low-frequency components (Qian 2001).

Given a mother wavelet function $\psi(t)$, the WT of a signal $x(t)$ is defined as

$$\xi(\tau, a) = \frac{1}{\sqrt{a}} \int_t x(t) \psi^* \left(\frac{t - \tau}{a} \right) dt \tag{4.8}$$

where:
  τ is the translation
  $a > 0$ is the scale parameter
  * denotes complex conjugation

When a is limited to dyadic scales (scales that are integer powers of 2), the WT can be implemented efficiently in a discrete form that is often termed as the stationary WT (SWT) (Mallat 2009).

In the underlying fall detection application, the WT is considered useful in capturing the high Doppler frequency components of the fall while protecting the low-frequency components in the data (Amin et al. 2015, Su et al. 2015). Similar to the spectrogram, the square magnitude of the WT is generally referred to as scalogram.

4.3 Radar Fall and Non-Fall Doppler Signatures

We shall gain some insights into the Doppler signatures of falls and non-falls. The Doppler signatures presented below come from two separate radars. The first set of signatures presented in Section 4.3.1 is from a CW radar, and the second in Section 4.3.2 is from a pulse-Doppler radar with a lower carrier frequency. The data from these two radars are processed for fall detection, and the results are presented in Section 4.5.

4.3.1 Doppler Signatures from a CW Radar

Figure 4.1 shows the spectrogram of selected fall and non-fall motion patterns (Wu et al. 2015). The first four patterns in Figure 4.1a–d, namely, fall backward, fall backward with arm motion, fall forward, and fall forward with arm motion, are collectively considered as falls, whereas the last four patterns in Figure 4.1e–h, namely, sit and stand, fast sit and

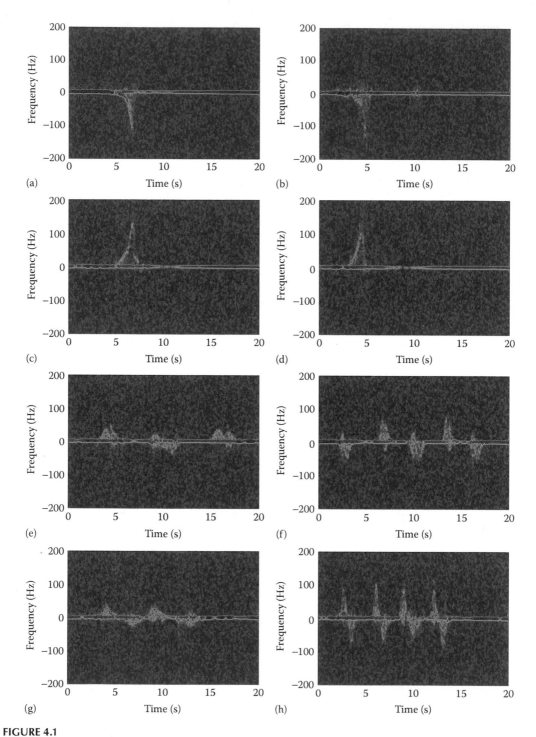

FIGURE 4.1
Doppler signatures of selected fall and non-fall activities. (a) Fall backward, (b) fall backward with arm motion, (c) fall forward, (d) fall forward with arm motion, (e) sit and stand, (f) fast sit and stand, (g) stand and sit, and (h) fast bend and stand up. (From Wu, Q. et al., *IET Radar Sonar Nav.*, 9, 164–172, 2015; Courtesy of the Institution of Engineering and Technology, London, U.K.)

stand, stand and sit, and fast bend and stand up, are collectively considered as non-fall motions. The data were collected using a CW radar, implemented using an Agilent E5071B RF network analyzer, in the Radar Imaging Lab at Villanova University, Villanova, Pennsylvania. A carrier frequency of 8 GHz was employed and the network analyzer was externally triggered at a 1 kHz sampling rate. The feed point of the antenna was positioned 1.15 m above the floor. Data were collected for eight different motion patterns using the two test subjects, with each experiment repeated 10 times (5 times each for the two test subjects). The motion patterns considered include (1) forward falling, (2) backward falling, (3) sitting and standing, and (4) bending over and standing up. Two different variations of each motion pattern were measured, one being a typical motion of such type, whereas the other demonstrating a high-energy form of the same motion to study the impact of such variations on the classification performance. The recording time for each experiment was 20 s (Wu et al. 2015). From Figure 4.1, it is observed that these different motion patterns show distinct Doppler signatures to permit classification, as will be discussed in Section 4.5.1.

4.3.2 Doppler Signatures from a Pulse-Doppler Radar

Next, we examine 21 types of falls, listed in Table 4.1, which are frequently incurred by seniors. The time-domain waveforms and the corresponding spectrograms of these falls are provided in Figure 4.2. It is instrumental to examine the Doppler signatures of human falls of a Doppler radar. The signatures presented here are obtained using an inexpensive commercially available pulse-Doppler radar that has a size and shape similar to a cup.

TABLE 4.1

Types of Human Falls

Fall Type Number	Types of Human Fall
1	Loose balance—forward fall
2	Loose balance—backward fall
3	Loose balance—left fall
4	Loose balance—right fall
5	Loss consciousness—forward fall
6	Loss consciousness—backward fall
7	Loss consciousness—left fall
8	Loss consciousness—right fall
9	Loss consciousness—straight-down fall
10	Trip and fall—forward fall
11	Trip and fall—sideways fall
12	Slip and fall—forward fall
13	Slip and fall—sideways fall
14	Slip and fall—backward fall
15	Reach-fall (chair)—forward fall
16	Reach-fall (chair)—left fall
17	Reach-fall (chair)—right fall
18	Reach-fall (chair) sliding—forward fall
19	Reach-fall (chair) sliding—backward fall
20	Lying on couch fall—upper body first fall
21	Lying on couch fall—hip first fall

The carrier frequency is 5.8 GHz, and the pulse repetition frequency is 10 MHz with a duty cycle of 40%. The radar produces an output proportional to the amount of Doppler frequency observed. The radar output is sampled at 960 Hz for processing.

The data were collected in a laboratory room at the University of Missouri, Columbia, Missouri, that simulates a home environment. The size of the room was $9 \times 8.2 \times 3$ m (length × width × height). The radar was mounted on the ceiling center facing down. The falls were performed by professional stunt actors who were trained to act and fall like elder people.

We observe from Figure 4.2 that the typical duration of a fall that is natural and unintentional is about 2–2.5 s long. Such duration appears to be independent of the age and physique of individuals in the data collected. Compared to the time-domain waveform, the spectrogram clearly reveals the signatures for different types of falls. The signatures are sensitive to the fall type and fall direction. The reach fall from chairs appears to have lower frequency content than the other falls.

For comparison purpose, Figure 4.3 illustrates the behavior of several non-fall activities. The time-domain signal returns are strong and have large magnitude. However, their spectral content is often <100 Hz, which is in contrast to the signals from falls that have spectral content often beyond 100 Hz. The non-fall spectrograms are distinguishable from those of falls.

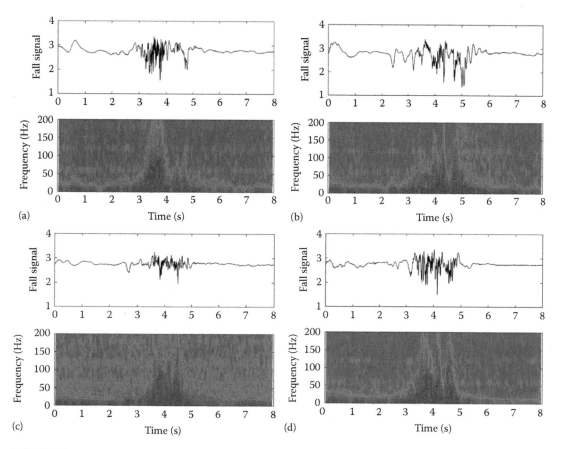

FIGURE 4.2
Time-domain signal and spectrogram of different falls. (a) Loose balance—forward fall, (b) loose balance—backward fall, (c) loose balance—left fall, (d) loose balance—right fall.   (*Continued*)

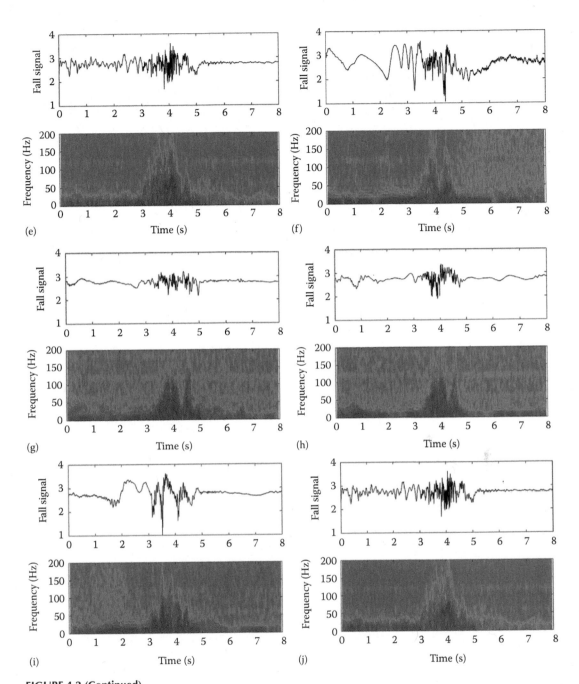

FIGURE 4.2 (Continued)
Time-domain signal and spectrogram of different falls. (e) Loss consciousness—forward fall, (f) loss consciousness—backward fall, (g) loss consciousness—left fall, (h) loss consciousness—right fall, (i) loss consciousness—straight-down fall, (j) trip and fall—forward fall. *(Continued)*

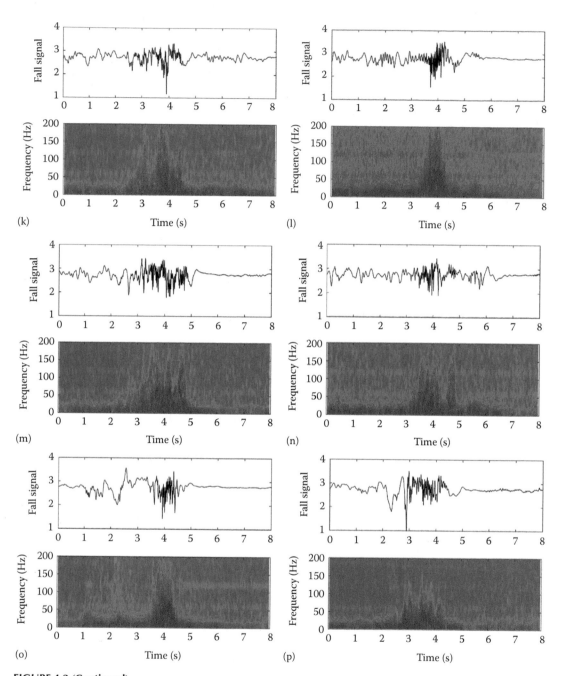

FIGURE 4.2 (Continued)
Time-domain signal and spectrogram of different falls. (k) Trip and fall—sideways fall, (l) slip and fall—forward fall, (m) slip and fall—sideways fall, (n) slip and fall—backward fall, (o) reach-fall (chair)—forward fall, (p) reach-fall (chair)—left fall.

(Continued)

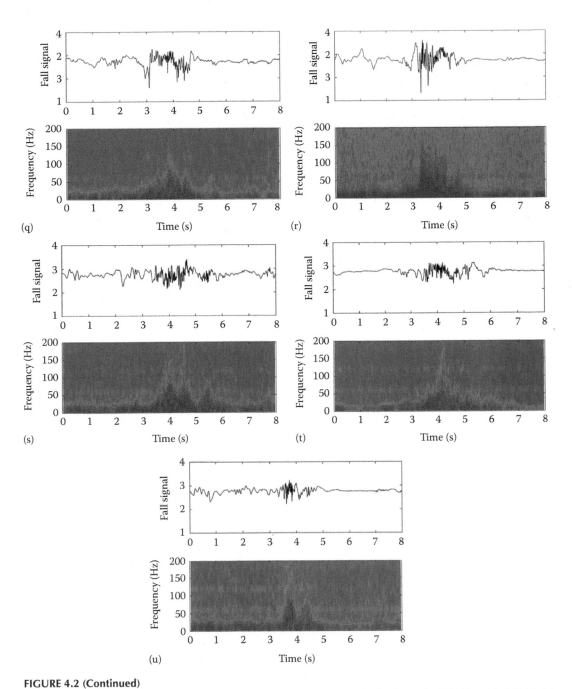

FIGURE 4.2 (Continued)
Time-domain signal and spectrogram of different falls. (q) Reach-fall (chair)—right fall, (r) reach-fall (chair) sliding—forward fall, (s) reach-fall (chair) sliding—backward fall, (t) lying on couch fall—upper body first fall, and (u) lying on couch fall—hip first fall.

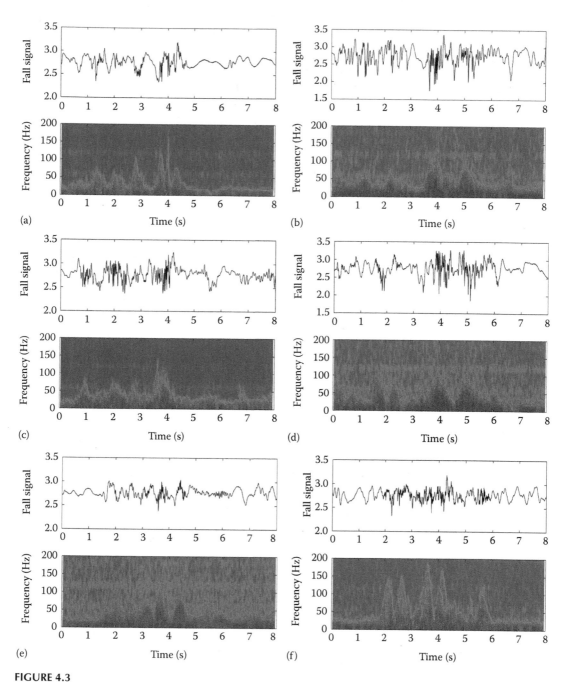

FIGURE 4.3
Time-domain signal and spectrogram of various non-fall motions. (a) Clapping hand strongly once and kicking three times, (b) standing up after a fall, (c) sleeping on a sofa, (d) bending over, (e) checking the watch and putting arm down, and (f) moving arms to warm up.

We would expect that a Doppler radar can offer relatively good performance for fall detection. From the spectrograms, the reach-falls and the couch-falls are more difficult to distinguish from the non-falls. Indeed, the experimental results presented in Section 4.5.2 confirm that the reach-falls and the couch-falls are the hardest to detect among others in Table 4.1. We would like to add that, although not included in Figure 4.3, motion coming from sitting or jumping could produce Doppler signatures close to that of human falls, due to the similarity of the motion dynamics. Distinguishing human fall from sitting or jumping motions will be challenging for narrowband radars. Because sitting and jumping have much smaller displacement of human body compared to a fall, incorporation of range information, which can be obtained from a wideband radar, would better distinguish them from falls (Erol et al. 2016).

4.4 Classification

4.4.1 Radar Fall Detection Framework

Doppler radar is a sensor that responds to any motion. In a home environment, various kinds of motions appear, typically caused by humans and pets. The radar response needs to be analyzed carefully for fall detection; otherwise, an exceedingly large number of false alarms would be the result.

The data processing blocks for fall detection is shown in Figure 4.4. The contiguous radar data is first applied to a prescreener that determines whether an important event may have occurred and, if so, its time location. Once an event is detected by the prescreener, a classification process is initiated to detect whether the event is a fall. More specifically, the transformed data, which is windowed around the identified event time location, is used to extract pertinent features; the window size should be sufficient wide to cover an entire fall activity (Su et al. 2015). Such features are then used by a classifier to perform fall versus non-fall classification.

A human fall spans certain time duration, typically 2–2.5 s. As such, there will be margin of errors in the time location of a possible fall event determined by the prescreener. The amount of errors depends on the prescreener used. Feature extraction should take the time location error into consideration when generating the features for classification.

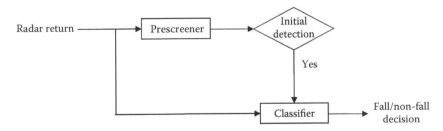

FIGURE 4.4
Block diagram of a radar fall detection system.

4.4.2 Prescreener

The purpose of the prescreener is to locate in time the activities in which falls may have occurred. The prescreener should be computationally efficient to operate in real time, and the prescreener should achieve nearly 100% detection of human falls. This may require a low prescreener detection threshold. Perhaps the simplest prescreener is the energy detector that makes the declaration of a possible fall activity based on the short-term energy of the radar signal return, because a fall creates high radar signal output. However, many daily activities would generate large radar returns as well. A prescreener with a smaller number of false alarms can relieve some burden of the classifier and provide better decision.

A power burst curve (also referred to as the energy burst curve), which represents the signal power within a specific frequency band as a function of time, can be utilized for prescreening (Liu et al. 2011, Wu et al. 2015). The frequency band chosen for prescreening should exclude the clutter-dominated near zero-frequency region but effectively captures human activities. An event is triggered for classification when the signal power in this specified band exceeds a certain level. The coefficients of wavelet decomposition at a given scale have also been used in the prescreening stage to identify the time locations where fall activities may have occurred (Su et al. 2015). The wavelet decomposition-based prescreener has a margin of errors in the fall location of about 0.125 s (Su et al. 2015). In the bathroom fall experiments performed, the prescreener reaches 90% detection of falls at a rate of 0.4 false alarms per minute and 100% detection at 2 false alarms per minute (Su et al. 2015). Such false alarm rate is too high, and feature extraction and classification must follow to improve performance.

4.4.3 Feature Extraction

Various features have been used in different studies. Most of these features are based on the time-varying Doppler signature and are obtained from the time–frequency analysis or the time–scale analysis.

In the work of Wu et al. (2015), pertinent features are obtained from the spectrogram and include extreme frequency magnitude, extreme frequency ratio, and the time span of event. These features are described as follows:

1. *Extreme frequency magnitude*: The extreme frequency magnitude is defined as $F = \max(f_{+\max}, -f_{-\min})$, where $f_{+\max}$ and $f_{-\min}$, respectively, denote the maximum frequency in the positive frequency range and the minimum frequency in the negative frequency range. As we described previously, critical falls often exhibit a high extreme frequency magnitude compared to other types of observed motions.

2. *Extreme frequency ratio*: The extreme frequency ratio is defined as $R = \max(|f_{+\max}/f_{-\min}|, |f_{-\min}/f_{+\max}|)$. For falls, due to the translational motion of the entire body, a high-energy spectrogram is concentrated in either the positive or the negative frequency, resulting in a high extreme frequency ratio. However, other types of motions, such as sitting and standing, often demonstrate high-energy content in both the positive and negative frequency bands because different body parts may undergo motions in opposite directions, thereby generating a low extreme frequency ratio.

3. *Time span of event*: This feature describes the length of time, in milliseconds, between the start and the end of an event, that is, $L = t_{\text{extrm}} - t_{\text{begin}}$, where t_{extrm}

denotes the time where the extreme frequency occurs and t_{begin} denotes the initiation time of the event. The latter is determined by the time when the magnitude of the frequency content of a signal passes a specific threshold. The different motion patterns being compared in this work generally show distinct time spans.

In addition to these features, other features have also been extracted from TFDs for classification of human activities (see, e.g., Kim and Ling 2009, Gurbuz et al. 2013 and references therein). These include torso Doppler frequency, total bandwidth of the Doppler signal, offset of the total Doppler, normalized standard deviation of the Doppler signal strength, period of the limb motion, shape of the spectrogram envelope, ratio of torso echoes to other echoes in the spectrogram, and Fourier series coefficients of spectrogram envelope. Nonparametric features derived from subspace representations of the TFDs have also been proposed. Effective and reliable fall detection often requires the combined use of multiple features. Once a set of features is extracted, a classification algorithm can be applied to determine whether an event is a fall or non-fall activity. By using deep learning methods, features may be automatically learned from the data (Jokanovic et al. 2016).

4.4.4 Classification Methods

A variety of classifiers have been employed for fall detection. Among them, the support vector machine (SVM) is commonly used (Kim and Ling 2009). Different classifiers, including *k*-nearest neighbor, are used to automatically distinguish falling from other activities, such as walking and bending down (Liu et al. 2011). The sparse Bayesian learning method based on the relevance vector machine improves fall detection performance over the SVM with fewer relevance vectors, and its effectiveness is demonstrated in (Wu et al. 2015). Hidden Markov model-based machine learning is used in the work of Wu et al. (2013) to characterize the signal spectrogram for fall detection.

However, the choice of employed features has been determined to have a greater impact on the classification performance than the specific classifier applied (see Gurbuz et al. 2013 and references therein).

4.5 Experimental Results

In this section, we present two sets of classification results, using the data from two different radars. The results of the first one are presented in Section 4.5.1 and correspond to the experiments described in Section 4.3.1. The spectrogram of the signal and a Bayesian classifier are used for classification. For the second set that corresponds to the study in Section 4.3.2, the results are presented in Section 4.5.2. The SWT is used to generate the time–scale features, which are used for classification utilizing both the SVM and the nearest neighbor (NN) algorithm.

4.5.1 Experiment 1—Bayesian Classifier on Spectrogram

Figure 4.5 shows the ground truth of three features defined in Section 4.4.3: the extreme frequency magnitude, the extreme frequency ratio, and the length of the event. Specifically, Figure 4.5a shows the 3D view of the three features, whereas their pairwise 2D plots are,

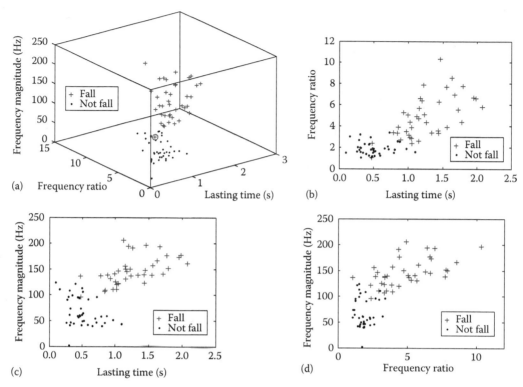

FIGURE 4.5
Ground truths of motion features in a three-dimensional view and three two-dimensional views. (a) 3D plot in the three-feature space, (b) frequency ratio versus lasting time, (c) frequency magnitude versus lasting time, and (d) frequency magnitude versus frequency ratio. (From Wu, Q. et al., *IET Radar Sonar Nav.*, 9, 164–172, 2015; Courtesy of the Institution of Engineering and Technology. With Permission.)

respectively, shown in Figure 4.5b–d. It is observed that these features generally provide a clear distinction between the fall and non-fall events, except one outlier fall event (marked with a circle in Figure 4.5a). Examination of the spectrogram of this outlier fall event shows that the signal is very weak, yielding low extreme Doppler frequency as well as a short length of event time.

The fall events exhibit larger extreme frequency magnitudes, higher extreme frequency ratios, and longer lengths of event time than the non-fall counterparts. These features, however, do not robustly classify the fall and non-fall activities based on a single feature alone.

We use fivefold cross-validation on the motion data. The entire sample set is randomly partitioned into five equal-sized subsets. Out of the five subsets, a single subset is retained as the validation data for testing the classifier, and the remaining four subsets are used as the training data. The cross-validation process is repeated 5 times, with each of the five subsets used exactly once as the validation data.

The Bayesian classification algorithm yields good classification results with one misclassification of the outlier fall event as described earlier, as shown in Figure 4.6a, where the misclassified result is marked with a circle. The probability of fall motion acquired in the classification is shown in Figure 4.6b. In this figure, the index of experiment is ordered with forward falling, forward falling with arm motion, backward falling, backward

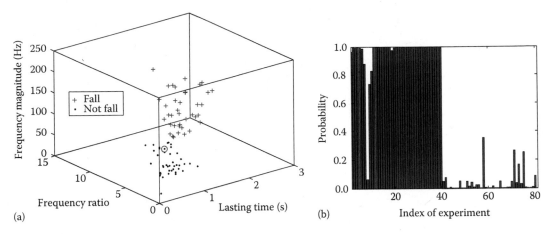

FIGURE 4.6
Classification results using Bayesian classifier. (a) Classification result and (b) probability distribution of fall motion. (From Wu, Q. et al., *IET Radar Sonar Nav.*, 9, 164–172, 2015. With Permission; Courtesy of the Institution of Engineering and Technology.)

falling with arm motion, sitting and standing, fast sitting and standing, bending over and standing up, and fast bending over and standing up (refer to Figure 4.1), each of 10 trials. We observe in Figure 4.6b that the fall events generally show a high probability of fall motion except the outlier event. For non-fall events, the normal sit-and-stand and bend-and-stand-up activities show consistently low probability of fall motion, whereas the fast sit-and-stand and bend-and-stand-up activities tend to exhibit a higher probability of fall motion, although their absolute value remains low for reliable classification.

4.5.2 Experiment 2—SVM on WT

We use a commercially available Doppler radar (RCR50 2013), and the Doppler radar has the same characteristics as described in Section 4.3.2. The data sampling rate is 960 Hz.

The dataset in this experiment was collected in a simulated home environment with the radar mounted in the ceiling. It contains 21 kinds of falls as listed in Table 4.1, each performed 5 times from a mix of one female and two male stunt actors aged between 30 and 46 years. It has eight kinds of non-fall activities (Su et al. 2015). The total number of falls is 105, and that of acted non-falls is 704. The total length of the entire dataset is 145 min. Please refer to the work of Su et al. (2015) for more details about the dataset.

Time–frequency analysis will be used for prescreening and classification. To maintain the balance between complexity and performance, the WT is used. It processes the data sequentially through successive filtering by a pair of the band-pass and low-pass filters and down-sampling by two. The pair of filters determines inherently the wavelet function. SWT is a reductant transform that generates the WT coefficients with the same length as the original data at each dyadic scale.

We select the pair of filters defined by the reverse bi-orthogonal 3.3 (rbio3.3) wavelet (Mallat 2009), which is found to produce the best results (Su et al. 2015) among the set of commonly used wavelet functions. Figure 4.7 shows the SWT coefficients from the lowest six dyadic scales of a data segment containing a fall in the middle with two non-fall activities at the beginning and the end, where D_i is the SWT coefficients at dyadic scale 2^i.

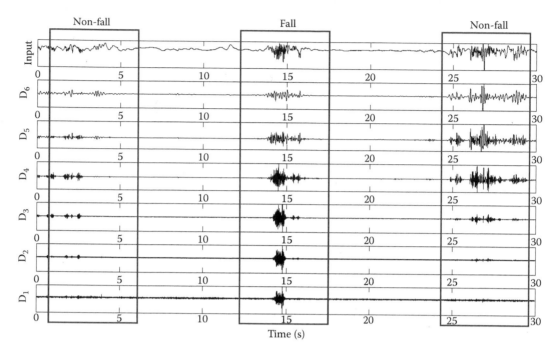

FIGURE 4.7
Stationary wavelet transform of a backscattered radar signal from a human.

The two non-fall motions have a high radar energy return. However, the SWT coefficients at lower scales provide much better distinction between the fall and the two non-falls, where the non-fall activities are significantly diminished.

The prescreener detection value at sample time t, denoted by $p(t)$ and expressed by

$$p(t) = \sum_{k=1}^{N} \left(w(i) D_2 \left(t + \frac{N}{2} - k \right) \right)^2 \tag{4.9}$$

is simply the running energy of the SWT coefficients at dyadic scale 4, where the running window has a duration of 0.5 s. In Equation 4.9, $w(i)$ is the Hamming window function, $N = 480$ is the number of samples in 0.5 s sampled at a rate of 960 Hz.

The features are the sequence of energies of the SWT coefficients over a 0.5 s time window with 50% overlap centered at the time position identified by the prescreener. Over 2.5 s, there are nine feature values for each scale. These feature values are normalized to sum to unity. The collection of the features from dyadic scales from 2 to 64 forms the feature vector of length 42 for classification (Su et al. 2015).

We applied two classifiers for fall versus non-fall classification: the SVM and the NN (Bishop 2006). For SVM, we applied leave-one-out cross-validation to obtain the fall detection performance. That is, the fall confidence of a given signature is obtained from a classifier that is trained using the rest of the signatures. For the NN, the confidence is the difference between the distance of the test point to the nearest member of the non-fall class and that of the fall class, where the distance was computed using the Manhattan distance. The dataset contains 105 falls and 704 acted non-falls that were performed by three

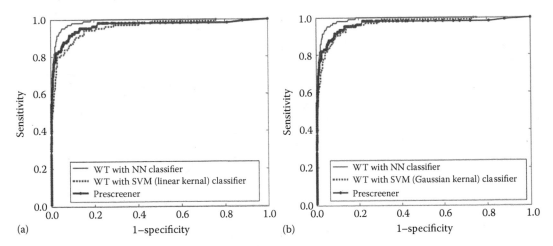

FIGURE 4.8
Performance of a fall detection system based on SWT features using NN and SVM classifiers with different kernels. (a) Linear kernel and (b) Gaussian kernel.

professionally trained stunt actors. The 105 falls contain 21 fall types as listed in Table 4.1; each type was repeated 5 times. Non-fall activities were walking, kicking, clapping, and bending over.

Figures 4.8a and b give the detection performances, respectively, using linear and Gaussian kernels for SVM. The sensitivity and specificity are defined as

$$\text{Sensitivity} = \frac{TP}{TP + FN}$$

$$\text{Specificity} = \frac{TN}{TN + FP}$$

(4.10)

where TP, FN, TN, and FP stand for true positive, false negative, true negative, and false positive, respectively. They have different values at a separate threshold, giving the performance of sensitivity versus (1 − specificity). In essence, sensitivity is the percentage of correct detection, whereas (1—specificity) is the fraction of false alarms over the total from the prescreener.

The results show that the NN classifier is able to reduce the number of false alarms from the prescreener by a factor of nearly 3 at 90% correct detection. The few hard to detect falls where the prescreener confidence values are low turn out to be the reach-falls and the couch-falls. The NN classifier is able to improve their detection considerably. The SVM does not perform as good as the NN, regardless of using either linear or Gaussian kernels. Another set of features using the Mel-frequency cepstral coefficients has also been tested for this dataset, and the performance is inferior to the features obtained from SWT (Su et al. 2015).

When the radar is not positioned in the ceiling or the radar has different center frequencies/characteristics, the performance of the SWT features may change and they may need to be adjusted accordingly. Indeed, Gurbuz et al. (2015) examined how the features vary with respect to the radar parameters and Gurbuz et al. (2016) illustrated the effect of the positioning of the radar in activity classification performance.

4.6 Fall Detection Based on Fused Radar and Passive Infrared Sensor Measurements

Passive infrared (PIR) motion sensors may supplement radar sensors to enhance fall detection and reduce false alarms (Liu et al. 2014, Erden and Cetin 2017). Such sensors have been placed at Tiger Place Apartment, Columbia, Missouri, since 2005 to report the absence or presence of the resident at a certain location in the home (Liu et al. 2014). An improved fall detection is achieved by fusing the radar sensor and the PIR motion sensor networks in the home of an older adult. Seven motion sensors are installed above the door of each room and the main facility area, such as kitchen, bathroom, and closet. However, one Doppler radar facing down to the floor (marked by a red cross) is placed on the ceiling. The detection range of the radar is about 6 m and the height of the room is about 3 m.

An event from the motion sensor network means someone is moving around the sensor. A fall is unlikely to occur if an event from a motion sensor is recorded immediately after it. Therefore, if motions are detected by the PIR motion sensors shortly after a potential fall detected by the radar, it is likely a false alarm. For example, Figure 4.9a shows that a fall is recognized by the radar with a confidence of 0.96, where the blue solid line gives the fall

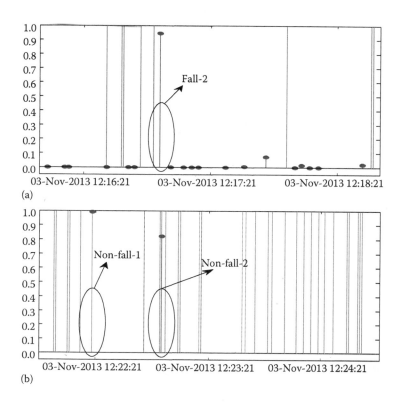

FIGURE 4.9
Examples of radar detected fall motions and their confidence (blue lines) along with detected motions by the PIR motion sensor (red lines). (a) Typical scenario showing a natural fall and (b) typical scenario showing non-fall motions only. (From Liu, L. et al., An automatic fall detection framework using data fusion of Doppler radar and motion sensor network, in *Proceedings of the Annual International Conference of IEEE Engineering in Medicine and Biology Society*, Chicago, IL, 2014, pp. 5940–5943. With Permission; Courtesy of IEEE.)

confidence generated by Doppler radar fall detection system, whereas the red dotted line presents the event activated by the sensor network. This plot shows a natural fall where no event is reported from the sensor network in the following 55 s. Figure 4.9b presents the typical false alarms from the senior daily activities: fast opening and closing the door near to ceiling radar (0.98 fall confidence), and fast turning around and shifting the walker direction (0.83 fall confidence). Multiple events from sensor network after non-fall-2 reflect that the resident is active or a visitor is in the apartment.

4.7 Conclusion

In this chapter, we described radar-based elderly fall detection techniques, and the discussions are mainly focused on the narrowband radar technology that generates rich Doppler signatures corresponding to different types of human motions, including falls. The Doppler signatures are highly nonstationary; therefore, the time–frequency and time–scale analyses are useful tools to reveal and extract the important features for motion classification and fall detection. Doppler signatures of fall and non-fall motions were depicted and compared. Key features that distinguish fall and non-fall motions were discussed. Fall detection performance evaluated using datasets collected from experiments and real-world operations was presented. In addition, motion sensor network using low-cost PIR sensors may supplement radar sensors to enhance fall detection and reduce false alarms.

Acknowledgment

The work of Y. D. Zhang was made possible by National Priorities Research Program (NPRP) Grant # NPRP 6-680-2-282 from the Qatar National Research Fund (a member of Qatar Foundation). The statements made herein are solely the responsibility of the authors. D. Ho expresses his thanks to the Agency for Healthcare Research and Quality for providing funding under grant No. R01HS018477 for collecting the human fall data. He also expresses his thanks to Mr. Bo-Yu Su for preparing the figures.

References

Amin, M. G., Y. D. Zhang, F. Ahmad, and K. C. D. Ho, Radar signal processing for elderly fall detection: The future for in-home monitoring, *IEEE Signal Processing Magazine*, 33(2): 71–80, 2016.

Amin, M. G., Y. D. Zhang, and B. Boashash, High-resolution time-frequency distributions for fall detection, in *Proceedings of SPIE Radar Sensor Technology Conference*, Baltimore, MD, April 2015.

Bagala, F., C. Becker, A. Cappello, L. Chiari, K. Aminian, J. M. Hausdorff, W. Zijlstra, and J. Klenk, Evaluation of accelerometer-based fall detection algorithms on real-world falls, *PLoS ONE*, 7(5): 1–9, 2012.

Bennett, T. R., J. Wu, N. Kehtarnavaz, and R. Jafari, Inertial measurement unit-based wearable computers for assisted living applications: A signal processing perspective, *IEEE Signal Processing Magazine*, 33: 28–35, 2016.

Bishop, S., *Pattern Recognition and Machine Learning*. Springer, New York, 2006.

Boashash, B., *Time-Frequency Signal Analysis and Processing: A comprehensive Reference*. Second version. Elsevier, Oxford, 2003.

Boashash, B., N. A. Khana, and T. Ben-Jabeura, Time-frequency features for pattern recognition using high-resolution TFDs: A tutorial review, *Digital Signal Processing*, 40: 1–30, 2015.

Chen, V. C., *The Micro-Doppler Effect in Radar*. Artech House, Norwood, MA, 2011.

Chen, V. C. and H. Ling, *Time-Frequency Transforms for Radar Imaging and Signal Analysis*, Artech House, London, 2002.

Cohen, L., *Time-Frequency Analysis*. Englewood Cliffs, Prentice Hall, NJ, 1995.

Erden, F. and E. Cetin, Infrared sensors for indoor monitoring, in M. G. Amin (ed.), *Radar for In-Door Monitoring: Detection, Localization, and Assessment*, CRC Press, Boca Raton, FL, 2017.

Erol, B., M. G. Amin, B. Boashash, F. Ahmad, and Y. D. Zhang, Wideband radar based fall motion detection for a generic elderly, in *Proceedings of Asilomar Conference on Signals, Systems, and Computers*, Pacific Grove, CA, November 2016.

Gurbuz, S. Z., B. Erol, B. Cagliyan, and B. Tekeli, Operational assessment and adaptive selection of micro-Doppler features, *IET Radar Sonar Navigation*, 9(9): 1196–1204, 2015.

Gurbuz, S. Z., B. Tekeli, M. Yuksel, C. Karabacak, A. C. Gurbuz, and M. B. Guldogan, Importance ranking of features for human micro-Doppler classification with a radar network, in *Proceedings of the 16th International Conference Information Fusion*, Istanbul, Turkey, July 2013, pp. 610–616.

Gurbuz, S. Z., C. Clemente, A. Balleri, and J. Soraghan, Micro-Doppler-based in-home aided and unaided walking recognition with multiple radar and sonar systems, *IET Radar Sonar Navigation*, 11(1): 107–115, 2017.

Igual, R., C. Medrano, and I. Plaza, Challenges, issues and trends in fall detection systems, *Biomedical Engineering Online*, 12(66): 1–24, 2013.

Jokanovic, B., M. G. Amin, and F. Ahmad, Radar fall detectors using deep learning, in *Proceedings IEEE Radar Conference*, Philadelphia, PA, 2016.

Kim, Y. and H. Ling, Human activity classification based on micro-Doppler signatures using a support vector machine, *IEEE Transactions on Geoscience Remote Sensing*, 47(5): 1328–1337, 2009.

Liu, L., M. Popescu, K. C. Ho, M. Skubic, and M. Rantz, Doppler radar sensor positioning in a fall detection system, in *Proceedings of the International Conference of IEEE Engineering in Medicine and Biology Society*, San Diego, CA, August 2012, pp. 256–259.

Liu, L., M. Popescu, M. Skubic, and M. Rantz, An automatic fall detection framework using data fusion of Doppler radar and motion sensor network, in *Proceedings of the Annual International Conference of IEEE Engineering in Medicine and Biology Society*, Chicago, IL, August 2014, pp. 5940–5943.

Liu, L., M. Popescu, M. Skubic, M. Rantz, T. Yardibi, and P. Cuddihy, Automatic fall detection based on Doppler radar motion signature, in *Proceedings of the International Conference on Pervasive Computing Technologies for Healthcare*, Dublin, Ireland, May 2011, pp. 222–225.

Mallat, S., *A Wavelet Tour of Signal Processing*. Academic Press, Burlington, MA, 2009.

Qian, S., *Introduction to Time-Frequency and Wavelet Transforms*. Prentice Hall, Englewood Cliffs, NJ, 2001.

RCR50 manual (April 1, 2013) [Online]. Available: http://www.interlogix.com/_/assets/library/1036806B_RCR50_inin.pdf. (accessed on May 19, 2017).

Sadigh, S., A. Reimers, R. Andersson, and L. Laflamme, Falls and fall-related injuries among the elderly: A survey of residential-care facilities in a Swedish municipality, *Journal of Community Health*, 29(2): 129–140, 2004.

Shingu, G., K. Takizawa, and T. Ikegami, Human body detection using MIMO-UWB radar sensor network in an indoor environment, in *Proceedings of 2008 International Conference on Parallel and Distributed Computing, Applications and Technologies*, Dunedin, New Zealand, December 2008, pp. 437–442.

Skolnik, M., *Introduction to Radar Systems, 3rd edition*. McGraw-Hill, New York, 2002.

Su, B. Y., K. C. Ho, M. J. Rantz, and M. Skubic, Doppler radar fall activity detection using the wavelet transform, *IEEE Transactions on Biomedical Engineering*, 62(3): 865–875, 2015.

Wu, M., X. Dai, Y. D. Zhang, B. Davidson, M. G. Amin, and J. Zhang, Fall detection based on sequential modeling of radar signal time-frequency features, in *Proceedings of IEEE International Conference on Healthcare Informatics*, Philadelphia, PA, September 2013, pp. 169–174.

Wu, Q., Y. D. Zhang, W. Tao, and M. G. Amin, Radar-based fall detection based on Doppler time-frequency signatures for assisted living, *IET Radar, Sonar and Navigation*, 9(2): 164–172, 2015.

5

Continuous-Wave Doppler Radar for Human Gait Classification

Fok Hing Chi Tivive, Abdesselam Bouzerdoum, and Bijan G. Mobasseri

CONTENTS

5.1 Introduction

Radar technology has commonly been used to estimate the speed and distance of a moving object. In recent years, it has been investigated as an alternative sensing modality for moving target classification. In contrast to optical imaging systems such as digital and infrared cameras, radars can work in all weather conditions and from a distance. Furthermore, radars have through-the-wall sensing capability and do not capture facial characteristics for identification; thus, they are less intrusive and are less likely to be considered to violate privacy. A modern Doppler radar detects not only the gross translation motion of the target but also local dynamics exhibited by moving parts attached to the target, for example, rotation of a helicopter blade, vibrations of an engine, or limb motion of a human. All these micro movements induce additional frequency modulations on the radar returns, thereby generating sidebands about the main Doppler frequency, which are referred to as the *micro-Doppler* (μ-D) signature [1]. Several studies have been conducted to analyze μ-D radar signatures of rigid and non-rigid moving targets [1–8]. Chen et al. formulated mathematical models and conducted experiments to investigate μ-D radar effects of targets under translation, rotation, and vibration motions [1,9]. Other researchers performed numerical simulations using a kinematic model or real radar data to analyze the μ-D signatures of nonrigid objects [3–8]. Many of these studies show that the μ-D signature reflects the kinetic motions of an object and provides a viable means for object identification. μ-D signals have been used to classify rigid targets, for example, to distinguish between a helicopter and an airplane [10], wheeled and tracked vehicles [11], different jet engines [12], and different ballistic targets [13].

Recently, the research focus has been diverted to the classification of human gait. Human gait refers to the walking manner achieved through the movement of the torso, legs, and arms. It is defined by the *gait cycle* consisting of two main phases: stance and swing. Stance phase is the period between heel strike and toe off of the same foot and consists of three phases: heel strike, mid stance, and toe off. Swing phase begins when the foot is no longer in contact with the ground and consists of an initial acceleration of the limb followed by deceleration to position the foot for heel strike. Human gait classification using μ-D signals has numerous potential civilian and military applications. For example, μ-D radar technology can be used for countering terrorism, conducting urban military operations, providing urban border security, rescuing hostages, and detecting human movement in a forest. This radar technology can also be used for in-home monitoring of the elderly to provide immediate assistance after a fall [14].

Apart from Doppler radar [15–28], there are other sensing modalities that have been adopted for human gait identification, namely, optical [29–32] and acoustic [33–35] sensors. With optical sensors, the gait is captured in the spatiotemporal space as a set of image sequences. The set of image sequences is then converted into the form of gait energy image [36,37], active energy image [38], and gait flow image [31,39]. The gait energy image is the average of binary silhouette images, whereas the active energy image is the aggregation of active regions, which are obtained by calculating the difference of two adjacent silhouette images. The gait flow image, however, is formed by determining the optical flow field from the silhouette images. For acoustic sensors, Altaf et al. [34] introduced the concept of the acoustic ghost profile, which is obtained from the temporal signal analysis of the sound of footsteps.

This chapter presents a μ-D feature extraction method for radar human gait classification. The local dynamics of human motions embedded in the Doppler signal are captured using a joint time–frequency (T–F) representation. The arm and leg motions, which are periodic, induce μ-D modulations near the torso frequency shift. Therefore, local T–F patches centered on the torso frequency shift are extracted for feature extraction, thereby introducing some tolerance to variations in target speed. Then, each patch is convolved with a set of log-Gabor filters to detect discriminative features, which are then compressed by a dimensionality reduction technique.

The remaining part of the chapter is organized as follows: Section 5.2 presents a brief description of the radar μ-D signal and the related work on the classification of human μ-D signatures. Section 5.3 describes the proposed μ-D feature extraction method. Section 5.6 presents the experimental results, and finally, Section 5.7 gives the concluding remarks.

5.2 Background and Related Work

This section presents the basic mathematical description of μ-D effect from a point scattering target, followed by the description of three T–F analysis techniques, namely Wigner–Ville distribution, short-time Fourier transform, and S-method used to analyze Doppler signals. Then, several existing methods for classifying human μ-D signals are reviewed.

5.2.1 Doppler Signal Model

When a point scattering target is moving toward a continuous-wave (CW) radar at a constant velocity $v(t)$, the carrier frequency of the radar signal is shifted according to the target velocity. Assuming that the transmitted signal is $s(t) = A\exp(j2\pi ft)$ and the initial range of the point scattering target is R_0 from the radar at time $t = 0$, the change in the distance between the radar and the point scattering target as a function of time can be expressed as

$$R(t) = R_0 + \int_0^t v(\tau)\mathrm{d}\tau \tag{5.1}$$

The received Doppler signal can be modeled as

$$x(t) = A\exp\left\{j2\pi f\left[t - \frac{2R(t)}{c}\right]\right\} \tag{5.2}$$

where c is the speed of light. The Doppler shift f_d induced by the motion of the point scattering target is given by

$$f_d = f\frac{2v(t)}{c} \tag{5.3}$$

A more complex target can be represented as a set of point scatterers; the backscattered signal is a superposition of radar returns from the point scatterers. To synthesize a μ-D signal for human gait, a person can be characterized by a set of M_s segments, where each segment consists of M_p points and moves at its own velocity [4,40]. Then, the received μ-D signal can be formulated as

$$x(t) = \sum_{k=1}^{M_s}\sum_{m=1}^{M_p} A_{k,m}(t)\exp\left[j\frac{4\pi l_{k,m}}{\lambda}\int_0^t \omega_k(\tau)\cos\beta_k(\tau)\mathrm{d}\tau\right] \tag{5.4}$$

where:

  $A_{k,m}$ is the amplitude of the mth point on the kth segment
  $\beta_k(\tau)$ is the instantaneous angle from the zenith of the kth segment
  $\omega_k(\tau)$ is the instantaneous angular velocity of the kth segment
  $l_{k,m}$ is the distance of the mth point along the kth segment

In practice, human locomotion is more complicated than the model given in Equation 5.4. Therefore, advanced modeling tools, such as electromagnetic wave scattering modeling and the computer animation software MAYA, have been employed to simulate the μ-D signals exhibited by different parts of a moving person [41,42].

5.2.2 T–F Representations

A T–F representation is often adopted to model the Doppler signal. Several T–F analysis methods, for example, Wigner–Ville distribution (WVD), short-time Fourier transform (STFT), and the S-method (SM), are often used to transform a one-dimensional (1D) signal into its corresponding two-dimensional (2D) T–F representation. These three methods are briefly reviewed next.

Let us consider a radar signal $x(t)$ reflected by a moving target. The WVD of the radar signal is given by

$$X_{\text{WVD}}(t,\omega) = \int_{-\infty}^{\infty} x\left(t+\frac{\tau}{2}\right)x^*\left(t-\frac{\tau}{2}\right)\exp(-j\omega\tau)\,d\tau \tag{5.5}$$

where * denotes the complex conjugate operator. The WVD, which is regarded as the basic quadratic T–F distribution, provides the best T–F resolution for mono-component signals. However, for a multicomponent signal such as the Doppler signal obtained from human gait, the existence of cross-terms renders the interpretation of the T–F representation difficult. Therefore, a kernel function is often used to suppress the cross-term interferences. A variant of the WVD that employs a time window $w(t)$ for smoothing in the frequency domain is the pseudo-WVD (PWVD), which is defined as follows:

$$X_{\text{PWVD}}(t,\omega) = \int_{-\infty}^{\infty} w\left(\frac{\tau}{2}\right)w^*\left(-\frac{\tau}{2}\right)x\left(t+\frac{\tau}{2}\right)x^*\left(t-\frac{\tau}{2}\right)\exp(-j\omega\tau)\,d\tau \tag{5.6}$$

The PWVD can be expressed in terms of the STFT as

$$X_{\text{PWVD}}(t,\omega) = \frac{1}{2\pi}\int_{-\infty}^{\infty} X\left(t,\omega+\frac{u}{2}\right)X^*\left(t,\omega-\frac{u}{2}\right)du \tag{5.7}$$

where $X(t,\omega)$ is given by

$$X(t,\omega) = \int_{-\infty}^{\infty} x(t+\tau)w(\tau)\exp(-j\omega\tau)\,d\tau \tag{5.8}$$

The spectrogram $X_{sp}(t,\omega)$ is another T–F representation that has commonly been used for depicting radar μ-D signature. It is the squared magnitude of the STFT:

$$X_{sp}(t,\omega) = \left|X(t,\omega)\right|^2 \tag{5.9}$$

Although the spectrogram can easily be implemented and does not have cross-terms, it generally gives a low-resolution T–F representation.

The SM, however, can achieve similar auto-term concentration as the WVD without producing any cross-terms. It is basically a modified WVD with a reduced integration extent [43]. The T–F distribution of the SM is given by

$$X_{sm}(t,\omega) = \frac{1}{2\pi}\int_{-\infty}^{\infty} Q(u)X\left(t,\omega+\frac{u}{2}\right)X^*\left(t,\omega-\frac{u}{2}\right)du \tag{5.10}$$

where:
 $Q(u)$ is a finite frequency window
 $X(t,\omega)$ is the STFT given in Equation 5.8

The SM can be written in discrete form as

$$X_{sm}(n,k) = \sum_{i=-N/2+1}^{N/2} Q(i)X(n,k+i)X^*(n,k-i) \tag{5.11}$$

where:
 n is the discrete-time index
 k is the discrete-frequency index
 N is the number of frequency samples

For a rectangular window, that is, $Q(i)=1$ for $|i| \le L$ and zero otherwise, the SM with L terms can be written as

$$X_{sm}(n,k) = \sum_{i=-L}^{L} X(n,k+i)X^*(n,k-i) \tag{5.12}$$

When $L = 0$, the SM behaves as the spectrogram, and when L spans the entire discrete Fourier domain, the SM is identical to WVD. The parameter L is chosen so that most of the auto-term energy is concentrated at the maximum value of the auto-term; it is usually set to a small value $\left(L \in [3,10]\right)$ [44].

5.2.3 Radar μ-D Feature Extraction Approaches

The key step in many μ-D classification approaches is the extraction of salient features from the radar signal. The feature extraction is performed in the time domain, frequency domain, or joint T–F domain. In the frequency domain, Bilik and Khomchuk applied speech processing techniques to extract three types of features: cepstrum,

Mel-frequency cepstrum coefficient (MFCC), and linear predictive coding (LPC) coefficients [19]. Then, the minimum divergence technique was applied to classify radar signals of four different classes: a person, a group of persons, a wheeled vehicle, a tracked vehicle, or a clutter. In the time domain, Smith et al. proposed a template-based method using dynamic time warping to classify short frame Doppler signals [45,46]. Molchanov et al. employed discrete cosine transform (DCT) and neural networks to discriminate between one, two, and three persons walking along the line of sight of the radar [17]. Fairchild and Narayanan decomposed a Doppler signal into a set of intrinsic mode functions (IMFs) using the empirical mode decomposition (EMD). Then, the energies of the IMFs were used as features to classify the following human activities: (1) breathing while standing still, (2) picking up an object, (3) standing up from a crouching pose, and (4) swinging arms. For efficient feature extraction, Javier and Kim extracted linear predictive coefficients directly from the Doppler signal [47]. The aforementioned methods, however, extract features that are localized either in time or in frequency.

There are techniques capable of extracting features localized in both time and frequency domains. Initially, a T–F analysis technique is employed to convert the signal into its corresponding T–F representation. Kim and Ling applied STFT to obtain the spectrogram where six types of features were defined and extracted: (1) the torso Doppler frequency, (2) the total bandwidth of the Doppler signal, (3) the offset of the total Doppler, (4) the bandwidth without μ-D, (5) the normalized standard deviation of the Doppler signal strength, and (6) the period of the limb motion [48]. For classification, they used support vector machines (SVMs) to discriminate seven human activities: running, walking, walking without moving arms, crawling, boxing, boxing while moving forward, and sitting relatively still. Bjorklund et al. exploited the periodicity of the μ-D frequencies by computing the cadence velocity diagram (CVD), which is obtained by applying Fourier transform on the spectrogram [22]. They extracted the cadence frequencies and the velocity profiles as features. Clemente et al. extracted translation and rotation invariant features by applying different orders of pseudo-zernike moments to the CVD [49]. Instead of STFT, Orovic et al. applied the Hermite SM to convert the Doppler signal into a high-resolution T–F representation and developed an envelope detection method to capture the evolution of the arm movements [21]. In [20] and [50], machine learning techniques were proposed to learn discriminative features from the spectrogram. In [20], a set of trainable filters arranged into a hierarchical network structure was optimized using Levenberg–Marquardt algorithm for feature extraction. In [50], an auto-encoder neural network was adopted to learn sparse features from the spectrogram. These machine learning methods produce good results when a large dataset is available for training. Another approach of feature extraction is to determine the subspace in which the μ-D modulations reside in. In [15], principal component analysis (PCA) was applied to learn several subspaces characterizing different arm motions. For feature extraction, input vectors were extracted along either the row or the column of the spectrogram. In [51] and [52], the authors extracted local patches from the T–F representation to account for the variations in target speed. They also applied a dimensional reduction technique to produce a low-dimensional feature vector. In [52], a set of 2D filters was employed for feature extraction before dimensionality reduction. More recently, centroid and SVD-based features have been extracted from multistatic Doppler data to detect and classify human motions at different aspect angles [53].

5.3 μ-D Feature Extraction

This section presents a method to extract features from the Doppler signal for classification of human gait, particularly the micro-motions of the arms while walking. First, the feature extraction method converts the Doppler signal into a T–F representation and extracts local patches. Then, the local patches, which are represented as small images, are convolved with a set of log-Gabor filters to detect motion energy at different scales and orientations. Finally, a dimensionality reduction technique is employed to produce a compact feature vector for classification. Figure 5.1 depicts the main steps of the μ-D feature extraction method, including the preprocessing and classification steps.

5.3.1 Contrast-Enhanced T–F Patch

When a person is walking toward or away from the radar, the arm and leg motions generate additional μ-D modulations around the human torso frequency. In addition, the torso frequency varies as a function of the target speed. Therefore, to cope with this issue, local patches centered on the torso frequency are extracted for classification. Because the torso induces the dominant frequency component, its location in the joint T–F representation, $X(n,k)$, can be easily determined by finding the main peak in the frequency profile, which can be expressed as

$$F_p(k) = \sum_{n=1}^{M} |X(n,k)| \tag{5.13}$$

where M is the number of time samples. The frequency profile is passed to a median filter, and the output is normalized by dividing by the maximum value to obtain $\tilde{F}_p(k)$. The location of the torso frequency is estimated as

$$k_t = \arg\max_k \tilde{F}_p(k) \tag{5.14}$$

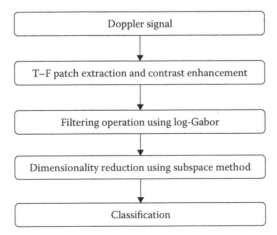

FIGURE 5.1
A schematic diagram of the μ-D feature extraction method.

Once the location of the torso frequency k_t is estimated, the height and width of the local patch can be determined as follows: Assuming that the T–F representation depicts positive μ-D modulations, the frequency index of the highest μ-D modulation, that is, the height of the main peak, is estimated by finding the smallest frequency index $k_0 \in (k_t, N/2)$ that satisfies the following condition:

$$\frac{\sum_{k=k_t}^{k_0} \tilde{F}_p(k)}{\sum_{k=k_t}^{\frac{N}{2}} \tilde{F}_p(k)} \geq \eta \tag{5.15}$$

where η is a fixed threshold chosen based on the noise level $(0 < \eta < 1)$. The vertical span of the patch is given by the frequency interval $[k_t - k_0, k_t + k_0]$. Then, a down-sampling or up-sampling operation is performed on the columns of the patch to fix the height to N_y so as to remove the variations in speed of the leg swing, or the arm swing when marching. Furthermore, the patch is aligned with respect to the main peak. Let n_i denote the time of the ith main peak, N_p denote the time duration between two consecutive main peaks, and N_x be the length of the input signal in samples. The horizontal span of patch is given by the time interval $\left[n_i + \left| \frac{N_p}{2} \right|, n_i + \left| \frac{N_p}{2} \right| + N_x - 1 \right]$. For feature extraction, the each local patch is considered as an image, and each element of the patch is considered as a pixel whose value is the magnitude, or absolute value, of the employed T–F representation.

When a person is far away from the radar, the induced μ-D signal is weak, due to signal path loss. Therefore, the Naka–Rushton equation [54] is applied to enhance the contrast of each extracted patch. This contrast enhancement technique not only increases the weak μ-D amplitudes but also suppresses the small amplitudes, which usually represent noise. Let $W(n,k)$ denote a local patch extracted from the T–F representation, where $k \in [1, N_y]$ and $n \in [1, N_x]$. The contrast-enhanced patch is computed as

$$\hat{W}(n,k) = \frac{W(n,k)^r}{W(n,k)^r + \mu^r} \tag{5.16}$$

where:
   μ is the mean value of the patch
   r is a constant that controls the slope of the function (here $r = 1$)

Figure 5.2 presents an example of a local patch extracted from a spectrogram of a walking person before and after contrast enhancement.

5.3.2 Log-Gabor Filtering

The contrast-enhanced patches are convolved with a set of log-Gabor filters to extract discriminative features, such as motion energy at different spatial frequencies (or scales) and orientations. Compared to their traditional Gabor counterparts, log-Gabor filters have neither direct current component nor bandwidth limitation [55]. Therefore, a small

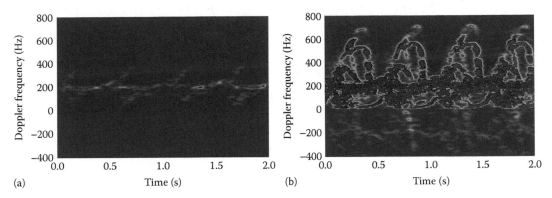

FIGURE 5.2
A local patch extracted from the spectrogram (a) before and (b) after contrast enhanced by the Naka–Rushton equation with $r = 1$.

set of filters is sufficient to cover the desired frequency spectrum. In the frequency domain, a log-Gabor filter is given by

$$H_{k,l}(f,\theta) = \exp\left\{ -\frac{\left[\log\left(\frac{f}{f_k}\right)\right]^2}{2\left[\log(\beta)\right]^2} \right\} \exp\left\{ -\frac{(\theta - \theta_l)^2}{2\sigma_\theta^2} \right\} \tag{5.17}$$

where:
  f_k is the center frequency at the kth scale
  θ_l is the lth orientation angle
  β is the bandwidth
  σ_θ is the angular bandwidth of the log-Gabor filter

$\beta = 0.75$ gives an approximate bandwidth of one octave, whereas $\beta = 0.55$ produces a bandwidth of two octaves. The angular bandwidth σ_θ is often set to 1.5 for even spectrum coverage [56]. Here, log-Gabor filters with bandwidth of two octaves ($\beta = 0.55$) were used for feature extraction. Figure 5.3 shows examples of the log-Gabor filters at four different orientations with a center frequency of 0.11 and a bandwidth of two octaves.

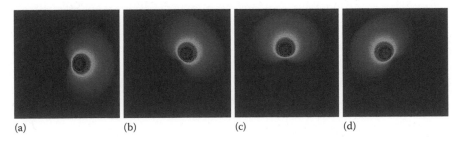

(a)   (b)   (c)   (d)

FIGURE 5.3
Examples of log-Gabor filters with a normalized center frequency of 0.11 at four different orientations ranging from 0° to 135°, with a step size of 45° in the 2D frequency domain. (a) $\theta = 0°$, (b) $\theta = 45°$, (c) $\theta = 90°$, and (d) $\theta = 135°$.

Because the log-Gabor is designed in the frequency domain, the filtering operation is performed in the same domain. Therefore, a 2D Fourier transform is applied to the contrast-enhanced patch $\hat{W}$, which is denoted by $\tilde{W}_F$. The feature map generated by the (k,l)-th log-Gabor filter is given by

$$Z_{k,l} = \left| \text{IFFT2}\left(\tilde{W}_F H_{k,l} \right) \right| \tag{5.18}$$

where IFFT2 denotes the 2D inverse fast Fourier transform. To achieve some degree of intensity and translation invariance, the feature map is first normalized as

$$\tilde{Z}_{k,l}(n,k) = \frac{Z_{k,l}(n,k)}{\sum_{l=1}^{N_\theta} Z_{k,l}(n,k)} \tag{5.19}$$

where N_θ is the number of orientations. Then, each feature map is partitioned into $R = r_1 \times r_2$ non-overlapping subregions, and the means of all the subregions are concatenated to form a mean vector $\mu_{k,l} = \left[\mu_1, \ldots, \mu_R\right]^T$. For a set of $S = N_f N_\theta$ log-Gabor filters (i.e., N_f scales and N_θ orientations), the mean vectors from all feature maps are arranged into a matrix $A \in \mathbb{R}^{R \times S}$:

$$A = \left[\mu_{1,1}, \ldots, \mu_{N_f, N_\theta}\right] \tag{5.20}$$

Finally, the elements in matrix A are normalized to the range $[0,1]$ before dimensionality reduction.

5.4 Dimensionality Reduction Techniques

The extracted features can be high dimensional, and not all of them are necessary to characterize the radar signal. Furthermore, the features can be correlated, and hence contain redundant information. In this section, four common dimensionality reduction techniques are investigated: PCA, 2D PCA (2DPCA) [57], two-directional 2D PCA ([2D]²PCA) [58], and two-directional 2D linear discriminant analysis ([2D]²LDA) [59]. The 2DPCA, (2D)²PCA, and (2D)²LDA methods are matrix based; they do not require a matrix-to-vector conversion to compute the covariance matrix. Therefore, they are more computationally efficient to compute the covariance matrix and the eigen-components. 2DPCA works only in the row direction of a 2D input, whereas (2D)²PCA considers both row and column directions simultaneously. Compared to 2DPCA, (2D)²PCA generates a much compact feature matrix and achieves better recognition accuracy [58]. Unlike the traditional LDA, (2D)²LDA does not suffer from the singularity problem for small training sets. More detailed descriptions of PCA, 2DPCA, (2D)²PCA, and (2D)²LDA are given in Sections 5.4.1 through 5.4.3.

5.4.1 Principal Component Analysis

PCA is an eigen-based technique to determine which features matter the most, or which ones best *explain* the data. Specifically, the PCA constructs an orthonormal basis by systematically maximizing the variance of the projected data on each basis vector.

The basis vectors have to obey three properties: they must be orthogonal to the previous basis vector, have a norm of one, and maximize the projected variance. These three conditions can be formulated as a constrained optimization problem. Central to the PCA is data reconstruction from projections. It follows that at the optimum basis vector, the minimum reconstruction error and the maximum projected variance occur simultaneously. It is shown in the following that the sought-after basis vectors are the eigenvectors of the data covariance matrix.

Let $x \in \mathbb{R}^{n \times 1} (n = RS)$ be the feature vector obtained by vectorizing the patch given in Equation 5.20, and let $X = \{x_1, x_2, \ldots, x_P\} \in \mathbb{R}^{n \times P}$ denote the centered data matrix containing feature vectors extracted from P local patches. The projection of the data along the candidate basis vector $u \in \mathbb{R}^{n \times 1}$ is computed as $u^T X$. The vector u that maximizes the variance of this projection is the first principal axis given by

$$\text{Var}(u^T X) = u^T \Sigma u \tag{5.21}$$

where $\Sigma = XX^T$ is the data covariance matrix. Let $\{e_1, e_2, \ldots, e_n\}$ be the n-dimensional eigenvectors of Σ and $\{\lambda_1, \lambda_2, \ldots, \lambda_n\}$ the corresponding eigenvalues. The variance of the projection $u^T X$ along $u = e_i$ is by definition λ_i. However, the variance along any other vector v is the weighted sum of all eigenvalues. This weighted sum will always be smaller than the largest eigenvalue; hence, the principal axis must be chosen to be the eigenvector corresponding to the largest eigenvalue if the variance of the projection is to be maximized. This choice of the basis function maximizes the variance and minimizes the reconstruction error simultaneously. Let the projection operator be defined by the matrix $\Phi = [e_1, e_2, \ldots, e_n]$. The projection of the data on Φ is given by

$$Y = \Phi^T X = \begin{bmatrix} \langle e_1, x_1 \rangle, & \langle e_1, x_2 \rangle, & \ldots & \langle e_1, x_P \rangle \\ \langle e_2, x_1 \rangle, & \langle e_2, x_2 \rangle, & \ldots & \langle e_2, x_P \rangle \\ \vdots & \vdots & \ldots & \vdots \\ \langle e_n, x_1 \rangle, & \langle e_n, x_2 \rangle, & \ldots & \langle e_n, x_P \rangle \end{bmatrix} \in \mathbb{R}^{n \times P} \tag{5.22}$$

The ith column of Y, y_i, is the projection of the ith feature vector on each eigenvector:

$$y_i = \begin{bmatrix} \langle e_1, x_i \rangle \\ \langle e_2, x_i \rangle \\ \vdots \\ \langle e_n, x_i \rangle \end{bmatrix} \tag{5.23}$$

As it stands, Equation 5.22 is an invertible transformation, and the transformed matrix Y is still of the same dimensions as X. Dimensionality reduction can be achieved by keeping only the l most significant eigenvectors. The reconstructed data is given by

$$\hat{X}_l = \Phi_l Y_l \tag{5.24}$$

where:
  Φ_l are the first l most significant eigenvectors
  Y_l are the projection of the data on these eigenvectors

5.4.2 Two-Directional 2DPCA

This matrix-based subspace technique generates two projection matrices to reduce the number of rows and columns of a 2D input simultaneously. Given the matrix A in Equation 5.20, the compressed features $D \in \mathbb{R}^{m_r \times m_c}$ ($m_r \leq R, m_c \leq S$), obtained from (2D)²PCA, can be written as

$$D = \Phi_r^T A \Phi_c \tag{5.25}$$

where Φ_r and Φ_c are the projection matrices with orthonormal components. Let $Y = A\Phi_c$ be the new features obtained from the projection matrix Φ_c. The optimal projection is obtained by maximizing the total scatter of the projected data, where the total scatter matrix is defined as the trace of the covariance matrix of the training samples. Therefore, the projection matrix Φ_c can be determined by maximizing the following criterion [58]:

$$
\begin{aligned}
J(\Phi_c) &= \text{trace}\left\{ E\left[(Y - E(Y))(Y - E(Y))^T \right] \right\} \\
&= \text{trace}\left\{ E\left[(A\Phi_c - E(A\Phi_c))(A\Phi_c - E(A\Phi_c))^T \right] \right\} \\
&= \text{trace}\left\{ \Phi_c^T G_c \Phi_c \right\}
\end{aligned} \tag{5.26}
$$

where $G_c = E\left[\{A - E(A)\}^T \{A - E(A)\} \right]$ denotes the image covariance matrix, which is an S-by-S nonnegative definite matrix. Suppose that the training set comprises P samples $\{A_1, \ldots, A_P\}$. The image covariance matrix G_c can be computed as

$$G_c = \frac{1}{P} \sum_{i=1}^{P} (A_i - \bar{A})^T (A_i - \bar{A}) \tag{5.27}$$

where $\bar{A}$ is the global mean given by

$$\bar{A} = \frac{1}{P} \sum_{i=1}^{P} A_i \tag{5.28}$$

The optimal solution of Equation 5.26 is a set of orthogonal projection vectors, $\{\phi_1, \ldots, \phi_{m_c}\}$, which are the eigenvectors of G_c corresponding to the first m_c largest eigenvalues. The number of eigenvectors m_c can be determined using the following condition:

$$\frac{\sum_{i=1}^{m_c} \lambda_i}{\sum_{i=1}^{S} \lambda_i} \geq \gamma \tag{5.29}$$

where:
λ_i denotes the ith eigenvalue of G_c
γ is a threshold

In 2DPCA, the matrix A is only projected onto Φ_c to obtain the feature matrix. For (2D)²PCA, another projection matrix Φ_r is determined to reduce the number of rows of the matrix A.

It can be computed by substituting the input A_i in Equations 5.27 and 5.28 with its transposed version. Thus, the image covariance matrix G_r can be expressed as

$$G_r = \frac{1}{P}\sum_{i=1}^{P}\left(A_i - \bar{A}\right)\left(A_i - \bar{A}\right)^T \tag{5.30}$$

The projection matrix Φ_r consists of the eigenvectors of the matrix G_r corresponding to the m_r largest eigenvalues, where m_r is estimated using a similar condition to Equation 5.29.

5.4.3 Two-Directional 2D Linear Discriminant Analysis

(2D)²LDA considers the class information when forming the projection matrices. Its principle is to find a linear transformation to maximize the between-class scatter and minimize the within-class scatter for both dimensions of an input matrix. Therefore, the compressed feature $D \in \mathbb{R}^{n_r \times n_c}$ is given by

$$D = \Psi_r^T A \Psi_c \tag{5.31}$$

where Ψ_r and Ψ_c are the projection matrices. Let K be the number of classes, N_i the number of training samples in the ith class, and P the total number of training samples, $P = \Sigma_{i=1}^{K} N_i$. If A_j^i denotes the jth sample of the ith class and $\bar{A}^i$ is the mean of all samples of the ith class, $\bar{A}^i = 1/N_i \Sigma_{j=1}^{N_i} A_j^i$, then the between-class and within-class scatter matrices for the row direction are given by, respectively,

$$G_{bc} = \frac{1}{P}\sum_{i=1}^{K}N_i\left(\bar{A}^i - \bar{A}\right)^T\left(\bar{A}^i - \bar{A}\right) \tag{5.32}$$

$$G_{wc} = \frac{1}{P}\sum_{i=1}^{K}\sum_{j=1}^{N_i}\left(A_j^i - \bar{A}^i\right)^T\left(A_j^i - \bar{A}^i\right) \tag{5.33}$$

where $\bar{A}$ is the global mean given in Equation 5.28. Similarly, the between-class and within-class scatter matrices for the column direction are expressed as, respectively,

$$G_{br} = \frac{1}{P}\sum_{i=1}^{K}N_i\left(\bar{A}^i - \bar{A}\right)\left(\bar{A}^i - \bar{A}\right)^T \tag{5.34}$$

$$G_{wr} = \frac{1}{P}\sum_{i=1}^{K}\sum_{j=1}^{N_i}\left(A_j^i - \bar{A}^i\right)\left(A_j^i - \bar{A}^i\right)^T \tag{5.35}$$

The projection matrices Ψ_r and Ψ_c are obtained by maximizing the following Fisher criteria, respectively:

$$J(\Psi_r) = \text{trace}\left(\frac{\Psi_r^T G_{br} \Psi_r}{\Psi_r^T G_{wr} \Psi_r}\right) \tag{5.36}$$

and

$$J(\Psi_c) = \text{trace}\left(\frac{\Psi_c^T G_{bc} \Psi_c}{\Psi_c^T G_{wc} \Psi_c}\right) \tag{5.37}$$

The bases of the projection matrices Ψ_r and Ψ_c are the eigenvectors of $G_{wr}^{-1}G_{br}$ and $G_{wc}^{-1}G_{bc}$ corresponding to the n_r and n_c largest eigenvalues, respectively. Similar condition to that given in Equation 5.29 can be used to determine the parameters n_r and n_c.

5.5 Classification

After the dimensionality reduction step, the compressed feature vector is sent to a classifier to assign the Doppler signal to one of the reference gait classes. Here, two classification methods are investigated: minimum distance classifier and SVMs.

5.5.1 Minimum Distance Classifier

The minimum distance classifier with the Mahalanobis distance metric is a simple technique to predict the label of a test sample. The Mahalanobis distance is a measure of the distance between a point and a distribution. It is a weighted distance in that it takes into account not only the distance to the mean but also the variance of the distributions. Assuming that there are K gait classes, a subspace for each gait class is created, and its corresponding projection matrix Φ_k with the top most significant eigenvectors is computed. The number of retained eigenvectors is controlled by thresholding the energy packed in the retained eigenvalues, that is, using similar condition as given in Equation 5.29. Let μ_k and Σ_k denote, respectively, the mean vector and covariance matrix associated with the kth subspace, and d be the low-dimensional feature vector obtained by vectorizing the features given in Equation 5.25 or 5.31 from a local patch. The low-dimensional feature vector is then projected on all K gait subspaces:

$$y_k = \Phi_k^T d, \ k = 1, 2, \ldots, K \tag{5.38}$$

The classification is performed by computing the distance between y_k and each gait subspace, which is given by

$$h(y_k, \Phi_k) = (y_k - \mu_k)^T \Sigma_k^{-1}(y_k - \mu_k), \ k = 1, 2, \ldots, K \tag{5.39}$$

The gait subspace with the smallest Mahalanobis distance to y_k is the unknown gait class:

$$\min_k h(y_k, \Phi_k), \ k = 1, 2, \ldots, K \tag{5.40}$$

The idea here is that a feature projects closer to its own subspace than to any other. Of course, there are outliers that will be accounted for in the error rates.

5.5.2 Support Vector Machines

SVMs are supervised learning models, which have been widely used in many pattern recognition problems, due to their good generalization capability. It can be briefly described as follows: Consider a training set $\{d_i, y_i\}_{i=1}^{P}$, where d_i is the ith low-dimensional feature vector and $y_i \in \{1, -1\}$ is the corresponding class label. The design of an SVM classifier involves the solution of the following optimization problem:

$$\min_{w,b,\xi} \left(\frac{1}{2} w^T w + C \sum_{i=1}^{P} \xi_i \right), \text{s.t. } y_i \left[w^T \Omega(d_i) + b \right] \geq 1 - \xi_i, \xi_i \geq 0 \tag{5.41}$$

where:
w is a weight vector
ξ_i denotes the margin error associated with the ith compressed feature vector
$\Omega(d_i)$ is a kernel operator that maps a feature vector d_i into a higher dimensional space so that the classification problem becomes linearly separable
C is a positive regularization constant to control the trade-off between the margin and the misclassification rate

The parameter C is often obtained through a cross-validation procedure using a validation set. The dual form of Equation 5.41 is given by

$$\max_{\alpha_i} \left[\sum_{i=1}^{P} \alpha_i - \frac{1}{2} \sum_{i,j=1}^{P} \alpha_i \alpha_j y_i y_j \mathcal{K}(d_i, d_j) \right], \text{ s.t. } 0 \leq \alpha_i \leq C \forall i, \sum_{i=1}^{P} \alpha_i y_i = 0 \tag{5.42}$$

where $\mathcal{K}(d_i, d_j) \equiv \Omega(d_i)^T \Omega(d_j)$ is the kernel function. Here, the linear kernel, $\mathcal{K}(d_i, d_j) \equiv d_i^T d_j$, is used to design the SVM classifier. The vector w in the primal problem is related to the variables $\alpha_i, i \in [1, P]$, in the dual problem as follows:

$$w = \sum_{i=1}^{P} y_i \alpha_i \Omega(d_i) \tag{5.43}$$

After the estimation of the vector of Lagrange multipliers α using a training set, the predicted label of a test sample d_t can be computed as

$$y_p(d_t) = \text{sgn} \left[\sum_{i=1}^{N_s} y_i \alpha_i \mathcal{K}(s_i, d_t) + b \right] = \text{sgn} \left[w^T \Omega(d_t) + b \right] \tag{5.44}$$

where:

$$b = \frac{1}{N_s} \sum_{i=1}^{N_s} w^T s_i - y_i \tag{5.45}$$

where:
sgn($\cdot$) denotes the signum function
s_i is the ith support vector
N_s is the number of support vectors

5.6 Experimental Results

Several radar human gait signatures have been investigated so far; they include a person or a group of people walking, running, jogging, and crawling. Here, three types of arm motions are classified: (1) walking with both arms swinging, (2) walking while carrying an object in one hand, and (3) walking while holding an object with both hands. Figure 5.4 shows the images of a subject walking toward the radar with different arm motions and their respective T–F representations. Classification of these radar human gait signatures is very useful for military and law enforcement agencies. For example, based on μ-D effects induced by the arm movement of a person, one can differentiate the combat mode of enemy soldiers. In a hostage situation, law enforcement officers can discern the hostage from the captor by identifying the person walking with no arm motion. Doppler signals backscattered from a person walking with these types of arm motions are used to test the μ-D feature extraction method. In Section 5.6.1, the experimental setup is described, followed by an evaluation of the standard PCA method in Section 5.6.2. Then, the different steps of the μ-D feature extraction method are investigated. Finally, different feature extraction methods are compared.

5.6.1 Experimental Setup

A 24 GHz frequency-modulated CW radar with an antenna beam width of 7° horizontal and 25° vertical was used for data acquisition. The radar was positioned at a height of 0.7 m from the ground in an outdoor environment. The Doppler data were acquired from 18 subjects (7 females and 11 males). Each subject walked toward the radar while performing the aforementioned arm movements and repeated each motion type 3 times. The radar signal was recorded at a sampling rate of 7812 Hz for 10 s. Overall, a set of 162 Doppler signals was selected to evaluate the feature extraction method.

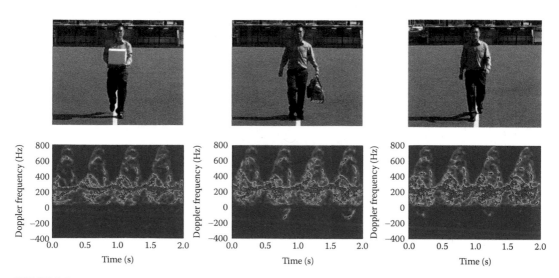

FIGURE 5.4
Images of a subject walking with different arm motions toward the radar and their T–F representations produced by the SM and enhanced by the Naka–Rushton equation.

5.6.2 Feature Extraction Using PCA

PCA was applied to extract features directly from the spectrogram, where time or frequency slices of the spectrogram were defined as input vectors [15]. A time slice is a snapshot of the spectrogram in time at selected frequencies. Each snapshot is, therefore, a measurement vector over a number of Doppler frequencies. A frequency slice is a snapshot of one Doppler frequency at multiple time points. Either one or both can be projected onto the gait subspaces, which are generated by PCA. In the experiment, the projected data were used to train the minimum distance classifier. To capture the variations of µ-D across the time axis, the time slices were used to form the three gait subspaces: no-arms swing, one-arm swing, and two-arms swing subspaces.

Figure 5.5a and b shows the magnitude of the sorted eigenvalues and the energy distribution for the no-arms swing subspace, respectively. The result in Figure 5.5b shows that PCA can capture 90% of the energy of the data with 53 eigenvectors. In other words, the time slice with a dimension of 1024 can be reduced to 53 features for classification. This consequently facilitates the training of the classifier and reduces the computational cost. After dimensionality reduction using PCA, each feature vector is assigned a label per Equation 5.40. The classification rate (CR) for no-arms swing as a function of the number of principal components is shown in Figure 5.6a. The crossover point is significant in that the number of times the no-arms gait data are classified to the no-arms gait class exceeds the other two labels. The number of principal components at the crossover point is 64. Further increasing the number of eigenvectors improves the CR. A majority vote scheme can be applied to assign the class label to the Doppler signal; thus, the unknown gait signal does not need to be correctly classified 100% of the time. Figures 5.6b and c show the classification results for one-arm and two-arms swings. The results show that at least 65 principal components are required to discriminate one class from the others, and the CR keeps improving when the gait subspaces are formed with more eigenvectors. This PCA-based method achieves good CR when considering long signal frames and assuming that there is insignificant variation in target speed. Therefore, to alleviate these issues, this chapter presents a method that extracts local T–F patches on the torso frequency shift as inputs. Then, log-Gabor filters and a matrix-based subspace method are used for feature extraction.

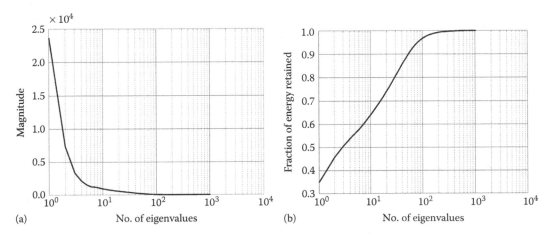

(a) No. of eigenvalues

(b) No. of eigenvalues

FIGURE 5.5
Sorted eigenvalues and energy distribution of the eigenvalues of the no-arms swing as a function of the number of eigenvalues: (a) Sorted eigenvalues and (b) energy distribution of the eigenvalues.

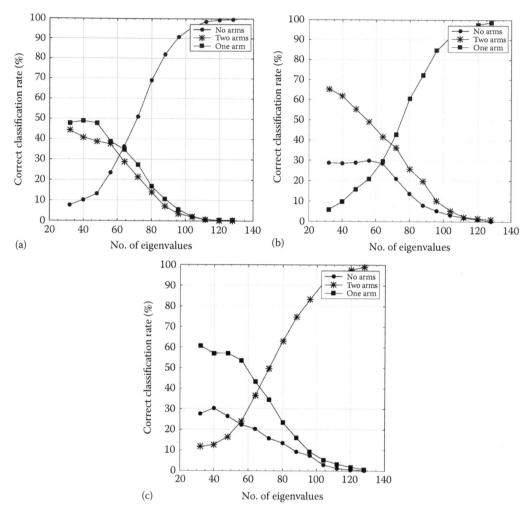

FIGURE 5.6
The CR as a function of the number of eigenvectors to form the (a) no-arms, (b) one-arm, and (c) two-arms subspaces.

5.6.3 T–F Distribution for μ-D Signature

The STFT and SM have commonly been used to depict the radar μ-D signature of human gait. The window lengths of these two T–F analyses control the resolution of the T–F representation, which in turn affects the performance of the feature extraction method. Therefore, different window lengths are investigated, ranging from 32 to 213 ms. The number of discrete Fourier transform (DFT) points and the overlap between consecutive windows are set to 2048 and 90%, respectively. Then a variable window length is used to form the T–F representation. For a window length smaller than 2048 samples, the signal is padded with zeros prior to T–F analysis. Initially, the input signal frame is set to 1 s, and the height of the local T–F patch is fixed at 128 samples (i.e., $N_y = 128$). Each local patch is contrast enhanced by the Naka–Rushton equation and convolved with the log-Gabor filters. The mean values of the non-overlapping subregions extracted from the feature maps of the log-Gabor filters are used as inputs for the SVM classifier. The number of

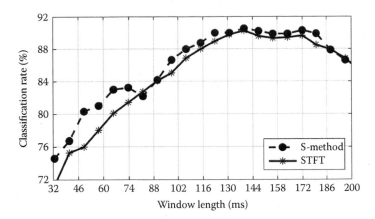

FIGURE 5.7
The effect of the window length of the T–F analysis method on the CR of the feature extraction method.

non-overlapping subregions extracted from each feature map is set to 8×8, that is, $r_1 = 16$ and r_2 is chosen to have a time duration of 125 ms. Figure 5.7 depicts the CR as a function of window length. In the feature extraction method, both STFT and SM achieve nearly similar CR and reach a peak CR with a window length of 139.3 ms. A much longer window reduces the CR, which may have been due to poor time resolution.

5.6.4 Log-Gabor Filter Set for Feature Extraction

The number of log-Gabor filters in the feature extraction method is determined experimentally for the given problem. To this end, a set of experiments is performed by varying the number of scales from three to five and the number of orientations from four to ten. For each combination of the number of scales and orientations, an SVM classifier is trained and the CR of the feature extraction method is computed. Figures 5.8a and b show the CR as a function of the number of scales and orientations for the STFT and SM, respectively. Increasing the number of scales markedly improves the CR of the feature extraction method. The CR reaches a steady state when the number of scales reaches four. For the number of orientations, the STFT obtains a peak CR of 90.0% with eight orientations, whereas the SM achieves a CR of 90.4% with nine orientations. Therefore, a set of 36 log-Gabor filters is used to conduct the remainder of the experiments.

5.6.5 Subspace Method for Dimensionality Reduction

After the conversion of the radar signal into T–F patches using the SM and the extraction of oriented motion energy features using the log-Gabor filters, $(2D)^2PCA$ and $(2D)^2LDA$ are applied to produce a low-dimensional feature vector. The CRs of these two dimensionality reduction methods as a function of the dimension of the feature vector are shown in Figure 5.9. The number of features can be further reduced without severely deteriorating the CR. Between the two matrix-based subspace methods, $(2D)^2LDA$ achieves better CR when the number of features is small. Further increasing the number of features improves the CR of the $(2D)^2PCA$ to the same level as the $(2D)^2LDA$. When the number of features is increased over 1024, the CR of $(2D)^2PCA$ is slightly higher than that of $(2D)^2LDA$. Therefore, $(2D)^2PCA$ is adopted for compressing the features before classification.

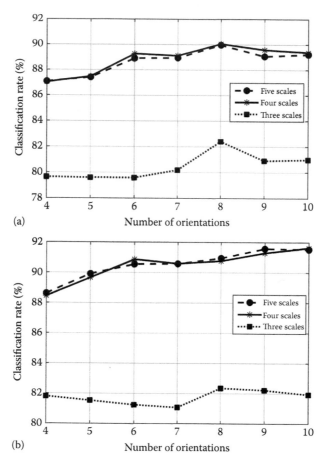

(a)

(b)

FIGURE 5.8
The effect of different number of log-Gabor filters on the CR of the feature extraction method in conjunction with (a) STFT and (b) SM.

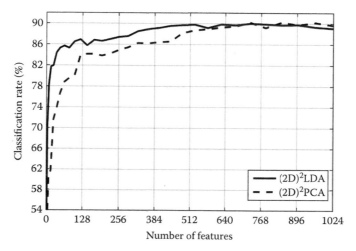

FIGURE 5.9
The CR of the matrix-based subspace dimensionality reduction methods as a function of the number of features.

5.6.6 Input Signal Length

The classification of µ-D signatures depends on the type of micro-motions exhibited by the moving object. A short signal frame may not contain sufficient arm and leg swings to differentiate the aforementioned arm movements, and a long signal frame can lead to feature redundancy and computational inefficiency. Therefore, in this subsection, different signal lengths varying from 0.5 to 3 s, with a step of 0.5 s are tested. For each input signal length, a new SVM classifier is trained. Figure 5.10 shows the CR of the feature extraction method as a function of the input signal length. The feature extraction method achieves a CR of 87.14% with an input length of 0.5 s, and increasing the signal length to 2 s improves the CR to 91.3%.

5.6.7 Comparison of Feature Extraction Methods

In comparison with other feature extraction methods, seven techniques were implemented and tested on the same dataset: (1) MFCC, (2) CVD, (3) EMD, (4) Gabor filters, (5) PCA, (6) 2DPCA, and (7) (2D)²PCA. In the MFCC-based method, a set of 40 triangular bandpass filters was employed to generate 64 Mel-scale cepstral coefficients. The window length of the analysis and the overlap between successive windows were set to 0.5 and 0.01 s, respectively. In the CVD-based method, the first three harmonic frequencies and their corresponding velocity profiles were used to produce the feature vector, which was then compressed using PCA. In the EMD-based method, each 2 s Doppler signal was decomposed into a set of IMFs, and the energies of the IMFs were used as features. In addition to log-Gabor filters, standard Gabor filters were used for extracting oriented motion energy features. Three eigen-based methods were implemented to extract features directly from the contrast-enhanced T–F patch. They employed PCA, 2DPCA, and (2D)²PCA to compute the eigen-components. The contrast-enhanced local patch was projected onto the subspace spanned by these eigen-components to form a low-dimensional feature vector. An SVM classifier was used with all feature extraction methods to predict the class of the Doppler signals.

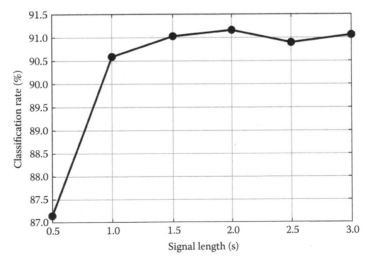

FIGURE 5.10
The effect of the length of the input Doppler signal on the CR of the feature extraction method.

TABLE 5.1

Classification Rates of Different Feature Extraction Methods
Using Sixfold Cross-Validation

Method	Classification rate (%)	Standard error (%)
Proposed method	91.3	0.40
Gabor filters	79.9	0.57
Eigen-based method with PCA	72.2	0.64
Eigen-based method with 2DPCA	70.6	0.65
Eigen-based method with (2D)²PCA	73.4	0.63
MFCC-based method	72.7	0.64
CVD-based method	62.3	0.69
EMD-based method	41.6	0.70

The CR of the feature extraction methods is obtained using sixfold cross-validation, that is, the dataset is divided into six parts where five parts are used for training and one part is kept for testing. Moreover, the subjects used for training are different from those used for testing. This procedure is repeated six times for different training sets and the final CR is computed as the percentage of correctly classified signals, which are aggregate across all six different test sets. Table 5.1 lists the CRs of different feature extraction methods. Among the eight feature extraction approaches, the proposed method achieves the highest CR. Substituting the log-Gabor filters with standard Gabor filters in the proposed method reduces the average CR from 91.3% to 79.9%. This shows that log-Gabor filters are more effective than Gabor filters in covering the feature spectrum. Gabor filters with more scales and orientations may improve the CR at the expense of increasing the size of the feature matrix and computational cost. The difference in CRs between the proposed method and the eigen-based method using (2D)²PCA shows that adding a filtering step enhances the classification performance by 17.9%. Also, the proposed method obtains 18.6% improvement in CR compared to MFCC and is 29.0% better than CVD. The EMD-based technique, which relies only on the energy of the extracted IMF, is least accurate with a CR of 41.6%.

5.7 Conclusion

This chapter presented a 2D feature extraction method for classifying μ-D human gait signatures. The proposed method applied joint T–F analysis, for example, SM, to transform a Doppler signal into a 2D representation from which a local patch centered on the torso frequency shift was extracted for classification. This local processing makes the proposed method robust against variations in target speed. In addition, subsampling and contrast-enhancement operations were performed on the local patches to reduce the influence of the leg swing speed and the signal propagation loss. Then, log-Gabor filters were applied to detect oriented motion energy features. The (2D)²PCA and (2D)²LDA methods were employed to reduce the dimension of the feature vector. In contrast to standard PCA and LDA, (2D)²PCA and (2D)²LDA do not require a matrix-to-vector conversion to compute the covariance matrix and do not suffer from singularity problem for small datasets. The proposed method was evaluated on real radar data and compared with seven different feature extraction methods. Experimental results showed that the proposed method can effectively classify human walking with different arm motions and achieve better CR than other existing feature extraction approaches.

References

1. V. C. Chen, F. Li, S. S. Ho, and H. Wechsler, Analysis of micro-Doppler signatures, *IEEE Proceedings of the Radar, Sonar and Navigation,* 150: 271–276, 2003.
2. V. C. Chen, *The Micro-Doppler Effect in Radar.* Boston, MA: Artech House, 2011.
3. T. Thayaparan, S. Abrol, E. Riseborough, L. Stankovic, D. Lamothe, and G. Duff, Analysis of radar micro-Doppler signatures from experimental helicopter and human data, *IET Radar, Sonar and Navigation,* 1: 289–299, 2004.
4. I. Bilik and J. Tabrikian, Radar target classification using doppler signatures of human locomotion models, *IEEE Transactions on Aerospace and Electronic Systems,* 43: 1510–1522, 2007.
5. S. S. Ram and H. Ling, Simulation of human microdopplers using computer animation data, in *IEEE Radar Conference,* Rome, Italy, 2008, pp. 1–6.
6. D. Tahmoush and J. Silvious, Stride rate in radar micro-Doppler images, in *IEEE International Conference on Systems, Man and Cybernetics,* San Antonio, TX, 2009, pp. 4218–4223.
7. S. S. Ram, C. Christianson, Y. Kim, and H. Ling, Simulation and analysis of human micro-Dopplers in through-wall environments, *IEEE Transactions on Geoscience and Remote Sensing,* 48: 2015–2023, 2010.
8. T. Damarla, M. Bradley, A. Mehmood, and J. M. Sabatier, Classification of animals and people ultrasonic signatures, *IEEE Sensors Journal,* 13: 1464–1472, 2013.
9. V. C. Chen, D. Tahmoush, and W. J. Miceli, *Radar Micro-Doppler Signatures: Processing and Applications,* The Institution of Engineering and Technology, London, vol. 34: IET, 2014.
10. P. Molchanov, A. Totsky, J. Astola, K. Egiazarian, S. Leshchenko, and M. Rosa-Zurera, Aerial target classification by micro-Doppler signatures and bicoherence-based features, in *Proceedings of the 9th European Radar Conference,* Amsterdam, the Netherlands, 2012, pp. 214–217.
11. Y. Li, L. Du, and H. Liu, Hierarchical classification of moving vehicles based on empirical mode decomposition of micro-Doppler signatures, *IEEE Transactions on Geoscience and Remote Sensing,* 51: 3001–3013, 2013.
12. S.-H. Park, Automatic target recognition using jet engine modulation and time-frequency transform, *Progress in Electromagnetics Research M,* 39: 151–159, 2014.
13. A. R. Persico, C. Clemente, C. Ilioudis, D. Gaglione, J. Cao, and J. Soraghan, Micro-Doppler based recognition of ballistic targets using 2D Gabor filters, in *Sensor Signal Processing for Defence,* Edinburgh: IEEE, 2015, pp. 1–5.
14. Y. D. Zhang and K. C. Ho, Continuous wave doppler radar for fall detection, in *Radar for Indoor Monitoring,* M. G. Amin, Ed. Boca Raton, FL: CRC Press, 2017.
15. B. G. Mobasseri and M. G. Amin, A time-frequency classifier for human gait recognition, in *Proceedings of the SPIE, Optics and Photonics in Global Homeland Security V and Biometric Technology for Human Identification VI,* Vol. 7306, Orlando, FL: SPIE, 2009, pp. 730628-1–730628-9.
16. M. B. Guldogan, F. Gustafsson, U. Orguner, S. Bjorklund, H. Petersson, and A. Nezirovic, Human gait parameter estimation based on micro-Doppler signatures using particle filters, in *IEEE International Conference on Acoustics, Speech and Signal Processing,* Prague, Czech Republic, 2011, pp. 5940–5943.
17. P. Molchanov, J. Astola, K. Egiazarian, and A. Totsky, Ground moving target classification by using DCT coefficients extracted from micro-Doppler radar signatures and artificial neuron network, in *Microwaves, Radar and Remote Sensing Symposium,* Kiev, Ukraine: IEEE, 2011, pp. 173–176.

18. P. D. Fairchild and M. R. Narayanan, Classification and modeling of human activities using empirical mode decomposition with S-band and millimeter-wave micro-Doppler radars, in *Proceedings of the SPIE Radar Sensor Technology XVI*, Vol. 8361, Baltimore, MD, 2012, pp. 83610X-1–83610X-15.
19. I. Bilik and P. Khomchuk, Minimum divergence approaches for robust classification of ground moving targets, *IEEE Transactions on Aerospace and Electronic Systems*, 48: 581–603, 2012.
20. F. H. C. Tivive, A. Bouzerdoum, and M. G. Amin, A human gait classification method based on radar doppler spectrograms, *EURASIP Journal on Advances in Signal Processing*, 2010: 1–12, 2010.
21. I. Orovic, S. Stankovic, and M. Amin, A new approach for classification of human gait based on time-frequency feature representations, *Signal Processing*, 91: 1448–1456, 2011.
22. S. Bjorklund, T. Johansson, and H. Petersson, Evaluation of a micro-Doppler classification method on mm-wave data, in *IEEE Radar Conference*, Atlanta, GA, 2012, pp. 0934–0939.
23. D. Tahmoush, Micro-range micro-Doppler for dismount classification, in *Proceedings of the SPIE, Radar Sensor Technology XVII*, Vol. 8714, Baltimore, MD, 2013, pp. 87141E-1–87141E-7.
24. P. O. Molchanov, J. T. Astola, K. O. Egiazarian, and A. V. Totsky, Classification of ground moving targets using bicepstrum-based features extracted from micro-Doppler radar signatures, *EURASIP Journal on Advances in Signal Processing*, 2013: 1–13, 2013.
25. S. Groot, R. Harmanny, H. Driessen, and A. Yarovoy, Human motion classification using a particle filter approach: multiple model particle filtering applied to micro-Doppler spectrum, *International Journal of Microwave and Wireless Technologies*, 5: 391–399, 2013.
26. J. Park, J. T. Johnson, N. Majurec, M. Frankford, K. Stewart, G. E. Smith, et al., Simulation and analysis of polarimetric radar signatures of human gaits, *IEEE Transactions on Aerospace and Electronic Systems*, 50: 2164–2175, 2014.
27. M. G. Amin, F. Ahmad, Y. D. Zhang, and B. Boashash, Human gait recognition with cane assistive device using quadratic time-frequency distributions, *IET Radar, Sonar and Navigation*, 9: 1224–1230, 2015.
28. R. Ricci and A. Balleri, Recognition of humans based on radar micro-Doppler shape spectrum features, *IET Radar, Sonar and Navigation*, 9: 1216–1223, 2015.
29. A. Kale, A. Sundaresan, A. N. Rajagopalan, N. P. Cuntoor, A. K. Roy-Chowdhury, V. Kruger, et al., Identification of humans using gait, *IEEE Transactions on Image Processing*, 13: 1163–1173, 2004.
30. X. Huang and N. V. Boulgouris, Human gait recognition based on multiview gait sequences, *EURASIP Journal on Advances in Signal Processing*, 2008: 1–8, 2008.
31. T. H. W. Lam, K. H. Cheung, and J. N. K. Liu, Gait flow image: A silhouette-based gait representation for human identification, *Pattern Recognition*, 44: 973–987, 2011.
32. T. Connie, M. K. O. Goh, and A. B. J. Teoh, A grassmannian approach to address view change problem in gait recognition, *IEEE Transactions on Cybernetics*, PP: 1–14, 2016.
33. K. Kalgaonkar and B. Raj, Acoustic Doppler sonar for gait recoginaton, in *IEEE Conference on Advanced Video and Signal Based Surveillance*, London, 2007, pp. 27–32.
34. M. U. B. Altaf, T. Butko, and B.-H. F. Juang, Acoustic gaits: Gait analysis with footstep sounds, *IEEE Transactions on Biomedical Engineering*, 62: 2001–2011, 2015.
35. Y. Qi, C. B. Soh, E. Gunawan, K.-S. Low, and R. Thomas, Assessment of foot trajectory for human gait phase detection using wireless ultrasonic sensor network, *IEEE Transactions on Neural Systems and Rehabilitation Engineering*, 24: 88–97, 2016.
36. J. Man and B. Bhanu, Individual recognition using gait energy image, *IEEE Transactions on Pattern Analysis and Machine Intelligence*, 28: 316–322, 2006.
37. Q. Yang and K. Qiu, Gait recognition based on active energy image and parameter-adaptive kernel PCA, in *Information Technology and Articial Intelligence Conference*, Chongqing, China, 2011, pp. 156–159.
38. E. Zhang, Y. Zhao, and W. Xiong, Active energy image plus 2DLPP for gait recognition, *Signal Processing*, 90: 2295–2302, 2010.
39. W. Kusakunniran, Recognizing gaits on spatio-temporal feature domain, *IEEE Transactions on Information Forensics and Security*, 9: 1416–1423, 2014.

40. S. S. Ram, S. Z. Gurbuz, and V. C. Chen, Modeling and simulation of human motions for micro-Doppler signatures, in *Radar for Indoor Monitoring*, M. G. Amin, Ed. Boca Raton, FL: CRC Press, 2017.
41. V. C. Chen, W. J. Miceli, and B. Himed, Micro-Doppler analysis in ISAR—Review and perspectives, in *International Radar Conference*, Bordeaux, France: IEEE, 2009, pp. 1–6.
42. G. Kirose, Animating a human body mesh with MAYA for Doppler signature computer modeling, *Army Research Laboratory, Report No. ARL-TN-0351*, Adelphi, MD, 2009.
43. L. Stankovic, A method for time-frequency analysis, *IEEE Transactions on Signal Processing*, 42: 225–229, 1994.
44. T. Thayaparan, L. Stankovic, I. Djurovic, S. Penamati, and K. Venkataramaniah, Intelligent target recognition using micro-Doppler radar signatures, in *Proceedings of the SPIE, Radar Sensor Technology XIII*, Vol. 7308, Orlando, FL: SPIE, 2009, pp. 730817-1–730817-11.
45. G. E. Smith, K. Woodbridge, and C. J. Baker, Template based micro-Doppler signature classification, in *European Radar Conference*, Manchester: IEEE, 2006, pp. 158–161.
46. G. E. Smith, K. Woodbridge, and C. J. Baker, Radar micro-Doppler signature classification using dynamic time warping, *IEEE Transactions on Aerospace and Electronic Systems*, 46: 1078–1096, 2010.
47. R. J. Javier and Y. Kim, Application of linear predictive coding for human activity classification based on micro-Doppler signatures, *IEEE Geoscience and Remote Sensing Letters*, 11: 1831–1834, 2014.
48. Y. Kim and H. Ling, Human activity classification based on micro-Doppler signatures using a support vector machine, *IEEE Transactions on Geoscience and Remote Sensing*, 47: 1328–1337, 2009.
49. C. Clemente, L. Pallotta, A. D. Maio, J. J. Soraghan, and A. Farina, A novel algorithm for radar classification based on doppler characteristics exploiting orthogonal pseudo-zernike polynomials, *IEEE Transactions on Aerospace and Electronic Systems*, 51: 417–430, 2015.
50. B. Jokanovic, M. Amin, and F. Ahmad, Radar fall motion detection using deep learning, in *IEEE Radar Conference*, Philadelphia, PA, 2016, pp. 1–6.
51. J. Li, S. L. Phung, F. H. C. Tivive, and A. Bouzerdoum, Automatic classification of human motion using Doppler radar, in *Proceedings of the International Joint Conference on Neural Networks*, Brisbane, Australia, 2012, pp. 1–6.
52. F. H. C. Tivive, S. L. Phung, and A. Bouzerdoum, Classification of micro-Doppler signatures of human motions using log-gabor filters, *IET Radar, Sonar and Navigation*, 9: 1188–1195, 2015.
53. H. Griffiths, M. Ritchie, and F. Fioranelli, Bistatic radar configuration for human motion detection and classification, in *Radar for Indoor Monitoring*, M. G. Amin, Ed. Boca Raton, FL: CRC Press, 2017.
54. K. I. Naka and W. A. H. Rushton, S-potentials from colour units in the retina of fish (cyprinidae), *The Journal of Physiology*, 185: 536–555, 1966.
55. D. J. Field, Relations between the statistics of natural images and the response properties of cortical cells, *Journal of the Optical Society of America A*, 4: 2379–2394, 1987.
56. J. Arrospide and L. Salgado, Log-gabor filters for image-based vehicle verification, *IEEE Transactions on Image Processing*, 22: 2286–2295, 2013.
57. J. Yang, D. Zhang, A. F. Frangi, and J.-Y. Yang, Two-dimensional PCA: A new approach to appearance-based face representation and recognition, *IEEE Transactions on Pattern Analysis and Machine Intelligence*, 26: 131–137, 2004.
58. D. Zhang and Z.-H. Zhou, (2D)^{2}PCA: Two-directional two-dimensional PCA for efficient face representation and recognition, *Journal Neurocomputing*, 69: 224–231, 2005.
59. S. Noushath, G. H. Kumar, and P. Shivakumara, (2D)^{2}LDA: An efficient approach for face recognition, *Pattern Recognition*, 39: 1396–1400, 2006.

6

Range-Doppler Processing for Human Motion Detection and Classification

David Tahmoush, Fauzia Ahmad, Anthony Martone,
Graeme E. Smith, and Zachary Cammenga

CONTENTS

6.1 Introduction

Human motion detection and classification in indoor environments are the primary objectives in many defense and security applications, ranging from through-the-wall radar imaging (TWRI) to hostage crisis [1–3]. These capabilities are also of fundamental importance in the emerging areas of remote patient monitoring and elderly assisted living, where radar technology is showing great potential [4–7]. Human activities are characterized by motions of the torso and limbs, breathing, and heartbeat. These attributes make humans distinguishable from other targets and clutter, even in inherently challenging indoor environments, and enable target detection to proceed based on changes in the phase of the scattered radar signals over time.

Indoor human motions, such as sitting, bending, picking an object, and head turning, typically do not generate fixed prolonged Doppler frequencies. This is because such human motions can be abrupt and highly nonstationary, producing a time-dependent phase whose rate of change may fail to translate into single or multicomponent sinusoids that can be captured by different individual Doppler filters. Rather, the corresponding wide spectrum of human motions can span the entire radar frequency band. In lieu of Doppler filters, time–frequency signal representations (TFRs) can be used to reveal the local frequency behavior of the multicomponent signal, including the component instantaneous frequencies (IFs) [8–10]. This constitutes the signal micro-Doppler signatures, which characterize the micro-motion dynamics of the indoor human activities.

A variety of published works have shown how human motion detection and classification can be achieved using micro-Doppler signatures [8–15]. Most of these works consider a continuous-wave (CW) signal. More recently, research has begun to consider wideband signals through a hybrid approach that combines high range resolution (HRR) observations and micro-Doppler signatures to provide a more informative radar human signature [16–25]. Fogle and Rigling [25] demonstrated the ability to decompose the scattering of a walking person into the individual scattering centers by separating the motion from the individual limbs. The decomposition depends on the use of both HRR and prior knowledge of the micro-Doppler signature for a walking person. The ability to separate the limbs of a person and track them in range and frequency underpins the use of a joint HRR micro-Doppler signature or a *wideband micro-Doppler signature* for extracting fine characteristic details of a target.

The wideband micro-Doppler signature may also be used to help extract target signatures in complex scenes with multiple, closely spaced targets. The traditional *narrowband* micro-Doppler analysis does not allow for a wide signal bandwidth. If multiple targets are present, the micro-Doppler signature of the targets would be overlaid in the time–frequency domain [26]. However, using a wideband signal, researchers have demonstrated the ability to separate the scatterings from the different targets resulting in the separation of their corresponding micro-Doppler signatures [17,23,24]. By separating the scattering from multiple humans, their micro-Doppler signatures can be independently estimated resulting in the ability to classify the individual targets separately rather than relying on a collective signature.

The move from narrowband to wideband micro-Doppler permits extraction of both structural information through HRR and micro-motions through the Doppler signature. As such, it requires the enhancement of the existing micro-Doppler theory and the development of new signal analysis methods that enable the use of the additional range information. The existing literature begins to approach this with individual studies developing parts of the theory relevant to hypothesis being considered [16–25]. However, a holistic theory that converts the narrowband micro-Doppler theory [27–29] to the wideband case has yet to be proposed. Conversely, signal transformations that create a three-dimensional (3D) data cube of range–time–frequency to help analyze and visualize the wideband micro-Doppler signature have been developed in [16,20,23]. These works show that by combining the wideband signal with micro-Doppler analysis, useful information can be extracted from the data.

Identifying moving humans in indoor environments comes with its own unique challenges, including lack of direct observations, obscuration from walls, and multiple-bounce pathways contributing to the received radar signals [30–35]. For the radar to perform well and achieve high performance, the system must be capable of incorporating environmental features and adapting dynamically to their time dependencies. Multipath can be used as an indirect means to view a target which is not in direct line of sight (LOS), thereby decreasing the obscuration effects from walls and increasing the sensing area in an indoor environment. Use of wideband signals provides range resolution capability, which permits separation of the multiple multipath reflections and inference of additional information about the human subject. Even when LOS is available, multipath returns can be used in conjunction with prior knowledge about the environment to confirm target detection [36]. Further, target information can be extracted from multipath returns to enhance tracking performance [37]. Standard data association techniques, such as the nearest neighbor standard filter, optimal Bayesian data association filter [38], or the probabilistic data association filter [39,40], need to be customized to accommodate multipath returns.

An alternative to pulse-Doppler processing for indoor moving target indication is change detection (CD) [41–44], wherein target detection is accomplished by subtraction of some data frames acquired over successive probing of the scene. CD is often favored for radar applications that require a high spatial resolution. In such applications, the moving target is typically represented by multiple range and cross-range resolution cells. The target in motion causes range migration, which is a transition from one resolution cell to another over a coherent processing interval (CPI). Range migration is often disruptive to pulse-Doppler processing due to straddle loss and results in reduced signal-to-noise ratio and range resolution [45]. An additional advantage of CD is its robustness to abrupt, complex, and nonuniform (i.e., nonlinear phase variation with time) human motions. These complex motions often cause phase discontinuities in pulse-Doppler processing that cannot be resolved into a single Doppler frequency shift [46].

In this chapter, we consider human motion detection and classification in indoor environments using wideband pulse-Doppler radars. Section 6.2 describes the wideband micro-Doppler theory from first principles and demonstrates the benefit of wideband micro-Doppler signatures for human gait estimation. Section 6.3 demonstrates the offerings of range-Doppler processing for elderly fall detection in assisted living applications. Section 6.4 considers motion detection, tracking, and classification in multipath-rich indoor environment focusing on hallways. Section 6.5 details a CD approach to human motion indication for TWRI applications. Real radar data of humans moving behind walls are used to demonstrate the effectiveness of the CD process and highlight the imaging artifacts that are manifested. Section 6.6 contains the concluding remarks.

6.2 Wideband Micro-Doppler with Application to Human Gait Estimation

The concept of micro-Doppler, as introduced for radar targets in [47: Chapter 2], considers a CW signal and treats complicated targets, such as humans, as a collection of point scatterers. The use of a continuous waveform eliminates the ability to separate these scatterers in range. As such, the target's micro-Doppler signature is the summation of the micro-Doppler responses of each scattering center. The case in which the waveform is wideband,

and the target can occupy multiple range bins and migrate between them during measurement, has not been fully considered in prior works and is the focus of this section.

6.2.1 Wideband Micro-Doppler Theory

A simple geometry of interest for the mathematical formulation of the micro-Doppler effect is the observation of a point scatterer that exhibits micro-motion as well as conventional Doppler. As shown in Figure 6.1, a stationary radar located at point O observes a point-like scatterer at point P on a target. The points are defined in a fixed coordinate system (X, Y, Z). The target exhibits a bulk velocity, v, defined in the fixed coordinate system, and a rotation with respect to the radar that is described in terms of a local coordinate system (x, y, z) as $\Omega = (\omega_x, \omega_y, \omega_z)^T$. The origin of the local coordinate system is placed at the center of rotation of the target, which is the point P_0 in the (X, Y, Z) system. Over time, the position of the point P_0 will change due to the bulk velocity v. The position of P relative to P_0 will also change due to the rotation of the vector r, which indicates the position of the point scatterer relative to the center of target motion in the local coordinates. Summation of these two motions causes the change in the range to the point P (described by $r(t)$), leading to a progressive phase change in the received signal and, ultimately, the Doppler shift.

For a CW radar with carrier frequency f_c, the return from the point scatterer P is given by

$$s(t) = \rho(x, y, z) \exp\left[j2\pi f_c \frac{2r(t)}{c} \right] = \rho(x, y, z) \exp\{ j\phi[r(t)] \} \qquad (6.1)$$

where:
$\rho(x, y, z)$ is the scatterer reflectivity described in the local coordinates (x, y, z)
$r(t)$ represents the scalar range to the target as a function of time
$\phi(\cdot)$ is the phase of the baseband signal

The Doppler frequency shift induced by the target's motion is defined as

$$f_D(t) = \frac{1}{2\pi} \frac{d\phi(r(t))}{dt} = \frac{2f_c}{c} \frac{dr(t)}{dt} \qquad (6.2)$$

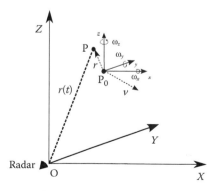

FIGURE 6.1
Geometry of a radar target with rotating parts.

Following the analysis in [47: Chapter 2], the Doppler frequency in Equation 6.2 takes the form

$$f_D = \frac{2f_c}{c}\left[v + \Omega \times r\right]\cdot n \tag{6.3}$$

where:

c is the speed of light

n can be approximated as the unit vector along the radar LOS, represented by vector $\overrightarrow{OP}$ in Figure 6.1, when the range to the target is much greater than the target displacement during the observation interval

The result of Equation 6.3 is very general and can be used to model a variety of motions, including rotation, vibration, and tumbling.

The use of a continuous waveform, however, limits the radar's ability to resolve targets in range. The micro-Doppler theory can be extended to include signal bandwidth through consideration of micro-motions when a pulsed waveform is used. A pulsed radar signal can be described mathematically as

$$s_{tx}(t) = \text{rect}\left(\frac{t - \frac{\tau}{2}}{\tau}\right)\exp\left(j2\pi f_c t\right)q(t) \tag{6.4}$$

where:

τ is the pulse duration

$q(t)$ captures the modulation (e.g., $q(t) = e^{j\alpha t^2}$ for a linear frequency modulation [LFM] signal [*chirp* signal] with a chirp rate of α), and the function rect($\cdot$) is defined as

$$\text{rect}(t) = \begin{cases} 1, & |t| < \frac{1}{2} \\ 0, & |t| \geq \frac{1}{2} \end{cases} \tag{6.5}$$

For a point scatterer, the received signal can be described as a time-shifted version of the transmitted signal, $s_{tx}(t)$. Let the time delay to the scatterer be represented by t_d. This time delay is dependent on the two-way distance between the point scatterer and the radar. If the target is moving, this distance changes for every consecutive pulse that is transmitted. The slow-time variable to help characterize the passage of time between pulses is the pulse repetition interval and is represented by t_s. Using slow time, the change in target range over consecutive pulses is given by $r(t_s)$. Then, t_d is expressed in terms of the slow-time change in range as

$$t_d(t_s) = \frac{2r(t_s)}{c} \tag{6.6}$$

Thus, the radar return from the point scatterer can be written in full by inserting the delay term into the transmitted signal expression as

$$s_{rx}(t,t_s) = \rho(x,y,z)s_{tx}(t - t_d(t_s)) \tag{6.7}$$

Combining Equations 6.4 and 6.7, the radar return is expressed as

$$s_{rx}(t,t_s) = \rho(x,y,z)\,\text{rect}\left(\frac{t-t_d(t_s)-\frac{\tau}{2}}{\tau}\right)\exp(j2\pi f_c(t-t_d(t_s)))q\,(t-t_d(t_s)) \tag{6.8}$$

Matched filtering [48], sometimes known as *dechirping*, of the received signal produces the output, $s_d\,(t,t_s)$, given by

$$s_d(t,t_s) = s_{rx}(t,t_s) * s_{tx}^*(-t) \tag{6.9}$$

where "*" denotes the convolution and the superscript "*" represents complex conjugation. Substituting Equations 6.4 and 6.8 in Equation 6.9 yields

$$s_d(t,t_s) = \rho(x,y,z)\,\text{tri}\left(\frac{t-t_d(t_s)-\frac{\tau}{2}}{\tau}\right)\exp\left(-j2\pi f_c t_d(t_s)\right)\tilde{Q}(t-(t_d(t_s))) \tag{6.10}$$

where $\tilde{Q}(\cdot)$ accounts for the convolution of the applied modulation [26] and the function tri$(\cdot)$, defined as

$$\text{tri}(t) = \begin{cases} 1-|t|, & |t| < 1 \\ 0, & \text{otherwise} \end{cases} \tag{6.11}$$

arises from the convolution of the rect$(\cdot)$ parts of the transmitted and received signals. It is noted that for an LFM modulation, $\tilde{Q}(\cdot)$ would assume the form of a *sinc* function.

From Equation 6.10, the phase of the return signal after matched filtering is given by

$$\phi(t_s) = 2\pi f_c t_d(t_s) \tag{6.12}$$

The micro-Doppler frequency f_D can now be determined from the signal phase as

$$f_D(t_s) = \frac{1}{2\pi}\frac{d\phi(t_s)}{dt_s} = \frac{2f_c}{c}\frac{dr(t_s)}{dt_s} \tag{6.13}$$

Following the analysis for the CW case, the Doppler frequency in this case can be shown to be equal to that in Equation 6.3. We observe from Equation 6.13 that the Doppler frequency is independent of t (also referred to as fast time) and only depends on the change in range over slow time. However, in the case of the pulsed signal, the magnitude of the matched filter output in fast time needs to be considered as well. From Equation 6.10, the magnitude can be determined as

$$|s_d(t,t_s)| = \left|\rho(x,y,z)\,\text{tri}\left(\frac{t-t_d(t_s)-\frac{\tau}{2}}{\tau}\right)\tilde{Q}(t-t_d(t_s))\right| \tag{6.14}$$

which represents the range–time surface with range corresponding to fast time, t, and the time represented by slow time, t_s. A theoretical joint range-time-frequency (JRTF) space can, thus, be created by mapping the range–time surface corresponding to $|s_d(t,t_s)|$ into the third dimension corresponding to f_D, that is,

$$\Upsilon((t,t_s),f_D) = \left(\left| \rho(x,y,z)\text{tri}\left(\frac{t - t_d(t_s) - \frac{\tau}{2}}{\tau} \right) \tilde{Q}(t - t_d(t_s)) \right|, \frac{2f_c}{c}[v + \Omega \times r] \cdot n \right) \qquad (6.15)$$

This function Υ is presented as the theoretical implementation of the ideal JRTF data cube produced when a wideband waveform is used. The JRTF data cube can be related to the narrowband formulation of the expected Doppler frequency shift by considering the limit as τ approaches infinity. The argument of tri$(\cdot)$ in Equation 6.15 approaches zero as τ approaches infinity, causing tri$(\cdot)$ to approach unity. Therefore, the range–time surface portion of Υ depends only on the modulation imparted on the signal represented by $\tilde{Q}$. For a pure tone CW signal, $|\tilde{Q}|$ is set to 1, and the expression for the JRTF data cube takes the form

$$\Upsilon_{\text{NB}}((t,t_s),f_D) = \left(|\rho(x,y,z)|, \frac{2f_c}{c}[v + \Omega \times r] \cdot n \right) \qquad (6.16)$$

This expression for the narrowband case shows that dependence on fast time is eliminated and the range–time surface assumes a constant, $\rho(x,y,z)$, whereas the expected micro-Doppler frequency remains unchanged compared with the wideband case.

6.2.2 Visualization of Wideband Micro-Doppler Signatures

The presented wideband micro-Doppler theory assumes that a target can be deconstructed into its individual point scatterers and then the JRTF data cube, Υ, can be evaluated for the motion of each scatterer separately. However, in practice, the motion of each individual scattering center is not known, and the motion of the various scatterers can overlap. Calculation of the JRTF data cube without prior information about the target should, therefore, be considered.

The JRTF signature can be estimated using conventional TFRs. One such representation is the short-time Fourier transform (STFT). The STFT permits the observation of frequency variations versus slow time and is a common technique for observing micro-Doppler signatures with the absence of observable cross terms [49,50]. With a narrowband system, the time–frequency representation provided by the STFT does not allow for range information. However, with a wideband system, targets that would otherwise be unresolvable with a narrowband system can now be resolved. Each sample in fast time provides range information about the target, and we would like a data transformation that retains this information. This process creates a TFR for each range bin in the target scene (Figure 6.2). The TFR for each range bin can then be considered simultaneously creating a data cube of information resulting in the JRTF signature. Through the use of the STFT, the JRTF data cube can be formed for complex scenes. This technique allows for further analysis of experimental results of targets consisting of multiple scattering centers.

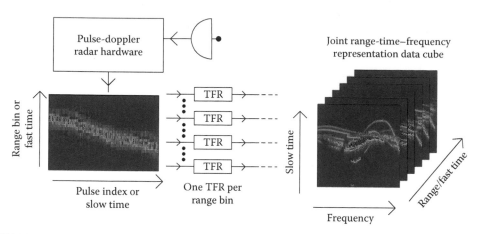

FIGURE 6.2
Calculation of the joint-range time–frequency data cube using a TFR.

6.2.3 Wideband Micro-Doppler of a Walking Human Target

To demonstrate the effect of wideband waveforms on the micro-Doppler signature, a simulation of a walking person is used. The radar and geometrical parameters employed in the simulation are listed in Table 6.1.

The human gait simulation is based on the human gait estimation model in References 47 and 51. The geometry setup consists of a single person walking toward the radar, as shown in Figure 6.3. With the wideband signal, an HRR profile is generated for each pulse. The series of HRR profiles is shown in Figure 6.4, where the color scale represents the uncalibrated power of the backscatter signal in dB. The range profiles show the target return observable in approximately 16 range bins over the course of the collection time. The general decrease in range over slow time shows that the motion of the person is toward the radar. This dominant response is attributed to the person's torso. Additional scattering, originating in front of and behind the main torso signature, is attributed to the limbs. In addition to the range profile, the traditional micro-Doppler signature can be generated by coherently summing range bins containing the target response [52] (Figure 6.5). The micro-Doppler signature shows multiple components corresponding to returns from different parts of the body without the use of wideband capabilities. The motion of the torso can be seen in high-intensity curves in the middle of the signature with an approximately sinusoidal shape. The limb motion is also visible in the lower intensity curves that surround the torso response. These curves are clearly periodic in nature, although their shape is no longer sinusoidal.

TABLE 6.1

Radar and Geometrical Parameters for
Simulation of a Walking Pedestrian

PRF	10 kHz
Carrier frequency	92.5 GHz
Bandwidth	1 GHz
Distance to target	7 m
Velocity	~1 m/s

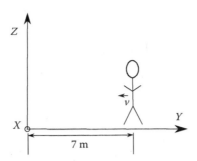

FIGURE 6.3
Geometry of walking person simulation.

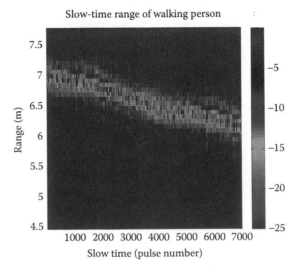

FIGURE 6.4
Range profile of person walking toward radar. The color scale is in dB normalized intensity.

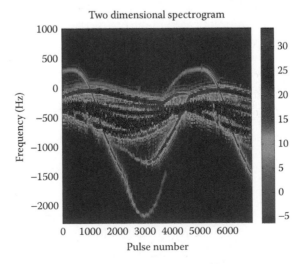

FIGURE 6.5
Micro-Doppler signature of person walking toward radar. The color scale is in dB intensity.

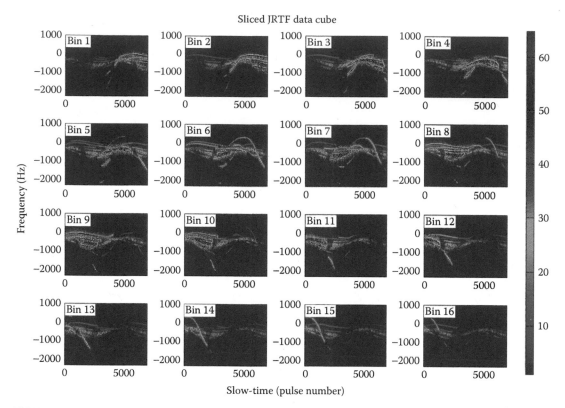

FIGURE 6.6
Spectrogram of each range bin contains scattering of walking person. The color scale is in dB intensity and common to all subplots.

With wideband capabilities, an STFT was used on the each of the 16 range bins that the target is present during the collection time. The results of the STFT show the time–frequency relationship for each specific range bin. With the increased complexity of the target's radar signature, it is impractical to identify a surface in the range–time–frequency data cube upon which the signature can be observed. Instead, the cuts from each range bin are shown in Figure 6.6, where *Bin 1* corresponds to the first range bin containing a target response that is closest to the radar, and the color scale shows the intensity of the frequency component in dB. It can be clearly observed that the frequency content in the first 2000 pulses in bins 13–16 is very similar. This is due to the presence of large radar cross section (RCS) scatterers whose sidelobes are present in adjacent range bins. This spreading of the scatterers in range was predicted by the $\mathrm{tri}(\cdot)$ and $\tilde{Q}$ functions of Equation 6.15. We can now see that the JRTF representation allows for the observation of the target's migration in range simultaneously with the micro-Doppler frequency information.

6.2.4 Discussion

Wideband micro-Doppler is a new form of target radar signature that combines the position information of the target structure with their micro-motions. The information estimated using wideband micro-Doppler techniques provides a greater insight into the structure of the target by breaking a complex target into a series of simple scatterers that can be separated in range. This provides the relative position of the scatterers exhibiting micro-motions within the structure of the target. The wideband micro-Doppler techniques also provide the ability of separating closely spaced targets before accessing their respective micro-Doppler signatures. The theoretical derivation of the JRTF data cube shows that the Doppler frequency is not dependent on range, so the 3D data cube can be represented as the two-dimensional (2D) range–time surface augmented in height by the Doppler frequency at each point in time. Using the STFT and HRR radar, the signature of a walking person is simulated, and the migration of the target can be viewed simultaneously with the micro-Doppler signature information.

6.3 Wideband Pulse-Doppler Radar-Based Elderly Fall Detection

Successful detection of a fall, locating its occurrence to, at least, room accuracy, and discriminating it from other gross motor activities with high classification rates would provide critical information to the first responders, leading to not only timely intervention and the most effective treatment but also a reduction in related medical expenses and subsequent burden on families that care for the senior relative remotely [7,53–59]. In this section, we use range information obtained from pulse-Doppler radar to distinguish between fall and non-fall motions with similar Doppler features. We also discuss the effects of the radar aspect angle on fall detection performance. A Kinect-based radar simulator is employed to demonstrate the merits of pulse-Doppler radar-based elderly motion monitoring.

6.3.1 Kinect-Based Wideband Pulse-Doppler Radar Simulator

Kinect-based wideband pulse-Doppler radar simulator employs the 3D position measurements obtained from the Kinect sensor in lieu of the time-varying range measurements typically obtained using radar [60]. The Kinect data for a human motion activity is first captured and stored in memory to create a database. Then, the Kinect time-varying 3D positions of different joints of the human body are computed per target position relative to the system and direction of motion. A smoothing function of low-pass filter characteristics is applied to the Kinect skeletal data to remove the high-frequency components associated with Kinect tracking errors of certain body joints. The processed 3D position measurements for each joint are used in Equation 6.8 to simulate the radar response of each joint. The resulting joint radar signals are superimposed to obtain the overall radar return corresponding to the considered human activity.

The range–slow time map and the micro-Doppler signature can be obtained after computation of the expected radar return for a specific activity. The micro-Doppler signature can be derived by adding all radar returns over all fast time/range samples during the observation period, and then applying STFT and its energetic representation, that is, spectrogram. Alternatively, summation of elements within the STFT-based JRTF data cube across the range/fast time dimension followed by its energetic representation results in the micro-Doppler signature. Likewise, summing of JRTF elements along the frequency axis leads to the range–slow time map.

6.3.2 Range-Spread as a Distinguishing Feature

A Kinect sensor was placed at approximately 2.5 m above the ground to capture the complete human body joints and the entire falling and sitting motions. Two subjects performed both sitting and falling motions at two different speeds: slow and fast. The sensor was tilted to capture the motions. This tilt angle due to Kinect's elevated position was compensated by applying a 3D rotation matrix, which results in the generation of the correct skeletal data. Figure 6.7a and b show the computed micro-Doppler signatures corresponding to a slow fall and a fast sitting motion, respectively. The upper and lower envelope frequencies for each signature are also superimposed. It is evident that the fast sitting and slow falling motions have similar Doppler signatures. The maximum negative Doppler frequency for falling is −160 Hz, whereas that for sitting is −157 Hz, resulting in an insignificant difference of 3 Hz. The highly overlapping nature of the falling and sitting time–frequency Doppler signatures will cause the Doppler-only-based classifier to become ineffective in discriminating falling and sitting motions.

Figure 6.7c and d depicts the range–slow time maps for slow falling and fast sitting motions, respectively. Despite the similarity of their Doppler signatures, the range extent

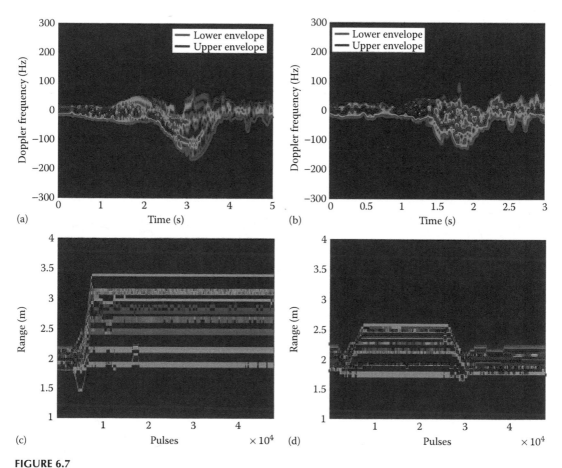

FIGURE 6.7
Micro-Doppler signatures and upper–lower envelopes of (a) falling and (b) sitting; Range-slow time maps of (c) falling and (d) sitting. (Reprinted from Erol, B. et al., Range Information for Reducing Fall False Alarms in Assisted Living, *Proceedings of IEEE Radar Conference*, 2016. With Permission.)

of falling is considerably higher than that of sitting. The latter is influenced by the type and the depth of the seat of the chair or the sofa. In the considered case, the range extent was found to be 1.60 m for falling, compared to 0.79 m for sitting. Clearly, there is a distinct difference in the two values, thus demonstrating the benefit of employing the motion range extent as a feature for discriminating falls from sitting.

6.3.3 Impact of the Aspect Angle on Fall Detection

The aspect angle between the radar LOS and the direction of target motion greatly influences the micro-Doppler signatures and range–slow time maps. The most distinctive signatures appear when target is moving toward or away from the radar, leading to the maximum Doppler spread. As the aspect angle increases, the Doppler effect also lessens. The resulting signatures have a compressed structure in frequency domain, which causes classification results to dramatically worsen. To investigate the effect of the aspect angle, a 3D rotation matrix was applied to the pre-processed Kinect skeletal data of falling and sitting to rotate the direction of the target motion to the desired aspect angle, prior to generating the corresponding radar return. Computed micro-Doppler signature and range–slow time map for falling are shown in Figure 6.8a and b, respectively, when the motion direction is orthogonal to the radar LOS. It is evident that the corresponding Doppler signature does not contain any micro and macro components of the motion. The same observation is also valid for the range–slow time map, which depicts a considerably compressed range extent compared to Figure 6.7c.

6.3.4 Classification Results

A database of different human motion articulations was constructed using the Kinect-based pulse-Doppler radar simulator for different aspect angles. Specifications of the simulated pulse-Doppler radar are provided in Table 6.2. The database contains 13 falling and 13 non-fall motion samples for each different aspect angle. Support vector machine is applied to discriminate falling from non-fall events for each aspect angle. The data are first separated into training and testing sets of sizes 60% and 40%, respectively. Samples for the training and testing sets were randomly selected. The random selection process

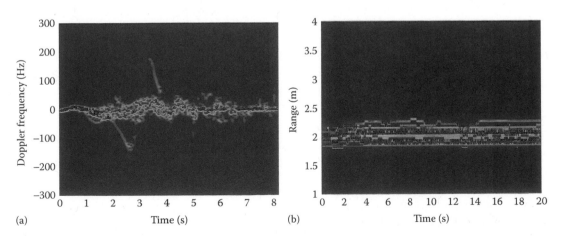

FIGURE 6.8
Kinect-based UWB simulator: (a) micro-Doppler signature and (b) range-slow time map of falling for 90°.

TABLE 6.2

Simulated Radar Parameters for Fall Detection

PRF	1 kHz
Carrier frequency	25 GHz
Bandwidth	2 GHz
Distance to target	2 m

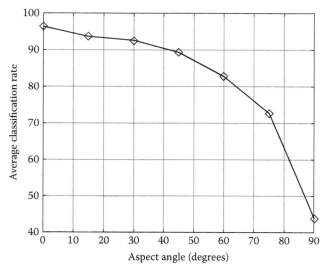

FIGURE 6.9
Dependency of average classification performance on aspect angle.

was repeated 1000 times for each aspect angle to follow the Monte Carlo approach and characterization of the classifier.

Classification accuracy of the maximum Doppler frequency and range extent features is plotted as a function of the aspect angle in Figure 6.9. As expected, the average classification performance is degraded toward 90° proving the detrimental effect of the higher aspect angles. The best classification performance achieved is 96% at 0°, whereas at 60°, the classification performance drops below 85%, and finally at 90° aspect angle, the classification performance degrades to as low as 45%. Consistent with the earlier discussions, the aspect angle is a critical factor affecting the classification rate.

6.3.5 Discussion

Significance of using the range information in distinguishing falls from sitting motions has been demonstrated. The fall exhibits approximately twice the range extent as that of sitting when the motion direction is along the LOS. This distinction leads to reduced false alarms. Because there are no guarantees on motion orientations, the observed detrimental effects of the aspect angle on fall detection necessitate the deployment of multi-radar monitoring system where blind spots due to transversal motions can be eliminated [61].

6.4 Range-Doppler Processing in Multipath Environment

Multipath returns can be treated as interference and mitigated [62,63]. However, there are scenarios in the indoor environment where the direct path is lost, and the only paths available utilize multipath. As such, use of radar for moving target indication in an indoor environment requires a functional understanding of the multipath environment. Measurements in a complex environment create multiple returns for each target based upon the number of reflections. In addition to a high number of material classes and surface properties, transmission and reflection properties of walls and objects are a function of both reflection angle and frequency. In this section, we report on the less-studied effect of indoor hallway-like environments on the multipath response of a moving human using real radar measurements.

6.4.1 Doppler Response in a Multipath Environment

The Doppler frequency shift, f_D, of a point scatterer moving with velocity v with respect to a stationary transmitter is given by

$$f_D = f_c \frac{2v}{c} \cos\theta \cos\varphi \tag{6.17}$$

where:
f_c is the frequency of the transmitted signal
θ is the angle between the subject motion and the beam of the radar in the ground plane
φ is the elevation angle between the subject and the radar beam

For complex objects, such as walking humans, the velocity of each body part varies over time. Additionally, the RCS of various body parts is a function of the aspect angle and frequency.

The typical Doppler response of a point target in a multipath environment changes due to the different paths that the transmitted and reflected signals can take. A diagram of possible paths in a hallway is shown in Figure 6.10. The Doppler response of a point target in a multipath environment becomes a function of range for the different lengths of the possible paths. At one particular range and for one particular set of incoming and outgoing paths, the Doppler can be expressed as

$$f_D = f_c \frac{v}{c} (\cos\theta_i \cos\varphi_i + \cos\theta_o \cos\varphi_o) \tag{6.18}$$

where:
θ_i and φ_i are the angles of the incoming wave
θ_o and φ_o are the angles of the outgoing wave

There are multiple paths that can exist in the target at the same range, and these combine coherently. When the target is extended in range and is not merely a point source, these interactions become more complex. Additionally, the particular radar will have a range resolution that will combine multiple paths together.

The response of a single target in a complex environment can be modeled as multiple targets in a simpler environment. This is illustrated in Figure 6.11, where three possible paths are mapped out for incoming or outgoing waves. Because of the different angles with

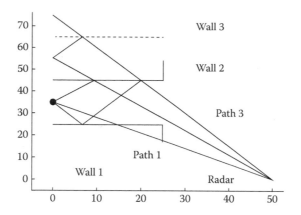

FIGURE 6.10
Possible paths within a hallway to a target. In this case, the direct path is blocked by wall 1. Wall 3 is a phantom wall, used to project the reflections from wall 1 after reflection from wall 2. For a perfect radar and a point source, the multiple reflections can look like multiple targets at ranges that would sometimes place them outside of the building.

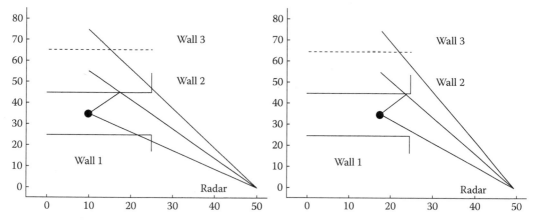

FIGURE 6.11
Examples of point targets moving in a hallway. On the left, the direct path is blocked by wall 1, whereas wall 2 reflects one indirect path. On the right, the direct path is finally available, and one indirect path is still available.

respect to the motion of the target, the Doppler response will vary depending on the paths that the radar energy takes. A simulated range-Doppler chip is provided in Figure 6.12, using the geometry of Figure 6.10 and a point reflector moving at 2 m/s along the hallway.

The Doppler response of a point target in a multipath environment can be considered as the summation over all possible paths combined coherently for a given geometry. For the hallway setup in Figure 6.10, the direct path does not give a response, but several other paths should provide returns. Considering that each reflection will introduce a loss that will be incorporated into the returned energy, and further simplifying to a ground scenario where $\varphi_i = \varphi_o = 0$, the set of Doppler responses from a point target takes the form

$$f_{D,r_1+r_2} = \sum_{r_1}\sum_{r_2} f_c \frac{v}{c}(\cos\theta_i + \cos\theta_o) \tag{6.19}$$

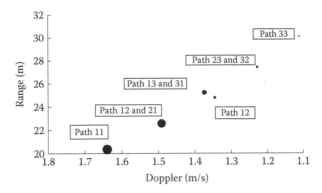

FIGURE 6.12
Range-Doppler response of a point target moving in a hallway. The size corresponds to the amplitude with attenuation due to reflections, whereas the Doppler and range assume the geometry from Figure 6.10.

where r_1 and r_2 represent the incoming and outgoing paths, respectively. Note that as one includes realistic target sizes and radar range resolutions, the phases will need to be included to combine the signals coherently. The different paths produce various angles reflected off the walls, and there is an increase in attenuation due to more reflections. The result of motion along the hallway is a *tail* of reflected signal with reducing frequency and amplitude extending in range from the target. A tail can be discrete trackable returns or a continuum of energy extending out in range depending on the extent of the subject and the resolution of the radar. A simulated tail is depicted in Figure 6.12, and the examples of measured tails can be seen in Figure 6.13.

The particular geometry of a hallway has a significant effect on the resulting ability to track moving targets. Range-Doppler chips of the person in motion along the hallway are shown in Figure 6.14, where multiple chips have been removed when there are no detections. The direct and multipath response from a person walking in a hallway is shown in Figure 6.15, which has time versus range maps that have been Doppler filtered and then integrated in Doppler. This leads to a 2D image that shows Doppler activity over range and time. The walker was close to the side, which means that there were times when the shadow of the near wall blocked the direct path and there was no indirect path. This shows up as gaps in the range–time image. This also means that the probable tail will be almost continuous, because the distance to the next reflection at the near wall is small, and two of these tails are expanded in Figure 6.15.

FIGURE 6.13
Examples of tailed targets moving in a hallway compared to a normal target (right) with direct response.

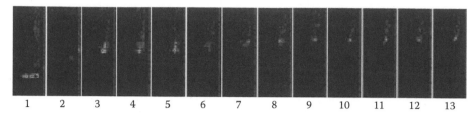

FIGURE 6.14
Range-Doppler chips of a person moving out from a hallway. The direct response is in chip 1, where there is significant micro-Doppler, and there is a significant amount of time where there is minimal response like chip 2, and then the reflected response in chips 3 through 13. This corresponds to the activity in Figure 6.15. The indirect path has a reduced return strength, which makes micro-Doppler characteristics harder to see. The tail characteristics vary with range. Chips 3 and 10 are expanded in Figure 6.13.

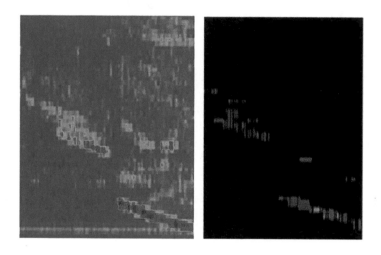

FIGURE 6.15
Range-time maps of the direct and multipath response from a human walking between and then emerging from a hallway (left) and the resulting tracks (right). This demonstrates the ability of a simple tracker in an indoor environment, but fails to connect the two tracks of the same target when they move from the non-LOS to direct response.

6.4.2 Tracking in Multipath Measurements

The tracker uses a simple constant false alarm rate (CFAR) approach in order to evaluate the tracking complication in an urban environment. A diagram of the general measurement setup with the radar and the hallways is shown in Figure 6.10. The non-LOS tracking from a subject in a hallway is illustrated in the right image of Figure 6.15, with a human emerging from between the two walls, shown in both the raw radar returns and the corresponding tracks. Individual chips out of range-Doppler maps used for the tracking are shown in Figure 6.14.

The position and tracking of non-LOS subjects are affected strongly by the geometry of the radar in relation to the hallway as well as the motion of the subject. In some cases, the tracking approach shows an inability to disambiguate two multipath tracks that are similar in range. This is especially clear in Figure 6.15, where the multipath tracks get combined into a single track due to the closeness of the target to the wall. For indoor environments when multipath tracks might be used for linking a subject to its previous multipath track, a more complex tracker is warranted.

6.4.3 Classification in Multipath Environments

Subject classification in multipath reflections based on micro-Doppler features has been determined to be hindered by the two-way attenuation loss, especially when the classification is heavily amplitude based or requires the detection of foot motion. An example of such a situation is provided in Figure 6.14, where the direct path has significant micro-Doppler, but the reflected paths are reduced in amplitude and lacking measurable micro-Doppler. The additional loss inhibits the measurement of lower RCS body parts used in a micro-Doppler classification scheme, though the obscuration and partial reflections are a more severe source of loss.

6.4.4 Discussion

The detection and tracking of non-LOS moving objects using multiple bounces was demonstrated. Although tracking using multipath returns is possible, there exists an ambiguity between real and multipath sources. Some of the difficulties in tracking the subjects through their multipath reflections were shown, including large simultaneous discontinuities in both range and Doppler. However, using direct and related multipath returns would allow a tracker to reduce the discontinuity to just Doppler and then range. Multipath classification is less predictable due to obscurations and signal attenuation. The use of amplitude characteristics is especially difficult in an indoor environment for classification, unless the attenuation can be calibrated using *a priori* knowledge of the environment.

6.5 Moving Target Indication via Change Detection

Detection of target position variations by subtraction of data or images at different time instants is the foundation of CD where details of motion profile are difficult to discern and become irrelevant to motion indication. In the CD paradigm, the radar probes a region of interest over multiple CPIs. Subtraction between any two consecutive CPI data frames effectively removes the stationary objects. The CD data are then processed by a focusing algorithm to enhance cross-range resolution [64], and the resulting CD images retain the moving target signature along with any image artifacts. Note that, due to the nature of the involved operations, CD can be applied before or after focusing. In ideal circumstances, the contribution of the imaging artifacts is negligible resulting in a sparse representation of the scene [46]. In non-ideal sensing scenarios, the subtraction does not adequately attenuate the stationary clutter resulting in a heavy contribution of imaging artifacts [43,44]. It has been shown that these imaging artifacts lead to false alarms and degrade motion tracking performance [64]. In this section, we present an analysis of CD and use both modeled and real data to demonstrate the effectiveness of the CD process.

6.5.1 Change Detection

Consider a multichannel monostatic radar system consisting of an M-element antenna array and a stepped-frequency signal with K frequency steps. Define the sampled frequency domain received data as $\{\gamma_{n,m}(1),\ldots,\gamma_{n,m}(K)\}$ for CPI n, where $n=\{1,\ldots,N\}$. A range profile is then estimated for each antenna element and CPI as follows:

$$R_{n,m} = [r_{n,m}(1),\ldots,r_{n,m}(K)] = \mathcal{F}^{-1}\{\Phi[\gamma_{n,m}(1),\ldots,\gamma_{n,m}(K)]\} \tag{6.20}$$

where:
   $\mathcal{F}^{-1}$ is the inverse Fourier transform operation
   $\Phi(\cdot)$ is a window function of size K
   $r_{n,m}(k)$ is the kth range bin in $R_{n,m}$

The range profiles for a given CPI are then processed by the back-projection algorithm [65] to focus the raw data into a sequence of N images, each of dimension $N_X \times N_Y$. The complex-valued image $I_n(i,j)$, corresponding to the nth CPI, is defined as

$$I_n(i,j) = \sum_{m=1}^{M} r_{n,m}(\kappa_{n,m}), i = 1,\ldots,N_X, j = 1,\ldots,N_Y \tag{6.21}$$

where $\kappa_{n,m} = \{1, \ldots, K\}$ denotes the range bin corresponding to the two-way distance between the mth antenna position and the image pixel (i,j).

Coherent CD is next applied to enhance the moving target signature and to remove the image artifacts. CD is defined for any two images I_{n_1} and I_{n_2} in the sequence of N images as [66]

$$\delta(i,j) = |I_{n_1}(i,j) - I_{n_2}(i,j)| \tag{6.22}$$

where $n_1 \neq n_2$, with I_{n_1} being the reference image. The magnitude of each pixel in the CD image $\delta(i,j)$ can be expressed as

$$|\delta(i,j)| = \sqrt{\alpha_{n_1}^2(i,j) + \alpha_{n_2}^2(i,j) - 2\alpha_{n_1}(i,j)\alpha_{n_2}(i,j)\cos(\zeta_{n_2}(i,j) - \zeta_{n_1}(i,j))} \tag{6.23}$$

where $\alpha_{n_a}(i,j)$ and $\zeta_{n_a}(i,j)$ denote the magnitude and phase, respectively, of the image $I_{n_a}(i,j)$ for $a = 1,2$. Typical implementation of Equation 6.22 requires that $n_1 = n_2 + 1$ so that sequential images are subtracted. Then, a progressive track of the moving target is established by letting $n_2 = \{1,\ldots,N-1\}$.

6.5.2 Modeling Scenario of a Brick Building

In this section, CD images are formed from a simulation of a person walking inside a brick building structure. The clutter objects are perfectly aligned in the model and are effectively subtracted using CD. The model is also used to investigate the imaging artifacts that persist after CD. The simulation implements a multichannel, stepped-frequency radar approach with a frequency range of 0.3–3 GHz with $K = 1623$ and a range resolution of 0.056 m. Near-field scattering simulations are computed using Xpatch, a physical optics-based code that uses a ray-tracing technique to calculate the scattering field from a target [67]. The array consists of 16 monostatic antennas distributed across a 2.01 m aperture. Simulated radar scattering data are computed for 23 moving target positions shown in Figure 6.16. The building structure consists of four brick walls and a ground plane (no ceiling). The walls are 8 in thick and are modeled as brick with a dielectric constant of 3.8–j0.24. The ground-plane has a dielectric constant

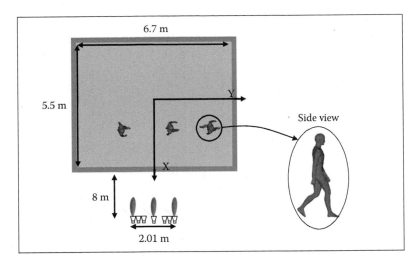

FIGURE 6.16
The CAD model of the building structure. $N = 23$ moving target positions are simulated.

of 6.8 (dry soil). The antenna array is located 8 m from the front wall. The target is moving parallel to the aperture (from left to right) with motion sampled every 0.15 m. The total distance covered by the moving target is 3.3 m.

A focused image of size 8×8 m^2 is formed for each moving target position using Equation 6.20 and comprises 400×400 pixels. A Hanning window is used to process the focused images. An example focused image for moving target position $n = 1$ is shown in Figure 6.17a using the dynamic range of 50 dB. The front wall is clearly visible and represented by two signatures separated by approximately 0.5 m in range. The first signature is formed by the edge of the wall closest to the radar, defined as the first edge, and the second signature is formed by the edge of the wall farthest from the radar, defined as the second edge. The back wall is also visible with two signatures. The moving target can be seen, albeit severely attenuated (approximately 35 dB)

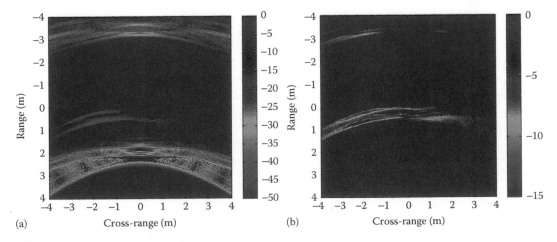

FIGURE 6.17
Example focus and change detection images of a simulated moving target walking inside a brick building structure. (a) Focused image, $n = 1$ and (b) CD image, $n_2 = 2$ and $n_1 = 1$.

compared to the wall reflections. Note that the moving target has two associated signatures approximately 0.4 m apart in range. A possible explanation for this second signature (further down range) is that a corner is formed between the ground plane and the human model. The rays that bounce between the ground plane and the human travel a longer distance than those reflected directly by the human, which creates a delayed moving target signature. This multiple-bounce phenomenon has been observed in [68] for front walls, back walls, and a stationary human using near-field Xpatch model data. *Ghost* artifacts are also present in the modeling data and formed when the multipath scattering adds coherently within the imaging area [67,69,70]. The ghost image appears as an extra copy of the target and may create a false alarm. Sidelobes are also noticeable throughout the image. Significant sidelobes extend in the cross-range direction and are dependent upon the employed cross-range processing scheme. Various apodization techniques have been presented to reduce sidelobes as part of classical synthetic aperture radar image formation algorithms [71]. These algorithms are well documented and are typically utilized by frequency domain image formation techniques.

Coherent CD is next applied to enhance the moving target signature. An example CD image is shown in Figure 6.17b for $n_2 = 2$ and $n_1 = 1$. A 15 dB dynamic range clearly indicates the moving target. The clutter objects (i.e., the walls) in Figure 6.17a are perfectly aligned in the model and are effectively subtracted in Figure 6.17b. It should be noted that, unlike the modeling scenario presented here, the alignment of the clutter is not perfect for real measurements due to inherent phase errors of the measurement system. As will be discussed in Section 6.5.3, these phase errors lead to imperfect clutter cancellation resulting in image artifacts in the CD image. Nevertheless, the model still shows evidence of persistent imaging artifacts that are present after CD—these include sidelobes and the target *shadow*. The shadow artifact occurs along the back wall and is due to the motion of the target. Modeling calculations indicate that such phenomenon is to be anticipated due to the multipath scattering characteristic to the indoor propagation environment.

6.5.3 Radar Scenarios of Wood and Cinderblock Building Structures

Real data measurements were collected by the U.S. Army Research Laboratory's ground-based synchronous impulse reconstruction (SIRE) radar [72]. The SIRE radar is an impulse-based, ultrawideband imaging radar, with a frequency range of 0.3–3 GHz. It employs two transmit antennas and 16 receiver antennas mounted in a wooden structure and attached to the top of a sport utility vehicle. The receive antennas are equally spaced across a linear 2 m long aperture. The transmitters are located at each end of the wooden structure, slightly above the receive array. The SIRE radar collects and processes data using Equation 6.20 with $K = 1351$. The equation is then used to form a sequence of images from the range profiles where each image is of size $8 \times 8 \ m^2$ and comprises 400×400 pixels. The SIRE radar must remain stationary during data collection so that the pixels of each image in the sequence of N images are properly aligned.

Data were collected using the SIRE radar for two scenarios: (1) a person randomly walking inside a wood building and (2) a person randomly walking inside a cinderblock building. The wood building consists of a room with minimal clutter (i.e., a plastic chair). The wall material is plywood and nailed into a framed structure of two-by-four lumber. The wood building schematic is shown in Figure 6.18a. Note that the radar is positioned 19° off the broadside angle to the wood wall. The off-broadside angle was

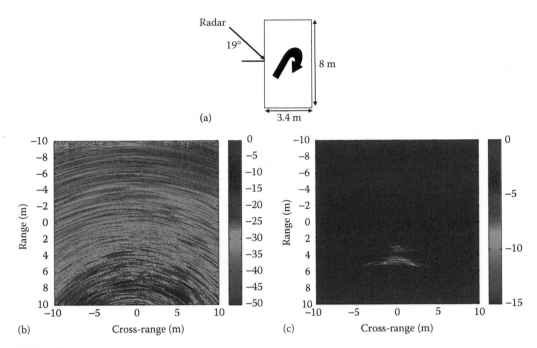

FIGURE 6.18
Example images of a moving target walking inside a wood building. (a) Wood building schematic, (b) focused image of wood building, and (c) CD image of wood building. The CD image clearly illustrates the moving target.

selected to reduce image artifacts present in the CD images due to large reflections from the wall [43]. The radar was positioned approximately 15.5 m from the front of the wall. The SIRE collected $N = 34$ frames of data and processed them to generate a 34-image sequence. The image for $n = 1$ is shown in Figure 6.18b, wherein the responses from the front wall and other clutter objects mask the moving target. The moving target is clearly identified using CD as shown in Figure 6.18c for a dynamic range of 15 dB. The front and back wall returns are attenuated.

The cinderblock building has four walls and four perpendicular support structures. The height and width of each piece of cinderblock is 0.203 and 0.191 m, respectively. Other properties of the cinderblocks are variable: (1) the length, (2) the number of hollow centers, (3) and the state of each hollowed center (i.e., some are empty, some are filled with cement, and some are partially filled with cement). Given the arbitrary properties of the cinderblock building, detection of moving targets placed inside is a highly challenging problem. The cinderblock building schematic is shown in Figure 6.19a where the radar is positioned 38° off the broadside angle to the front of the cinderblock wall. The SIRE again collected $N = 34$ frames of data and processed them. The image for $n = 1$ is shown in Figure 6.19b, whereas the CD image is shown in Figure 6.19c. Clearly, the moving target, shadow, and sidelobes are visible in the CD image. Image artifacts are also visible from the front wall, that is, the wall flash. The wall flash is a phenomenon caused by phase errors in the measurement data. The magnitude of the wall flash is on the same order as the moving target signature and leads to false alarms [43].

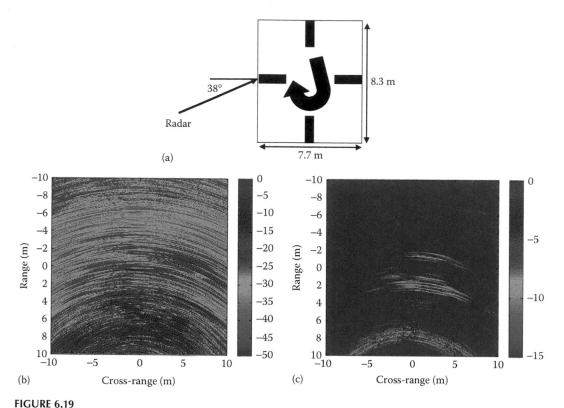

FIGURE 6.19

Example images of a moving target walking inside a cinderblock building structure. (a) Cinderblock structure schematic, (b) focused image of cinderblock structure, and (c) CD image of cinderblock structure. The CD image clearly illustrates the moving target in the presence of image artifacts.

6.5.4 Discussion

CD was shown to effectively subtract stationary clutter within the focused images. Analysis of the modeling scenario illustrates image artifacts (sidelobes and the shadow) that persist after CD, a result validated by the wood building measurement scenario. The image artifacts are more significant in the cinderblock measurement scenario, where the wall flash is evident in addition to the sidelobes and the target shadow. Detection and tracking of the moving target is therefore challenging in the presence of these image artifacts. The current literature highlights several techniques to mitigate the image artifacts for improved detection and tracking. These techniques include sparsity-driven techniques [46], CFAR approach for false alarm mitigation [73], multiple-input multiple-output processing for shadow effect mitigation [74], noncoherent CD for sidelobe mitigation [66], 3D processing for clutter mitigation [75], and techniques for wall clutter mitigation [76,77].

6.6 Conclusion

This chapter addressed range-Doppler processing for human motion detection and classification in indoor environments. First, wideband micro-Doppler theory was developed to enable joint extraction of structural information and micro-motions. The wideband micro-Doppler theory was applied to human gait estimation problem, and its superiority over traditional micro-Doppler signature was demonstrated for the case when the target occupies multiple range bins and migrates between them during measurement, Second, it was shown that augmentation of Doppler-only features with motion-range extent for elderly fall detection enables robustness against other similar motions, such as sitting, which would serve as confusers for Doppler-only-based classifiers. Third, detection, tracking, and classification of human motion in hallways were addressed in the presence of multipath propagation. It was demonstrated that multipath can be used as an indirect means to view a target that is not in radar's direct LOS. Finally, the CD paradigm was discussed and its performance was validated for the TWRI application using both modeled and real data. For each of the aforementioned approaches, their respective requirements, assumptions, and offerings were highlighted along with brief mention of some open problems.

References

1. M. G. Amin, *Through-the-Wall Radar Imaging*, CRC Press, Boca Raton, FL, 2011.
2. M. G. Amin and F. Ahmad, Through-the-wall radar imaging: Theory and applications, in N. D. Sidiropoulos, F. Gini, R. Chellappa, and S. Theodoridis (Eds.), *Academic Press Library in Signal Processing, vol. 2: Communications and Radar Signal Processing*, Elsevier, Oxford, UK, pp. 857–909, 2014.
3. F. Ahmad, T. Dogaru, and M. G. Amin, Ultrawideband sense through the wall radar technology, in J. D.Taylor (Ed.), *Advanced Ultrawideband Radar: Targets, Signals and Applications*, CRC Press, Boca Raton, FL, 2017.
4. M. G. Amin, Y. D. Zhang, F. Ahmad, and K. C. Ho, Radar signal processing for elderly fall detection: The future for in-home monitoring, *IEEE Signal Process. Mag.* 33(2): 71–80, 2016.
5. F. Ahmad, R. M. Narayanan, and D. Schreurs (Eds.), Special issue on application of radar to remote patient monitoring and eldercare, *IET Radar, Sonar, and Navig.*, 9(2): 115–190, 2015.
6. F. Ahmad, A. E. Cetin, K. C. Ho, and J. E. Nelson (Eds.), Special section on signal processing for assisted living, *IEEE Signal Process. Mag.*, 33(2): 25–94, 2016.
7. B. Y. Su, K. C. Ho, M. J. Rantz, and M. Skubic, Doppler radar fall activity detection using the wavelet transform, *IEEE Trans. Biomedical Eng.*, 62(3): 865–875, 2015.
8. Y. Kim and H. Ling, Human activity classification based on micro-Doppler signatures using a support vector machine, *IEEE Trans. Geosci. Remote Sens.*, 47(5): 1328–1337, 2009.
9. M. Otero, Application of a continuous wave radar for human gait recognition, *Proceedings of the SPIE*, vol. 5809:538–548, 2005.
10. I. Orovic, S. Stankovic, and M. Amin, A new approach for classification of human gait based on time-frequency feature representations, *Signal Process*, 91(6): 1448–1456, 2011.
11. C. Karabacak, S. Z. Gurbuz, A. C. Gurbuz, M. B. Guldogan, G. Hendeby, and F. Gustafsson, Knowledge exploitation for human micro-Doppler classification, *IEEE Geosci. Remote Sens. Lett.*, 12(10): 2125–2129, 2015.
12. B. G. Mobasseri and M. G. Amin, A time-frequency classifier for human gait recognition, *Proceedings of the SPIE*, vol. 7306:730628, 2009.
13. T. Thayaparan, L. Stankovic, and I. Djurovic, Micro-Doppler human signature detection and its application to gait recognition and indoor imaging, *J. Franklin Inst.*, 345(6): 700–722, 2008.
14. R. M. Narayanan and M. Zenaldin, Radar micro-Doppler signatures of various human activities, *IET Radar, Sonar, Navig.*, 9(9): 1205–1215, 2015.
15. M. G. Amin, F. Ahmad, Y. D. Zhang, and B. Boashash, Human gait recognition with cane assistive device using quadratic time-frequency distributions, *IET Radar, Sonar, Navig.*, 9(9): 1224–1230, 2015.

16. G. E. Smith, F. Ahmad, and M. G. Amin, Micro-Doppler processing for ultra-wideband radar data, in *Proceedings of the SPIE*, Bellingham, Washington, vol. 8361, pp. 83610L-1–83610L–10, 2012.

17. T. Sakamoto, P. Molchanov, Y. He, A. Yarovoy, F. Le Chevalier, and P. Aubry, Range-Doppler surface: A tool to analyse human target in ultra-wideband radar, *IET Radar, Sonar Navig.*, 9(9): 1240–1250, 2015.

18. Y. Luo, S. Li, T. Soon Yeo, Q. Zhang, and C. Qiu, Micro-Doppler feature extraction for wideband imaging radar based on complex image orthogonal matching pursuit decomposition, *IET Radar, Sonar Navig.*, 7(8): 914–924, 2013.

19. Y. He, F. Le Chevalier, and A. G. Yarovoy, Range-Doppler processing for indoor human tracking by multistatic ultra-wideband radar, in *Proceedings of the 13th International Radar Symposium*, pp. 250–253, 2012.

20. Z. A. Cammenga, G. E. Smith, and C. J. Baker, High range resolution micro-Doppler analysis, in *Proceedings of the SPIE*, vol. 9461, Bellingham, WA, pp. 94611G, 2015.

21. Z. A. Cammenga, G. E. Smith, and C. J. Baker, Combined high range resolution and micro-Doppler analysis of human gait, in *Proceedings of the IEEE International Radar Conference*, pp. 1038–1043, 2015.

22. D. Tahmoush, Wideband radar micro-Doppler applications, in *Proceedings of the SPIE*, vol. 8734, Bellingham, WA, pp. 873403, 2013.

23. T. Sakamoto, T. Sato, P. J. Aubry, A. G. Yarovoy, and S. Member, Texture-based automatic separation of echoes from distributed moving targets in UWB radar signals, *IEEE Trans. Geosci. Remote Sens.*, 53(1): 352–361, 2015.

24. K. Saho, H. Homma, T. Sakamoto, T. Sato, K. Inoue, and T. Fukuda, Accurate image separation method for two closely spaced pedestrians using UWB Doppler imaging radar and supervised learning, *IEICE Trans. Commun.*, E 97.B(6): 1223–1233, 2014.

25. O. R. Fogle and B. D. Rigling, Micro-range/micro-Doppler decomposition of human radar signatures, *IEEE Trans. Aerosp. Electron. Syst.*, 48(4): 3058–3072, 2012.

26. I. Bilik, J. Tabrikian, and A. Cohen, GMM-based target classification for ground surveillance Doppler radar, *IEEE Trans. Aerosp. Electronic Syst.*, 42(1): 267–278, 2006.

27. V. C. Chen and H. Wechsler, Micro-Doppler effect in radar: Phenomenon, model, and simulation study, *IEEE Trans. Aerosp. Electron. Syst.*, 42(1): 2–21, 2006.

28. V. Chen, *The Micro-Doppler Effect in Radar*, Artech House, Norwood, MA, 2011.

29. V. C. Chen, Analysis of radar micro-Doppler signature with time-frequency transform, in *Proceedings of the 10th IEEE Workshop on Statistical Signal and Array Processing*, pp. 463–466, 2000.

30. J. Durek, *Multipath Exploitation Radar Industry Day Presentation*, DARPA, Herndon, VA, 2009.

31. P. R. Barbosa, E. K. P. Chong, S. Suvarova, and B. Moran, Multitarget-multisensor tracking in an urban environment: A closed-loop approach, in *Proceedings of the SPIE*, vol. 6969, Bellingham, WA, pp. 69690W, 2008.

32. T. Trueblood, Multipath exploitation radar for tracking in urban terrain, Master's thesis, Arizona State University, Tempe, AZ, 2009.

33. P. Setlur, M. Amin, and F. Ahmad, Multipath model and exploitation in through-the-wall and urban radar sensing, *IEEE Trans. Geosci. Remote Sens.*, 49(10): 4021–4034, 2011.

34. M. Leigsnering, F. Ahmad, M. G. Amin, and A. M. Zoubir, Multipath exploitation in through-the-wall radar imaging using sparse reconstruction, *IEEE Trans. Aerosp. Electronic Syst.*, 50(2): 920–939, 2014.

35. M. Leigsnering, M. Amin, F. Ahmad, and A. M. Zoubir, Multipath exploitation and suppression for SAR imaging of building interiors: An overview of recent advances, *IEEE Signal Process. Mag.*, 31(4): 110–119, 2014.

36. M. Mertens and M. Ulmke, Precision GMTI tracking using road constraints with visibility information and a refined sensor model, in *Proceedings of the IEEE Radar Conference*, pp. 1–6, 2008.

37. J. L. Krolik, J. Farrell, and A. Steinhardt, Exploiting multipath propagation for GMTI in urban environments, in *Proceedings of the IEEE Radar Conference*, pp. 65–68, 2006.

38. Y. Bar-Shalom, *Tracking and Data Association*, Academic Press, San Diego, CA, 1987, pp. 157–190.

39. Y. Bar-Shalom, F. Daum, and J. Huang, The probabilistic data association filter, *IEEE Control Syst. Mag.*, 29(6): 82–100, 2009.
40. T. Kirubarajan and Y. Bar-Shalom, Probabilistic data association techniques for target tracking in clutter, *Proc. IEEE*, 92(3): 536–557, 2004.
41. R. Radke, S. Andra, O. Al-Kofahi, and B. Roysam, Image change detection algorithms: A systematic survey, *IEEE Trans. Image Process.*, 14(3): 294–307, 2005.
42. L. Novak, The effects of SAR data compression on coherent and non-coherent change detection, in *Proceedings of the International Radar Conference*, IEEE, Bordeaux, France, pp. 1–6, 2009.
43. A. Martone, K. Ranney, and R. Innocenti, Automatic through the wall detection of moving targets using low-frequency ultra-wideband radar, in *Proceedings of the IEEE International Radar Conference*, pp. 49–43, 2010.
44. M. G. Amin and F. Ahmad, Change detection analysis of humans moving behind walls, *IEEE Trans. Aerosp. Electronic Syst.*, 49(3): 1410–1425, 2013.
45. D. Mooney and W. Skillman, Pulse-Doppler radar, in M. Skolnik (Ed.) *Radar Handbook,*McGraw-Hill, New York, 1970, pp. 23–24.
46. F. Ahmad and M. Amin, Through-the-wall human motion indication using sparsity-driven change detection, *IEEE Trans. Geosci. Remote Sens.*, 51(2): 881–890, 2013.
47. V. Chen, D. Tahmoush, and W. Miceli, *Radar Micro-Doppler Signature Processing and Applications*, IET, U.K, 2014.
48. B. Keel, Fundamental of pulse compression waveforms, in M. A.Richards, J. A. Scheer, and W. A. Holm (Eds.), *Principles of Modern Radar, Volume 1—Basic Principles*, SciTech Publishing, Raleigh, NC, 2010, pp. 774–834.
49. L. B. Almeida, The fractional Fourier transform and time-frequency representations, *IEEE Trans. Signal Process.*, 42(11): 3084–3091, 1994.
50. V. C. Chen, Doppler signatures of radar backscattering from objects with micro-motions, *IET Signal Process.*, 2(3): 291–300, 2008.
51. P. van Dorp and F. C. A. Groen, Human walking estimation with radar, *IEE Proc.—Radar, Sonar Navig.*, 150(5): 356, 2003.
52. G. E. Smith, K. Woodbridge, and C. J. Baker, Naïve Bayesian radar micro-Doppler recognition, in *Proceedings of the IEEE International Radar Conference*, Adelaide, Australia, pp. 111–116, 2008.
53. R. Igual, C. Medrano, and I. Plaza, Challenges, issues and trends in fall detection systems, *Biomed. Eng. Online*, 12(66): 66, 2013.
54. L. R. Rivera, E. Ulmer, Y. D. Zhang, W. Tao, and M. G. Amin, Radar-based fall detection exploiting time-frequency features, in *Proceedings of the IEEE China Summit and International Conference Signal Information Processing*, pp. 713–717, 2014.
55. Q. Wu, Y. D. Zhang, W. Tao, and M. G. Amin, Radar-based fall detection based on Doppler time-frequency signatures for assisted living, *IET Radar, Sonar, Navig.*, 9(2): 164–172, 2015.
56. J. Hong, S. Tomii, and T. Ohtsuki, Cooperative fall detection using Doppler radar and array sensor, in *Proceedings of the IEEE 24th International Symposium on Personal, Indoor, and Mobile Radio Communications*, pp. 3492–3496, 2013.
57. A. Gadde, M. G. Amin, Y. D. Zhang, and F. Ahmad, Fall detection and classifications based on times-cale radar signal characteristics, in *Proceedings of the SPIE*, Baltimore, MD, vol. 9077, pp. 907712, 2014.
58. L. Liu, M. Popescu, M. Skubic, M. Rantz, T. Yardibi, and P. Cuddihy, Automatic fall detection based on Doppler radar motion signature, in *Proceedings of the 5th International Conference on Pervasive Computing Technologies for Healthcare*, Dublin, Ireland, pp. 222–225, 2011.
59. F. Wang, M. Skubic, M. Rantz, and P. E. Cuddihy, Quantitative gait measurement with pulse-Doppler radar for passive in-home gait assessment, *IEEE Trans. Biomedical Eng.*, 61(9): 2434–2443, 2014.
60. B. Erol, C. Karabacak, and S. Z. Gurbuz, A Kinect-based human micro-Doppler simulator, *IEEE Aerosp. Electronics Syst. Mag.*, 30(5): 6–17, 2015.
61. S. Tomii and T. Ohtsuki, Learning based falling detection using multiple Doppler sensors, *AIT*, 30(2A): 3343, 2013.

62. B. Krach and R. Weigel, Markovian channel modeling for multipath mitigation in navigation receivers, in *Proceedings of the European Conference on Antennas and Propagation*, Berlin, Germany, pp. 1441–1445, 2009.

63. B. D. Rigling, Urban RF multipath mitigation, *IET Radar, Sonar, Navig.*, 2: 419–425, 2008.

64. A. Martone, K. Ranney, and R. Innocenti, Through the wall detection of slow moving personnel, *Proc. SPIE*, 7308: 73080Q-1–73080Q-12, 2009.

65. J. McCorkle, Focusing of synthetic aperture ultra-wideband data, in *Proceedings of the IEEE International Conference on Systems Engineering*, Dayton, OH, pp. 1–5, 1991.

66. A. Martone, K. Ranney, and C. Le, A non-coherent approach for through the wall moving target indication, *IEEE Trans. Aerosp. Electronic Syst.*, 50(1): 193–206, 2014.

67. T. Dogaru, L. Nguyen, and C. Le, *Computer Models of the Human Body Signature for Sensing through the Wall Radar Applications*, ARLTR-4290, U.S. Army Research Laboratory, Adelphi, MD, September 2007.

68. T. Dogaru, L. Nguyen, and C. Le, *Synthetic Aperture Radar Images of a Simple Room Based on Computer Models*, ARLTR-5193, U.S. Army Research Laboratory, Adelphi, MD, May 2010.

69. C. Le, L. Nguyen, and T. Dogaru, Radar imaging of a large building based on near-field Xpatch model, in *Proceedings of the IEEE International Symposium on Antennas and Propagation Society*, Toronto, Canada, pp. 1–4, July 2010.

70. T. Dogaru, and C. Le, Synthetic aperture radar techniques for through-the-wall radar imaging, in M. Amin (Ed.), *Through-the-Wall Radar Imaging*, CRC Press, Boca Raton, FL, 2010.

71. H. Stankwitz, R. Dallaire, and J. Fienup, Nonlinear apodization for sidelobe control in SAR imagery, *IEEE Trans. Aerosp. Electronic Syst.*, 31(1): 267–279, 1995.

72. M. Ressler, L. Nguyen, F. Koenig, D. Wong, D, and G. Smith, The Army Research Laboratory (ARL) synchronous impulse reconstruction (SIRE) forward-looking radar, in *Proceedings of the SPIE*, vol. 6561, Bellingham, WA, pp. 656105-1–656105-12, 2007.

73. N. Anwar and M. Abdullah, A novel technique for multiple targets detection in through-the-wall radar imaging, in *Proceedings of the International Conference on Science Technology*, Pathum Thani, Thailand, pp. 70–75, November 2015.

74. J. Hu, Y. Song, T. Jin, B. Lu, G. Zhu, and Z. Zhou, Shadow effect mitigation in indication of moving human behind wall via MIMO TWIR, *IEEE Geosci. Remote Sens. Lett.*, 12(3), 2015, pp. 453–457.

75. G. Gennarelli, R. Solimene, F. Soldovieri, and M. Amin, Three-dimensional through-wall sensing of moving targets using passive multistatic radars, *IEEE J. Sel. Topics Appl. Earth Observ. Remote Sens.*, 9(1), 2016, pp. 141–148.

76. F. Fioranelli, S. Salous, I. Ndip, and X. Raimundo, Through-the-wall detection with gated FMCW signals using optimized patch-like and vivaldi antennas, *IEEE Trans. Antennas Propag.*, 63(3), 2015, pp. 1106–1117.

77. F. Ahmad, J. Qian, and M. Amin, Wall clutter mitigation using discrete prolate spheroidal sequences for sparse reconstruction of indoor stationary scenes, *IEEE Trans. Geosci. Remote Sens.*, 53(3), 2015, pp. 1549–1557.

7

Tracking Humans in the Indoor Multipath Environment

Traian Dogaru, Christopher Sentelle, Gianluca Gennarelli, and Francesco Soldovieri

CONTENTS

7.1 Introduction

When tracking is mentioned in the context of radar processing, the applications that first come to mind involve fast and highly maneuverable targets (such as aircrafts or missiles), or precision weapon guidance systems. From this perspective, it may not be immediately obvious that tracking could also be an effective technique in indoor sensing radar systems designed to detect moving humans. Nonetheless, as demonstrated in this chapter, the performance of a radar device for indoor sensing applications can greatly benefit from the inclusion of well-designed tracking algorithms.

In sensing scenarios involving indoor human personnel, the targets move with fairly low velocities, in patterns that are relatively predictable over the typical subsecond scan interval. However, one major challenge to radar detection in such environments is the presence of strong multipath propagation and scattering. This phenomenon has a great impact on the radar detection performance, especially by increasing the false alarm rate. When multipath returns from valid targets are counted as positive detections, the radar processor may indicate the wrong number of targets inside the region of interest (ROI). A more severe type of error occurs when multipath returns from targets or clutter outside the ROI appear as valid detections inside the ROI, although that region may be empty.

In these scenarios, the tracker logic can be a very effective tool in discriminating the direct path from the multipath target returns, and thus significantly improve the radar detection performance.

The material in this chapter assumes the reader is familiar with the basic concepts of radar tracking (for an introduction to the topic, as well as more advanced treatments, one should consult [1–4]). Because the track filtering poses less of a challenge in the case of human targets, all the examples in this chapter use a more or less conventional Kalman filter approach. However, the multipath propagation and scattering mechanisms characteristic to this environment require an advanced treatment of the data-to-track association problem.

Multiple investigations published in the literature over the last decade explore the radar detection of moving targets inside buildings. Although the techniques discussed in this chapter were developed specifically for through-the-wall radar (TWR) systems placed outside the building, they are equally applicable to scenarios where both the sensor and the target are placed inside the same room or building. Two possible approaches are typically employed for the moving target radar detection. One implementation, characteristic to radar systems with low-to-medium bandwidth, processes the target in the range-Doppler space, with a possible postdetection azimuth estimation step where multiple channel measurements are available. The other, typically associated with ultrawideband waveforms and antenna arrays, is based on the formation of high-resolution radar images (the target is processed in down- and cross-range). The second approach is often coupled with a change detection scheme to separate the stationary clutter from the moving targets [5–8]. Surprisingly though, very few of these studies explicitly incorporate tracking algorithms in the radar processing chain. References [9–11] demonstrate the effectiveness of tracking techniques for indoor sensing of human targets in fairly benign scenarios. However, the trackers implemented in these studies use relatively simplistic data association algorithms, such as the nearest neighbor (NN) or the global NN methods [3]. In this chapter, we illustrate two modern implementations of a tracking system as applied to TWR problems: the multiple hypothesis tracking (MHT) and the joint probabilistic data association (JPDA).

The dimensionality of the state vector available for track characterization is limited by the number of coordinates measured by the radar system. Section 7.2 discusses the implementation of a tracker where the range, the range rate, and the azimuth of the potential targets are provided at the detector output via range-Doppler processing. The tracker associated with this radar employs the MHT algorithm. In Section 7.3, we present an alternative approach, where the target detection and localization is performed in high-resolution two-dimensional (2D) images. The tracker employed by this radar implements the JPDA algorithm. Section 7.4 summarizes the findings in this chapter and suggests other possible approaches to indoor radar tracking of human personnel.

7.2 Radar Tracking System Operating in the Range-Doppler Space

7.2.1 Description of the L-3 Sensing TWR Radar Tracking System

Indoor monitoring by radar systems is closely related to the sensing through-the-wall (STTW) radar technology, which has attracted a large amount of interest within the community since the early 2000s. Government agencies, industries, and academia from several countries have funded both fundamental research and practical system development, contributing to the implementation of new concepts and the understanding of performance limits of these sensors [12,13]. Several prototypes were developed under U.S. Army programs between 2005 and 2013 [14]. In this section, we describe one of these systems, developed by L-3 Communications Security and Detection Systems for through-the-wall detection of indoor personnel.

The L-3 STTW radar system is a soldier-borne, handheld device (Figure 7.1) providing the ability to interrogate and detect both moving and stationary (i.e., breathing) human targets within buildings, behind doors, or beyond other visible obstructions at standoff distances. Here, we limit our discussion to the operation of the L-3 radar system under moving target scenarios, with particular emphasis on the tracking algorithm ability to mitigate the false alarms produced by multipath propagation and scattering.

The L-3 STTW system (Patent No. US20150301167A1) is a fully coherent S-band radar employing a stepped frequency continuous wave with an effective radiated power equivalent to that of a standard WiFi device (<100 mW) and runs off of lithium AA batteries. It is lightweight (<5 lbs), compact (<10 in. in the longest dimension), and highly ruggedized. The multichannel interferometer design supports 2D target location and tracking, and is easily extendable to full three-dimensional (3D) tracking. The antenna architecture supports up to 180° of field of view (FOV) in azimuth and >160° FOV in elevation. Additional guard channels are employed to adaptively suppress potential sources of false alarms arising from behind the sensor, for example, motion of friendly forces behind the sensor. Multiple targets can be detected, located, and tracked on the other side of a wall constructed from a variety of materials, including adobe, concrete block, reinforced concrete, wood structures, and brick, to name a few, at standoff ranges of >20 m.

The L-3 STTW device employs a multiple hypothesis tracking (MHT) (Patent No. US009229102B1) system [15], which is built from the ground up for the TWR application and tuned to allow efficient operation, in terms of both memory and computational complexity,

FIGURE 7.1
Depiction of the L-3 STTW sensor.

within the embedded application. A fully coupled, extended Kalman tracking filter is employed as part of the MHT system for tracking targets in full 3D Cartesian coordinate space.

The Kalman tracking filter forms recursive, least-squares estimates of the target trajectory using a user-specified dynamic model. The tracking filter is highly customizable with general applicability to a wide range of problems and is ideal within the MHT framework because it conveniently provides a prediction covariance estimate that can be used to compute probabilities. The extended Kalman filter (EKF) [3] solves the following set of equations at each time step k:

$$\hat{x}_{k|k} = \hat{x}_{k|k-1} + \mathbf{K}_k \left(y_k - \mathbf{H}\hat{x}_{k|k-1} \right)$$

$$\mathbf{K}_k = \mathbf{P}_{k|k-1}\mathbf{H}^T \left(\mathbf{H}\mathbf{P}_{k|k-1}\mathbf{H}^T + \mathbf{R} \right)^{-1}$$

$$\mathbf{P}_{k|k} = \left(\mathbf{I} - \mathbf{K}_k\mathbf{H} \right)\mathbf{P}_{k|k-1} \tag{7.1}$$

$$\hat{x}_{k+1|k} = \Phi\hat{x}_{k|k} + f_{k+1|k}$$

$$\mathbf{P}_{k+1|k} = \Phi\mathbf{P}_{k|k}\Phi^T + \mathbf{Q}$$

where:
  $\hat{x}$ is the state vector
  $\mathbf{K}$ the Kalman gain matrix
  y is the measurement vector
  $\mathbf{H}$ is the measurement matrix that maps the state vector into the measurement space and
        represents the Jacobian of the nonlinear mapping function in the EKF
  $\mathbf{Q}$ represents the user-specified process noise
  $\mathbf{P}$ is the process noise covariance matrix estimate
  Φ is the state transition matrix
  $\mathbf{R}$ is the user-specified measurement noise covariance matrix
  f represents the contribution from an external forcing function

The first three equations in the set perform a filtering step, whereas the last two equations perform the prediction step.

The prediction error covariance matrix $\mathbf{S}$ can be derived from the process noise and measurement covariance matrices as $\mathbf{S} = \mathbf{H}\mathbf{P}\mathbf{H}^T + \mathbf{R}$ and provides the error ellipsoid surrounding the predicted location in the measurement coordinate space. This is a useful quantity for detection gating as well as for computing a tracking score, discussed later.

In terms of a specific through-the-wall application, the user is free to choose the coordinate system, dynamic model, and process noise parameters. The EKF is commonly employed for converting between a polar coordinate system (range, Doppler, azimuth) and the Cartesian coordinate system, and works by linearizing the nonlinear transfer function between coordinate systems, where $\mathbf{H}$ becomes the Jacobian of the mapping between the measurement space and the state space. The dynamic model, along with process noise, are best chosen based upon the expected target trajectory and must account for target unpredictability. This is a challenging task for the through-the-wall application where targets can exhibit highly nonlinear trajectories within short intervals. A trade must be made between using higher order dynamic models (constant acceleration or beyond), which can become unstable if not tuned correctly and lower order dynamic models where loss in track must be contended with in highly nonlinear trajectory events. Regardless of the chosen filter design, additional coasting logic and filter reset logic is likely necessary to achieve satisfactory performance.

The MHT system was found to be an important method for TWR due to its ability to resolve multitarget scenarios as well as aid in rejecting multipath. The MHT is a method for solving the data association problem. Between each coherent processing interval (CPI), the system must associate new detections with existing tracks. In a commonly employed NN data association approach, each detection is assigned to the closest track, and each remaining, nonassociated detection is assigned to a new track. However, ambiguities may exist such as shown in Figure 7.2. Here, there are two tracks (T1 and T2) along with two detections (1 and 2). The typical NN approach first assigns detect 2 to track T1, and because detect 1 is outside of track T2, a new track is formed. However, as an alternative, it may be more likely that detect 1 belongs to track T1 and detect 2 to track T2. The MHT algorithm, instead of forcing a fixed assignment, maintains a set of hypotheses consisting of all possible assignments and defers the final association decision until sufficient data are obtained to make a final determination based upon computed probabilities for a set of hypotheses.

In the track-oriented MHT scheme, a tree structure is maintained containing all of the possible data-to-track associations over several scans. Each node of the tree represents a specific track-to-data association where the parent node track has been updated according to the corresponding data association. At each scan, a leaf node is chosen from the current tree, replicated, and updated according to each possible data association, which includes the coasting case where the track is not updated with an observation. A new root node is created for newly formed tracks. Figure 7.3 depicts an example where there are initially two tracks, T1 and T2. In the leftmost tree, the parent track, T1, has been replicated and updated using O1 (observation 1) to create T3 and O2 to create T4, and updated with no data association (O0) to create T5. The observations O1 and O2 are also used to create entirely new tracks T9 and T10. During the next scan, this same process is repeated for the new set of observations and tracks T3 through T10.

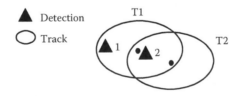

FIGURE 7.2
Example detection scenario showing possible ambiguities in the detect-to-track association problem.

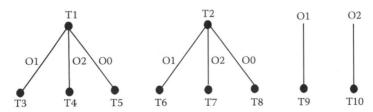

FIGURE 7.3
Example multiple hypotheses formed in the data association problem.

This process results in a family of trees with newly formed tracks creating the root node for each tree as shown in Figure 7.3, where there are four trees. The tracks within each family are related in that they share a set of common observations because they share at least the root node. Therefore, a compatible set of tracks, those not sharing a common observation, must not contain more than one track from a single family. This becomes important for hypothesis formation, discussed later, where a set of compatible tracks is chosen for each hypothesis. It is generally assumed that each tree represents a single target; however, scenarios involving potentially multiple targets that are not clearly resolvable within the measurement space are also possible.

A tracking score is maintained for each tracking filter corresponding to the likelihood ratio associated with a track being related to a true target, defined as follows [3]:

$$LR^j = \frac{p(D|H_j)P_0(H_j)}{p(D|H_0)P_0(H_0)} \tag{7.2}$$

where:

$p(D|H_0)$ is the probability density function (PDF) associated with data D belonging to the false alarm case

$p(D|H_j)$ is the PDF for a given detection being associated with track j

The quantities $P_0(H_0)$ and $P_0(H_j)$ are the prior probabilities of false alarm and detection, respectively. Therefore, H_0 represents the false alarm null hypothesis, whereas H_j corresponds to the alternative hypothesis of data D being associated with track j.

In terms of the Kalman tracking filter, the likelihood ratio (assuming a normal distribution) becomes [3]

$$LR_k^j = \frac{V_C \exp\left[-\frac{1}{2}\left(y_{i,k} - \hat{y}_k^j\right)^T S_k^{j-1}\left(y_{i,k} - \hat{y}_k^j\right)\right]}{(2\pi)^{M/2}\sqrt{\left|S_k^j\right|}} \tag{7.3}$$

where:

$\left|S_k^j\right|$ is the determinant of the prediction error covariance matrix associated with the jth Kalman filter at time step k

M is the measurement dimension

$y_{i,k}$ is the ith measurement vector at time step k

$\hat{y}_k^j$ is the jth Kalman filter prediction vector

V_C represents the measurement volume in which a true detection and false alarm event are independent (typically based upon radar resolution).

The total likelihood ratio includes the information associated with all updates to the track from previous scans and is computed after K updates as follows [3]:

$$LR_{tot}^j = L_0 \prod_{k=1}^K LR_k^j \tag{7.4}$$

where:

    L_0 is the initial likelihood ratio based upon prior probabilities for a newly formed track, defined as $L_0 = P_0(H_0)/P_0(H_1) = P_D\beta_{NT}/\beta_{FT}$

    P_D is the probability of detection

    β_{NT} is the new target density

    β_{FT} is the false target density

Note that it is possible to include other measures such as the target signal-to-noise ratio (SNR) within the likelihood ratio expression [3]. Generally, most systems work with the log likelihood ratio to update the track score based upon data association. For a newly formed track, the track score is initialized to $L_0^j = \ln(L_0)$ and is then updated during each CPI according to

$$L_k^j = L_{k-1}^j + \begin{cases} \ln\left(LR_k^j\right), & \text{track update} \\ \ln(1-P_D), & \text{no track update} \end{cases} \tag{7.5}$$

The track score is typically used in determining whether a track is valid and can be displayed. A two threshold system is employed, with the track becoming active when the track score crosses a minimum threshold and inactive once crossing below a lower threshold. These thresholds are typically chosen empirically so as to balance rejection of spurious false alarms while providing clean target trajectories with minimal latency (or minimal detection times).

In the track-oriented MHT implementation, a new set of hypotheses is formed from existing tracks and associated track scores at each scan. This is done by finding all combinations of sets of compatible tracks that can explain the current observations. In the example given earlier, one possible set of hypotheses might include {T3, T7}, {T5, T8, T9, T10}, {T3, T8, T10}. Each hypothesis assumes each observation is only used once. Once the set of hypotheses is enumerated, the hypothesis likelihood score simply becomes the sum of the individual track scores comprising the set of tracks within the hypothesis. The hypothesis probability is then computed for m = 1, ..., M_k hypotheses as follows:

$$p(H_{m,k}) = \frac{\exp\left(\sum_{v \in H_{m,k}} L_k^v\right)}{1 + \sum_{l=1}^{M_k} \exp\left(\sum_{v \in H_{l,k}} L_k^v\right)} \tag{7.6}$$

Once the probability of a hypothesis is computed, the probability associated with a specific track can be computed by summing all of the probabilities for all of the hypotheses that include that track. The computed hypothesis probabilities are predominantly used for determining which tracks to present for display. A simple, yet effective technique, in some applications, is to present the track with the highest probability from each family tree. Only those tracks meeting the track score criterion discussed earlier are displayed. Other approaches include presenting a probability-weighted sum of each track within a family tree and using a similarly weighted track score to determine when the target should be displayed.

Clearly, the size of the hypothesis tree can grow exponentially, and, therefore, intelligent pruning of the hypothesis tree is essential to successful implementation on an embedded real-time system. Numerous methods can be employed for tree pruning. First, the number of data-to-track associations can be limited by considering only those pairings where the observation falls within the tracking gate, which can be derived from the prediction error covariance matrix **S**. Second, tracks with low or negative track scores or data-to-track associations that will result in low track scores can be eliminated from consideration. Additional criteria are possible for eliminating consideration of track-to-data associations.

The predominant tree pruning method typically relies on the hypothesis probabilities in an N-scan pruning approach. The idea here is to maintain only N scans worth of data or, equivalently, an N-level tree structure. At each scan, the track with the highest probability, determined from the overall hypothesis probabilities, is selected from each family tree, and the root node, going N scans back, is selected as the new parent node for the family. All tracks within the family not sharing the new parent node are eliminated. The number of scans to employ in the N-scan approach must be determined for the specific application and is often best chosen empirically.

In general, there are numerous methods in addition to those just described for managing the size of the tree. A set of highly efficient and successful heuristics is employed within the L-3 sensor, which are capable of maintaining minimal tree sizes without sacrificing performance in the TWR application and allow for smooth and nearly false alarm-free display of target trajectories.

7.2.2 Multipath Scattering Analysis for Moving Targets

In this section, we focus on the multipath response of a target moving inside the room. The presence of *ghost images* of targets in the context of TWR is a well-known phenomenon and has been extensively documented in the literature [13,16]. In [16], where only stationary targets are analyzed, it is shown that the expected ghost image locations are the projection point(s) of the target on the back wall(s), as well as multiple mirror image points behind the back walls, placed at equal range intervals representing the distance between the target and the back wall. In this section, where we consider moving targets, we investigate the phenomenology of ghost images with respect to both their location in the range/Doppler (R/D) maps and their dynamic evolution in time.

The number of possible multipath propagation scenarios inside a four-wall room is very large; however, because the radar signal loses power with each successive reflection, as well as with the total length of the propagation path, only low-order multi-bounce paths (made of two or three reflections) are typically detected by the radar receiver. In Figure 7.4, we consider three such scenarios that were encountered in our data analysis. Notice that scenario 1 (Figure 7.4a) involves two bounces (one from the target and one from the back wall), whereas scenarios 2 and 3 (Figure 7.4b and c, respectively) involve three bounces (one from the target, one from the front/back wall, and another one from the back wall). For reference, we call the direct scattering from the target scenario 0.

Because the L-3 STTW radar receiver uses R/D maps combined with interferometric bearing estimation in the track formation processing, the measurement variables available are the range (R), azimuth angle (ϕ), and Doppler frequency, which is proportional to the range rate (V). Based on these, the tracking algorithm estimates the state variables, which are typically the Cartesian spatial coordinates and their time derivatives (or velocity components). To each valid track, including those generated by multipath returns, corresponds a separate measurement and state vector. Throughout Section 7.2, the track index is denoted as a subscript, running in our multipath scenarios from 0 to 3. If we assume a single moving target in the scene, all the track measurement and state vectors ultimately depend on the target state vector $\begin{bmatrix} x_T & y_T & v_x & v_y \end{bmatrix}^T$. In Figure 7.5,

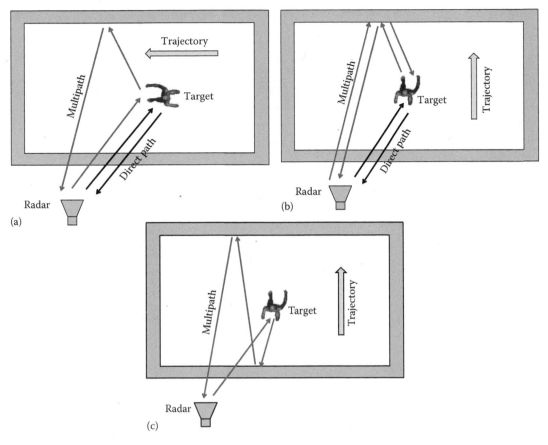

FIGURE 7.4
Three multipath propagation scenarios encountered in TWR measurements for a moving target in a four-wall room: (a) mechanism 1, (b) mechanism 2, and (c) mechanism 3.

we can establish the following equations for the measurement variables of the four target return mechanisms as a function of the target state variables:

$$R_0 = \sqrt{x_T^2 + y_T^2} \tag{7.7}$$

$$\phi_0 = \arctan\left(\frac{x_T}{y_T}\right) = \arcsin\left(\frac{x_T}{R_0}\right) \tag{7.8}$$

$$V_0 = \frac{dR_0}{dt} = \frac{x_T v_x}{\sqrt{x_T^2 + y_T^2}} + \frac{y_T v_y}{\sqrt{x_T^2 + y_T^2}} \tag{7.9}$$

$$R_2 = \sqrt{x_T^2 + \left(2D - y_T\right)^2} \tag{7.10}$$

$$\phi_2 = \arctan\left(\frac{x_T}{2D - y_T}\right) = \arcsin\left(\frac{x_T}{R_2}\right) \tag{7.11}$$

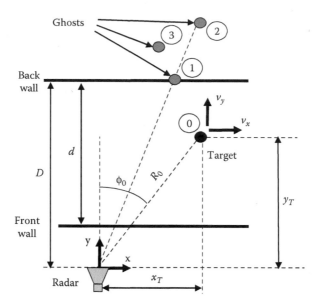

FIGURE 7.5
Diagram for deriving the range and range-rate equations of the target responses associated with the propagation mechanisms in Figure 7.4.

$$V_2 = \frac{x_T v_x}{\sqrt{x_T^2 + (2D - y_T)^2}} - \frac{(2D - y_T)v_y}{\sqrt{x_T^2 + (2D - y_T)^2}} \tag{7.12}$$

$$R_1 = \frac{R_0 + R_2}{2} \tag{7.13}$$

$$\phi_1 = \arctan\left(\frac{x_T}{2D - y_T}\right) = \arcsin\left(\frac{x_T}{R_2}\right) \tag{7.14}$$

$$V_1 = \frac{V_0 + V_2}{2} \tag{7.15}$$

$$R_3 = \frac{R_0 + \sqrt{x_T^2 + (2d + y_T)^2}}{2} \tag{7.16}$$

$$\phi_3 = \arctan\left(\frac{x_T}{2d + y_T}\right) \tag{7.17}$$

$$V_3 = \frac{V_0 + \dfrac{x_T v_x}{\sqrt{x_T^2 + (2d + y_T)^2}} + \dfrac{(2d + y_T)v_y}{\sqrt{x_T^2 + (2d + y_T)^2}}}{2} \tag{7.18}$$

Most of the tests performed with the L-3 STTW radar involved targets walking along radial (parallel to the y-axis) or transversal (parallel to the x-axis) trajectories. For these particular trajectories, we can establish some simple approximations to the measurement

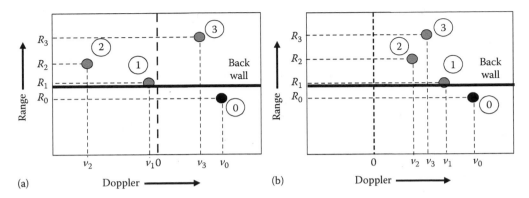

FIGURE 7.6
Diagram showing the typical location of the direct and multipath target responses in the R/D maps for (a) radial target trajectory and (b) transversal target trajectory.

vector equations. Thus, for radial motion ($v_x = 0$) and small cross-range displacements ($x_T \cong 0$), we have $R_0 \cong y_T$; the response 1 appears approximately along the back wall ($R_1 \cong 0$), 2 at a range symmetric to R_0 with respect to the back wall ($R_2 \cong 2D - R_0$), and $R_3 \cong R_0 + d$. At the same time, $V_0 \cong v_y$, $V_1 \cong 0$, $V_2 \cong -v_y$, $V_3 \cong v_y$. For transversal motion ($v_y = 0$), the range-rate expressions are more complicated, but, as a general rule, we can establish that the range rates of all four responses have the same sign as v_x and $|V_0| > |V_1| > |V_2|, |V_3|$. Schematic diagrams describing typical locations of the target responses in the R/D maps for the two types of motion are shown in Figure 7.6 (note that the diagrams are dynamic, with the location of the target multipath responses changing in time as a function of the direct response location). Notice that all the equations established in this section neglect the additional delays of the radar waves induced by propagation through the wall materials.

Besides the propagation mechanisms described previously, multipath target return scenarios involving reflections from the side walls, or other large scatterers inside the building (such as interior walls or metallic cabinets), may create sizeable responses. However, it is expected that these propagation paths appear only intermittently within the allowable FOV (120°), and as such, they are typically rejected by the MHT because of the low tracking scores.

A general formulation of multipath propagation involving successive reflections from flat surfaces such as the walls of a building can be obtained via the Householder transform [17,18]. In the following, we describe the positions of the target and its mirror images as vectors in a 2D space (the generalization to a 3D geometry is straightforward), while the radar is placed in the coordinate origin. Let $r_T = [x_T \quad y_T]^T$ be the target position vector and $r_A = [x_A \quad y_A]^T$ be the apparent position vector of its mirror image upon a wall reflection (note that we use the subscript T to denote the target and the superscript T to denote the matrix transpose). Then, r_A can be obtained from r_T by applying the operator $\Re$ given by

$$\Re : r_T \to r_A = \Gamma r_T + 2q = \left(I - 2uu^T\right) r_T + 2q \tag{7.19}$$

In this equation, $\Gamma = I - 2uu^T$ is the Householder matrix [18], with u representing the outward-looking normal unit vector to the wall, and q the vector distance from the origin to the wall. For example, in Figure 7.5, we have $u = [0 \quad 1]^T$ and $q = [0 \quad D]^T$ for the back wall and $u = [0 \quad -1]^T$ and $q = [0 \quad D-d]^T$ for the front wall. A one-way propagation path from the radar to the target (or vice versa), involving M wall reflections, indexed by $m = 1, \ldots M$, can be characterized by successively applying the operators $\Re_m$ to the original target vector position r_T:

$$\Re_M \Re_{M-1} \ldots \Re_1 : r_T \rightarrow r_A = \Gamma_M \left(\Gamma_{M-1} \ldots \left(\Gamma_1 r_T + 2q_1 \right) + \ldots 2q_{M-1} \right) + 2q_M \quad (7.20)$$

To derive the vector describing the round trip of the radar wave from the transmitter to the target (*transmitted path*) and from the target back to the receiver (*received path*), we take into account the following: (1) the transmitted and received paths may be different, as shown by the propagation mechanisms 1 and 3 introduced earlier in this section, and (2) the interferometric processing for azimuth estimation in the receiver means that the received path alone dictates the direction of the vector r_A. Using the subscripts t and r to denote the transmitted and received paths, respectively, the vector r_A obtained via a propagation mechanism made of M reflections in the transmitted path (indexed by $m = 1, \ldots, M$) and N reflections in the received path ($n = 1, \ldots N$), can be written as

$$r_A = \frac{1}{2} \frac{r_r}{\|r_r\|} \left(\|r_t\| + \|r_r\| \right) \quad (7.21)$$

From an algorithmic point of view, the vectors r_t and r_r are computed recursively starting from r_T and applying Equation 7.20. The link between the position vector $r_A = [x_A \ y_A]$ and the measurement variables R, ϕ, and V for a given propagation path is straightforward: the range represents the norm of r_A, the azimuth represents the angle made by the vector with the y–axis, and the range rate represents the range derivative with respect to time. Their equations are as follows:

$$R_A = \frac{1}{2} \left(\|r_t\| + \|r_r\| \right) \quad (7.22)$$

$$\phi_A = \arctan \left(\frac{x_r}{y_r} \right) \quad (7.23)$$

$$V_A = \frac{1}{2} \left(\frac{r_t^T \Gamma_t}{\|r_t\|} + \frac{r_r^T \Gamma_r}{\|r_r\|} \right) \dot{r}_T \quad (7.24)$$

with $\dot{r}_T = [v_x \quad v_y]^T$, $\Gamma_t = \prod_{m=1}^{M} \Gamma_{tm}$, and $\Gamma_r = \prod_{n=1}^{N} \Gamma_{rn}$ In the commonly encountered TWR scenario where the reflections along a multipath occur only between two parallel and opposite walls (as in all the examples considered earlier in this section), all the Γ_{tm} and Γ_{rn} matrices in the last equation have the same expression ($\Gamma_{tm} = \Gamma_{rn} = \Gamma$ for any m and n), and the matrix products simplify as

$$\prod_{m=1}^{M} \Gamma_m = \begin{cases} \Gamma & \text{for } M \text{ odd} \\ I & \text{for } M \text{ even} \end{cases} \quad (7.25)$$

As an experimental confirmation of the theoretical formulas presented in this section, Figure 7.7 shows the R/D maps obtained for three different CPIs, for two test scenarios where one target walks on either radial or transversal trajectory. Notice that mechanism 1 cannot be distinguished in the radial trajectory case (it coincides roughly with the location

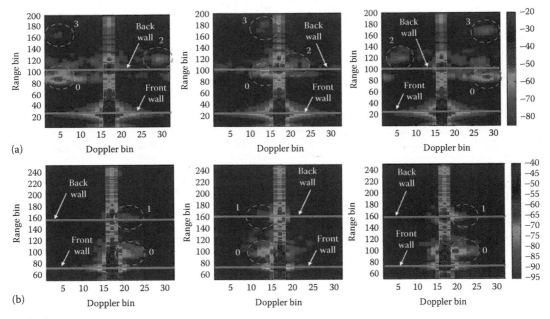

FIGURE 7.7
R/D maps obtained with the L-3 radar in TWR experiments (one moving target), for three different CPIs, showing the direct (0) and the multipath (1 through 3) target responses, for (a) radial target trajectory and (b) transversal target trajectory.

of the back wall in both the range and Doppler). In the transversal trajectory case, mechanisms 2 and 3 are too weak to appear in the R/D maps. Moreover, because the range rate is fairly small for transversal trajectories, the responses associated with all types of propagation mechanisms are relatively close to the zero-Doppler line and may be more difficult to detect than in the case when a radial motion component is present.

The analysis presented in this section is employed in a post-tracking ray-tracing algorithm that predicts the multipath propagation in order to reject the tracks that are consistent with ghost targets. To make this analysis possible, the radar must first detect and estimate the ranges to the reflecting walls or other large scattering surfaces. This is most easily done for the front and back walls of the room; however, the exact wall orientation is difficult to discern, and other large surfaces such as side walls may be difficult to detect. Incorrect, insufficient, or inaccurate information concerning the large scattering surfaces may lead to wrong decisions in the track classifier resulting in either rejection of valid targets or the inability to reject ghost targets. An alternative approach to discriminate the multipath returns, based on the dynamic evolution of each target track, is presented in Section 7.2.3.

Figure 7.8 presents the time evolution of the tracker output (range only), obtained during field experiments [14], for the three major propagation mechanisms described previously, with one target moving inside the four-wall room. The trajectory in Figure 7.8a is transversal to the radar line of sight (multipath mechanism 1 is prevalent in this case) and radial in Figures 7.8b and c (where multipath mechanisms 2 and 3 are prevalent). In all cases, the direct track (index 0, round markers in Figure 7.8) appears correctly between the front and back walls. The multipath tracks (triangular markers in Figure 7.8) are not displayed on the user screen, because they are located behind the back wall. Notice that some tracks appear as intermittent (the target is *lost* over certain intervals), and there is a short latency time of about 3 s before first establishing the tracks.

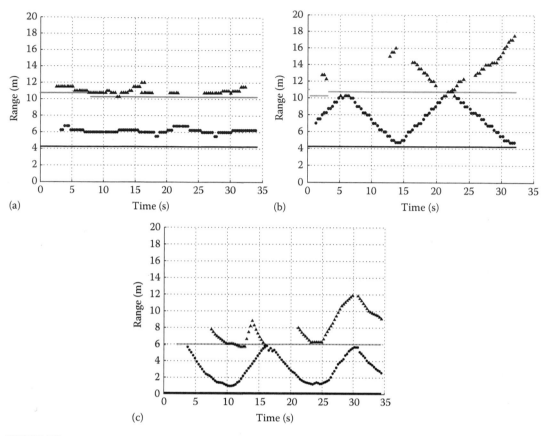

FIGURE 7.8
Range versus time plots of the tracker output for three one-mover scenarios, representing various multipath scattering mechanisms: (a) mechanism 1 under transversal target trajectory; and (b) mechanism 2 and (c) mechanism 3, both under radial target trajectories. The straight lines represent the estimated ranges to the front wall (black) and back wall (gray), respectively. Round markers describe the direct tracks, whereas triangular markers describe the multipath tracks.

7.2.3 Dynamic Evolution of the Multipath Target Tracks

The time evolution of the multipath tracks in Figure 7.8, as well as the linear nature of the Householder transform discussed in Section 7.2.2, suggests that a strong correlation exists between the variables characterizing the direct and multipath tracks generated by the same target. When we consider the measurement variables (range, azimuth, and range rate), we notice that the linear dependence between the tracks generated by one target is valid only in an approximate sense. Nevertheless, because we expect a large cross-covariance of these variables between tracks associated with the same target, we propose using it as a metric to group together all the tracks generated by that particular target. Subsequently, this track sorting procedure can be used to discriminate the direct from the multipath tracks.

As a validation of this approach to discriminate direct from multipath tracks, we consider a simulation scenario involving two targets moving back and forth on orthogonal trajectories—one radial and one transversal. The radar is at 4 m away from the front wall, whereas the room is 8 m deep. Target A moves from 5 to 11 m down-range at 3 m cross-range, with a velocity of 0.6 m s^{-1}. Target B moves at 8 m down-range between −3 and 3 m cross-range, with a velocity of 0.75 m s^{-1}. The observation interval spans 160 CPIs at 0.25 s per CPI. The radar measurement errors are accounted for by adding independent white Gaussian noise sequences to each track, with $\sigma_R = 0.25$ m for range, $\sigma_V = 0.1$ ms^{-1} or range rate, and $\sigma_{AZ} = 2.0°$ for azimuth. We consider the multipath mechanisms 2 and 3 for target A and 1 for target B, and apply the equations developed in Section 7.2.2. We employ a fully coupled EKF [3] to predict the next CPI radar data for each track. The state vector of the Kalman filter is represented in the Cartesian coordinate space and uses a constant velocity dynamic model. The resulting raw radar tracks, as well as the Kalman filter tracker output (produced by direct and multipath scattering), are shown in Figure 7.9.

Table 7.1a and b shows the covariance of the range and range rate, respectively, between all the track pairs that can be formed with the data previously described, using directly the radar measurements. Table 7.2a and b depict the same covariance estimates when Kalman filtering is employed in the tracker. In all the cases presented here, we consider the absolute value of the correlation coefficient [19], normalized such that it always produces a number between 0 and 1. In general, the tracks of the ghosts associated with a specific target are strongly correlated with that target's direct track, but they do not correlate with tracks generated by another independent target. We also note that the correlation between the two tracks generated by target B (moving on transversal trajectory) is not as large as for the other correlated track pairs. This demonstrates the large impact of both range and range-rate measurement errors on this procedure as applied to transversal target tracks. Nevertheless, in all cases, the track correlation is improved by the presence of the Kalman filter in the tracking algorithm (the improvement is more notable for target B). This highlights the potential importance of employing Kalman filtering prior to forming correlation estimates when given noisy radar measurements.

In general, scenarios involving more complex dynamic models than those presented here, such as accelerated motion, curved trajectories, or even random walk-type motion patterns, are expected to maintain the high correlation due to the linear relationship suggested by Equation 7.20. Several factors can act to reduce the effectiveness of using the correlation metric for multipath discrimination, including nonlinearity associated with the norm in Equations 7.22 and 7.24, inherent errors within the radar measurements and signal processing algorithms, as well as Kalman tracking errors. In particular, this work shows that the Kalman tracking performance directly affects the accuracy of the method, and, therefore, careful attention must be given to tuning the Kalman filter to the appropriate dynamic model for optimal performance. Another issue to consider is the possibility for this procedure to generate false positives: for instance, pairs of independent target tracks, originating from separate targets, can show strong correlation with one another, as well as with all the multipath tracks that they generate. A simple example is that of two targets moving along parallel radial trajectories, in the same or opposite directions, with the same absolute velocity. From this discussion, it is apparent that a combination of ray-tracing analysis and track correlation estimation should offer a robust solution of eliminating the multipath returns in a TWR scenario.

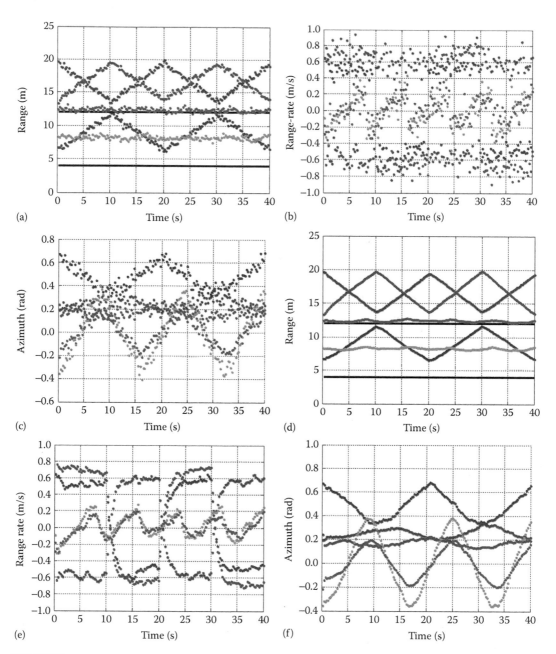

FIGURE 7.9
Radar tracks obtained in the simulation of two targets moving back and forth in a room on orthogonal trajectories, showing (a) and (d) track range versus time, (b) and (e) track range rate versus time, and (c) and (f) azimuth versus time. The plots (a) through (c) represent direct (unfiltered) measurement data, whereas (d) through (f) represent the Kalman tracking estimates. The colors represent the following: black—front and back wall; red—track A_0; green—track A_2; blue—track A_3; cyan—track B_0; magenta—track B_1.

TABLE 7.1

Covariance Matrix of (a) the Range and (b) the Range Rate for Five Different Tracks Obtained from Raw Data in the Simulation Scenario

(a)						(b)					
Track	A_0	A_2	A_3	B_0	B_1	Track	A_0	A_2	A_3	B_0	B_1
A_0	1.0	0.974	0.975	0.016	0.074	A_0	1.0	0.971	0.977	0.081	0.119
A_2	0.974	1.0	0.977	0.013	0.049	A_2	0.971	1.0	0.988	0.057	0.099
A_3	0.975	0.977	1.0	0.025	0.043	A_3	0.977	0.988	1.0	0.061	0.098
B_0	0.016	0.013	0.025	1.0	0.246	B_0	0.081	0.057	0.061	1.0	0.628
B_1	0.074	0.049	0.043	0.246	1.0	B_1	0.119	0.099	0.098	0.628	1.0

Note: The shaded cells indicate the track pairs that should be correlated.

TABLE 7.2

Covariance Matrix of (a) the Range and (b) the Range Rate for Five Different Tracks Obtained in the Simulation Scenario When Kalman Filtering Is Used

(a)						(b)					
Track	A_0	A_2	A_3	B_0	B_1	Track	A_0	A_2	A_3	B_0	B_1
A_0	1.0	0.998	0.991	0.146	0.055	A_0	1.0	0.993	0.995	0.137	0.217
A_2	0.998	1.0	0.994	0.156	0.037	A_2	0.993	1.0	0.996	0.114	0.189
A_3	0.991	0.994	1.0	0.132	0.038	A_3	0.995	0.996	1.0	0.125	0.199
B_0	0.146	0.156	0.132	1.0	0.798	B_0	0.137	0.114	0.125	1.0	0.901
B_1	0.055	0.037	0.038	0.798	1.0	B_1	0.217	0.189	0.199	0.901	1.0

Note: The shaded cells indicate the track pairs that should be correlated.

7.3 Radar Tracking System Operating in the High-Resolution Image Domain

7.3.1 Multiple Extended Target Tracking for TWR

In this section, based on a recently published paper [20], we present an alternative procedure to address the problem of indoor target tracking by TWRs. Whereas the radar system described in Section 7.2 relied on target processing in the R/D space, the approach illustrated in this section employs high-resolution images to detect and localize the targets. As previously emphasized in this chapter, tracking multiple targets in the indoor environment requires a careful treatment of the data-to-track association problem. The radar processing scheme presented in this section focuses less on the target–wall interactions but takes into account the multipath returns created by target-to-target multiple scattering mechanisms.

The tracking algorithm presented in this section is capable to deal with the data association problem in multitarget scenarios by exploiting the JPDA method [21] and the extended target tracking (ETT) paradigm [22,23]. Based on the fact that human targets are imaged as spots (groups of pixels) that produce detection clouds, the ETT procedure obtains an estimate of target sizes jointly with their location and velocity. Notably, the additional information related to object size can be effectively exploited to enhance the data association and thus improve the overall system performance.

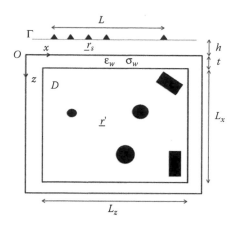

FIGURE 7.10

Geometry of the TWR scenario. Moving targets (circles) and static objects (rectangles) are located in an enclosed room and probed by a TWR. (From Gennarelli, G. et al., *IEEE Trans. Geosci. Remote Sens.*, 53, 6482–6494, © 2015 IEEE.)

For the sake of simplicity, let us consider the 2D scenario depicted in Figure 7.10. The Cartesian coordinate system relevant to this scenario involves down-range (z) and cross-range (x). Moving targets (circles) and static objects (rectangles) are located in an enclosed room D. The room walls have thickness t and are made of a homogeneous and nonmagnetic dielectric material characterized by the relative permittivity ε_w and the electric conductivity σ_w. The scene is probed by an antenna array deployed along the line Γ with length L and placed at a standoff distance h from the front wall interface at $z = 0$. Each antenna is sequentially turned on and off. In the *on* phase, one antenna radiates a wideband electromagnetic (EM) pulse with bandwidth $B = [\omega_{min}, \omega_{max}]$ and collects the radar echo so that a multi-monostatic/multifrequency measurement configuration is achieved. The position of each radiating element is denoted as $r_s = x_s\hat{x} - h\hat{z}$, whereas $r' = x'\hat{x} + z'\hat{z}$ is a generic point in the investigation domain D. The antennas are modeled as filamentary electric currents directed along the y-axis, such that the EM fields involved in the scattering problem are scalar.

When activated, each antenna in the array records the total field $E_{tot}(r_s, t_f)$, where t_f represents the fast-time coordinate. This operation is repeated periodically at times $t_s = k\Delta$, $k = 1,...,K$, where t_s is the slow-time coordinate, Δ is the temporal discretization step between two successive data frames, and K is the total number of frames recorded over the interval $[0, T_{st}]$. The entire radar signal processing chain for ETT is represented in Figure 7.11.

The *first* processing step consists in time gating of the total field dataset $E_{tot}(r_s, t_f, k)$, in order to filter out the stationary clutter related to the direct coupling signal and the front wall reflection. The scattered field data so obtained are transformed into the frequency domain, yielding the dataset $E_s(r_s, \omega, k)$, $k = 1,...,K$.

The *second* stage of the data processing chain is concerned with the image formation process, that is, a tomographic image of the TWR scene is achieved from the frequency domain scattered field data. In the following, it is assumed that the targets are metallic (perfectly conducting). Specifically, by denoting with τ_p the contour of object p, with $\tau = \bigcup_{p=1}^{P} \tau_p$ the union of all target contours, and invoking the physical optics (PO)

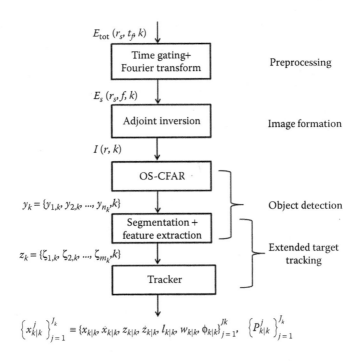

FIGURE 7.11
Block diagram of the TWR signal processing chain for multiple extended target tracking. (From Gennarelli, G. et al., *IEEE Trans. Geosci. Remote Sens.*, 53, 6482–6494, © 2015 IEEE. With Permission.)

approximation [24], the scattered field E_s at a slow-time index k, produced by currents induced over τ, is expressed by the integral equation:

$$E_s(r_s,\omega,k)=-j\omega\mu_0\int_\tau G(r_s,r',\omega)E_i(r_s,r',\omega)\gamma(r',k)d\tau \tag{7.26}$$

where:

$G(r_s,r',\omega)$ is the Green's function of the scenario
E_i is the incident field in D
γ is an unknown distribution function supported over τ

The above-mentioned expression can be synthetically formulated by adopting the following operator notation:

$$E_s = \mathcal{L}\gamma + \nu \tag{7.27}$$

where:

$\mathcal{L}:U \to V$ is a linear projection operator mapping the unknown space U onto the data space V
ν is a disturbance term that has been added to model the effect of clutter and noise

Solving the linear inverse problem described by Equation 7.27 consists of determining the unknown object function γ when E_s is known. This problem is ill-posed [25] and a regularization scheme must be applied in order to find a stable solution in the presence of noisy data. For the sake of simplicity and computational efficiency, the inversion is accomplished by resorting to the adjoint operator, $\mathcal{L}^*$:

$$\gamma = \mathcal{L}^* E_s \tag{7.28}$$

which essentially corresponds to a migration or matched filtering inversion scheme [26]. The magnitude $I(r,k)$ of the estimated function γ, normalized to its maximum value, is denoted as the tomographic image from this point on.

According to the block diagram in Figure 7.11, the *third* stage of the data processing deals with the object detection. The tomographic images obtained by adjoint inversion are given as input to the ordered statistic constant false alarm rate (OS-CFAR) detector originally introduced in [27]. As well known, CFAR detectors estimate the clutter/noise power by adapting the threshold in order to achieve a desired probability of false alarm P_{FA}. The square amplitude of each pixel in the image (cell under test) is compared to a threshold depending on the estimated noise power and desired P_{FA}. The noise power is evaluated by considering the cells in a reference window. A guard window is used to exclude cells close to the cell under test because they contain both clutter/noise and target energy. In OS-CFAR detection, the observations in the reference window are sorted to form a sequence in ascending numerical order $\{I_{(1)}, I_{(2)}, \ldots I_{(Nc)}\}$, where Nc is the number of cells in the reference window. The qth order statistic $I_{(q)}$ is chosen as the representative of the interference level, and the threshold Th is set as a multiple of this value, that is,

$$Th = \alpha I_{(q)} \tag{7.29}$$

By assuming that the real and imaginary parts of noise contribution in the complex value tomographic image are independent and identically distributed Gaussian variables, the constant α is evaluated by inverting the following equation [27]:

$$P_{FA} = q \frac{Nc!}{q!(Nc-q)!} \frac{(q-1)!(\alpha+Nc-q)!}{(\alpha+Nc)!} \tag{7.30}$$

According to [27], the 75th percentile (i.e., $I_{(0.75Nc)}$) is selected to determine Th.

At each time scan k, the OS-CFAR provides the set of detections $Y_k = \{y_{1,k}, y_{2,k}, \ldots, y_{n_k,k}\}$, where $y_{i,k} \in D$ and n_k are the position of the ith detection and the total number of detections, respectively. Note that the subscript k means that the abovementioned parameters vary for each scan time k.

Starting from the set Y_k, a segmentation operation is performed to group the detections that are possibly generated by the same targets. The adopted procedure exploits the group connectivity of detections, that is, the spatial relation of a detection with its neighbors. Two detections $y_{i,k}$ and $y_{j,k}$ are said to be *eight-connected* if $y_{j,k} \in N_8(y_{i,k})$, where $N_8(x)$ denotes the eight neighbors of x taken along the horizontal, vertical, and diagonal directions. When this condition is verified, the two detections are considered part of the same cluster.

With regard to the modeling of target shape, the common ellipsoidal representation is adopted [28]. Starting from the segments, the parameters characterizing the elliptical model are estimated using the normalized second central moments [29]. After this stage, a new set of measurements $Z_k = \{\zeta_{1,k}, \zeta_{2,k}, \ldots, \zeta_{m_k,k}\}$ is obtained, where m_k is the number of objects detected at time k. The entry $\zeta_{i,k} = [\zeta_{i,k}^{(p)}, \zeta_{i,k}^{(s)}, \zeta_{i,k}^{(\phi)}]^T$ contains the estimated parameters of the ellipse that fits the ith object. Specifically, $\zeta_{i,k}^{(p)} \in D$ is the position of the center of the object, $\zeta_{i,k}^{(s)} \in \mathbb{R}^2$ is the object size constituted by the major and minor ellipse axes, and $\zeta_{i,k}^{(\phi)}$ is the object orientation with respect to the x-axis.

The ETT procedure represents the *fourth* stage of the data processing chain (see Figure 7.11). At each slow-time scan k, the tracker receives as input the set Z_k to produce its outcomes. The ETT procedure is based on the JPDA rule [21]. Section 7.3.2 provides a detailed description of this stage, by first reviewing the motion and measurement models, and then the ETT algorithm itself.

7.3.2 Target Motion and Measurement Models

The target state vector x_k at time k is defined in Cartesian coordinates as the composite vector obtained by adding the target's size and orientation to the position and velocity, that is,

$$x_k = [x_k, \dot{x}_k, z_k, \dot{z}_k, l_k, w_k, \phi_k]^T \tag{7.31}$$

where:

x_k, z_k and $\dot{x}_k, \dot{z}_k$ are the position and velocity components along x and z, respectively
l_k and w_k represent the length and width, respectively
ϕ_k is the target orientation

Generally, ETT procedures assume that the target orientation is the same as the motion orientation, that is, $\phi_k = \arctan(\dot{z}_k/\dot{x}_k)$. In TWR tracking, this assumption does not seem to be always valid—thus, target and motion orientations are treated independently.

The motion of targets is described by the nearly constant velocity model [21], that is,

$$x_k = \mathbf{F} x_{k-1} + \mathbf{\Gamma} w_k \tag{7.32}$$

$$\mathbf{F} = \begin{bmatrix} \mathbf{I}_2 \otimes \tilde{\mathbf{F}} & \mathbf{O}_{4\times3} \\ \mathbf{O}_{3\times4} & \mathbf{I}_3 \end{bmatrix} \tag{7.33}$$

$$\mathbf{\Gamma} = \begin{bmatrix} \mathbf{I}_2 \otimes \tilde{\mathbf{\Gamma}} & \mathbf{O}_{4\times3} \\ \mathbf{O}_{3\times2} & \mathbf{I}_3 \end{bmatrix} \tag{7.34}$$

$$\tilde{\mathbf{F}} = \begin{bmatrix} 1 & T_s \\ 0 & 1 \end{bmatrix} \tag{7.35}$$

where:
$\tilde{\mathbf{\Gamma}} = [T_s^2/2, T_s]^T$
$\mathbf{I}_d$ is the identity matrix with size d
$\mathbf{O}_{r\times c}$ is the null matrix with r rows and c columns
T_s is the sampling time
$\otimes$ is the Kronecker product
the vector w_k accounts for target acceleration and other unmodeled dynamics and is assumed to be Gaussian with covariance matrix

$$\mathbf{Q} = \text{diag}(\sigma_v^2, \sigma_v^2, \sigma_l^2, \sigma_w^2, \sigma_\phi^2) \tag{7.36}$$

where:

diag($\cdot$) represents a diagonal matrix
$\sigma_v^2, \sigma_l^2, \sigma_w^2, \sigma_\phi^2$ are the variances of the additive acceleration, length, width, and orientation, respectively

The measurement ζ_k produced by a target at time k is expressed as

$$\zeta_k = \mathbf{H}x_k + n_k \tag{7.37}$$

$$\mathbf{H} = \begin{bmatrix} \mathbf{I}_2 \otimes \tilde{\mathbf{H}} & \mathbf{O}_{2\times3} \\ \mathbf{O}_{3\times4} & \mathbf{I}_3 \end{bmatrix} \tag{7.38}$$

where:

$\tilde{\mathbf{H}} = [1,0]$
n_k is the noise vector, assumed Gaussian, with zero mean and covariance matrix

$$\mathbf{R} = \text{diag}(\sigma_{n,x}^2, \sigma_{n,z}^2, \sigma_{n,l}^2, \sigma_{n,w}^2, \sigma_{n,\phi}^2) \tag{7.39}$$

where:

$\sigma_{n,x}^2$ and $\sigma_{n,z}^2$ are the variances along Cartesian axes
$\sigma_{n,l}^2, \sigma_{n,w}^2$, and $\sigma_{n,\phi}^2$ are the variances of sizes and orientation, respectively

7.3.3 Multitarget Tracking Procedure

The tracking algorithm is based on the JPDA technique, which is a Bayesian approach that associates by probabilistic weights all the validated measurements to the tracks. The track management is based on the *M-out-of-N* logic, meaning that a track is confirmed if at least M positive detections are declared out of N scans. The filtering stage is carried out using a Kalman filter [21]. The updated (predicted) target state and its covariance matrix at time k are denoted by $x_{k|k}^j (x_{k|k-1}^j)$ and $\mathbf{P}_{k|k}^j (\mathbf{P}_{k|k-1}^j)$. Assume that, at frame k, a set of J_k tracks is active/preliminary $T_k = \{T_1(k), T_2(k), ..., T_{J_k}(k)\}$, where $T_j(k)$ identifies the jth track. For each $j = 1, 2, ..., J_k$, a validation gate region G_k^j is constructed. Because the target-originated measurements are Gaussian, distributed around a predicted measurement $\zeta_{k|k-1}^j = \mathbf{H}x_{k|k-1}^j$ of target j, the gate is defined as

$$G_k^j = \{\zeta : (\zeta - \zeta_{k|k-1}^j)^T (\mathbf{S}_k^j)^{-1} (\zeta - \zeta_{k|k-1}^j) < \rho\} \tag{7.40}$$

where $\mathbf{S}_k^j = \mathbf{H}\,\mathbf{P}_{k|k-1}^j \mathbf{H}^T + \mathbf{R}$ is the innovation covariance and the threshold ρ determines the gating probability P_G, that is, the probability that a measurement originated by target j is correctly validated.

The track management consists of the following steps:

- *Track initiation*: A measurement is associated with the track $T_j(k)$ when it falls in its gate region. Each unassociated measurement is named initiator and produces a tentative track. After the detection of an initiator, a gate is set up, and, if a detection falls in the gate, this track becomes a preliminary track; otherwise, it is dropped. For each preliminary track, the JPDA can be initialized and used to set up a gate for the next sampling time. Starting from the third scan, a logic of M detections out of N scans is used for the subsequent gates. If, at the end of the process (scan $N+2$), the logic requirement is satisfied, the track becomes a confirmed or active track; otherwise, it is discarded.

- *Track termination*: A confirmed track is terminated if one of the following conditions is met: (1) no detection has been validated in the past N^* most recent sampling times, (2) the target's track uncertainty has exceeded a given threshold, and (3) the target has reached an unfeasible maximum velocity v_{max}.

- *Track update*: For each active and preliminary track, the target state is updated according to the measurement-to-track association rule of the JPDA [21], whereas the target state prediction follows directly from the motion model.

- *Data association*: A validation matrix is established for all confirmed and preliminary targets. This matrix is populated with validated measurements falling inside the gate, as well as the case of no measurement. All feasible joint association events are constructed in this way. Each measurement originates from one target, or is a false alarm. Each target generates, at most, one measurement with detection probability P_D. The probabilities of the joint events are evaluated under the assumptions: (1) target-originated measurements have a Gaussian distribution around the predicted location of the corresponding target measurement and (2) false alarms are distributed in the surveillance region according to a Poisson point process of parameter λ (the clutter density), assumed uniformly distributed in the gating region. The marginal association probabilities of target j with measurement i, $\beta_{i,k}^j$ are obtained by summing over all joint events where the marginal event of interest occurs [21].

- *Update and prediction*: The jth target state $x_{k|k}^j$ and its covariance $\mathbf{P}_{k|k}^j$ are updated by averaging the updates for all validated measurements with the association probabilities [21]:

$$x_{k|k}^j = x_{k|k-1}^j + \mathbf{K}_k^j v_k^j \tag{7.41}$$

$$\mathbf{P}_{k|k}^j = \mathbf{P}_{k|k-1}^j - \left(\sum_{i=1}^{m_k^j} \beta_{i,k}^j \right) \mathbf{K}_k^j \mathbf{S}_k^j (\mathbf{K}_k^j)^T + \mathbf{K}_k^j \left(\sum_{i=1}^{m_k^j} \beta_{i,k}^j v_{i,k}^j (v_{i,k}^j)^T - v_k^j (v_k^j)^T \right) (\mathbf{K}_k^j)^T \tag{7.42}$$

where:

$v_k^j = \sum_{i=1}^{m_k^j} \beta_{i,k}^j v_{i,k}^j$ is the combined innovation

$v_{i,k}^j = \zeta_{i,k}^j - \zeta_{k|k-1}^j$ is the innovation for the ith validated measurement

$\mathbf{K}_k^j = \mathbf{P}_{k|k-1}^j \mathbf{H}^T (\mathbf{S}_k^j)^{-1}$ is the Kalman gain

m_k^j is the number of validated measurements

$\zeta_{i,k}^j$ is the ith validated measurement

The predicted state $x_{k+1|k}^j$ and its covariance $\mathbf{P}_{k+1|k}^j$ are evaluated according to the standard Kalman filter prediction step.

7.3.4 Numerical Validation

A multitarget numerical experiment is reported in the following to assess the effectiveness of the tracking procedure described previously. The radar signals are generated for the indoor scenario depicted in Figure 7.12 by using the forward solver GPRMAX2D [30], which is based on the finite-difference time-domain method. The room walls are made

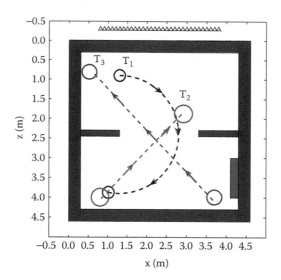

FIGURE 7.12
Multitarget TWR scenario. Dashed line depicts the target trajectories. (From Gennarelli, G. et al., *IEEE Trans. Geosci. Remote Sens.*, 53, 6482–6494, © 2015 IEEE. With Permission.)

of concrete with relative permittivity $\varepsilon_w = 4$ and conductivity $\sigma_w = 0.01$ S/m. The external walls have thickness $t = 0.3$ m, and the inner sizes of the room are $L_x = L_z = 4$ m. Two internal walls with length 1 m and thickness 0.15 m are located in the room at a distance of 2 m from the front wall.

Moreover, a static target (of size 0.2 m × 1 m), resembling a piece of furniture and having a relative permittivity equal to 6, is placed nearby the right lateral wall (blue rectangle in Figure 7.12). Three metallic cylinders (T_1, T_2, T_3) having different sizes are simultaneously moving in the room. T_1 has a radius of 0.15 m, and its center moves from (1.3, 0.9) m to (1.0, 3.9) m describing a uniform circular motion with velocity along the tangent to trajectory, equal to 1.0 m s^{-1}. T_2 has a radius of 0.225 m and its center moves from (0.8, 4.0) to (2.9, 1.9) m following a linear path with a constant speed of 0.6 m s^{-1}. T_3 has a radius of 0.187 m and moves from (3.7, 4.0) to (0.5, 0.8) m along a line with a speed of 0.9 m s^{-1}. As noticed in Figure 7.12, the targets get close together as they approach the center of the room. This scenario is useful for testing the performance of the ETT procedure, because target-to-target interactions are expected to produce ghost objects at locations where no physical target exists, with a consequent increase in the number of false alarms.

The scene is probed by a 3 m-long array, which is placed at a distance $h = 0.5$ m from the front wall. Each antenna radiates a Ricker pulse with a center frequency of 1.5 GHz and records the radar echo over the fast-time window [0, 40.0] ns. The antenna spacing is uniform and fixed at 0.1 m. The room is discretized into square pixels with size 0.05 m. The time-domain raw signals are preprocessed by a gating operation over the time window [0, 10] ns to filter direct coupling and the main front wall echoes. The remaining part of the signals are transformed into the frequency domain by considering the interval [1.0, 2.0] GHz with a step of 0.025 GHz. Moreover, the frequency-domain data are corrupted with additive white Gaussian noise by assuming an SNR of 10 dB. Each dataset is processed to form a tomographic image every 0.1 s (i.e., frame rate of 10 Hz). The slow-time observation window [0, 5] s, corresponding to a sequence of 51 images, is considered for the application of the ETT procedure.

TABLE 7.3

Tracker Parameters

Parameter	Value	Specification
T_s	0.1 s	Sampling time
σ_v	0.9 m s^{-2}	Process noise
σ_l	0.975	Process noise
σ_w	0.016	Process noise
σ_ϕ	0.074	Process noise
$\sigma_{n,x}$	0.02 m	Standard deviation x
$\sigma_{n,z}$	0.02 m	Standard deviation z
$\sigma_{n,l}$	0.05 m	Standard deviation length
$\sigma_{n,w}$	0.05 m	Standard deviation width
$\sigma_{n,\phi}$	$\pi/3$ rad	Standard deviation orientation
P_D	0.9	Detection probability
P_G	1	Gate probability
λ	10^{-5}m^{-2}	Clutter density
ρ	5^2	Gate threshold
$v_{\max}$	5 m s^{-1}	Maximum velocity
M/N	5/6	Track initialization logic
M^*/N^*	8/8	Track termination logic

Source: Gennarelli, G. et al., *IEEE Trans. Geosci. Remote Sens.*, 53, 6482–6494, © 2015 IEEE.

As for object detection, the OS-CFAR detector has been set by considering a P_{FA} = 1e−5 and the 75th percentile for threshold evaluation. The reference window was fixed to 13 × 13 pixels, and a guard window of 7 × 7 pixels was chosen. The tracker parameters are summarized in Table 7.3.

Figure 7.13 illustrates the tomographic images obtained from the data recorded at t_s = 0, 1, 2, 3, 4, and 5 s, respectively. Notice that the targets are always correctly localized, but the amplitude of each spot is variable, depending on their down-range. This phenomenon is a peculiar feature of the adjoint inversion, which can be explained in the framework of the singular value decomposition expansion [25]. Furthermore, as predicted, the presence of false targets among the true objects is evident in the images corresponding to t_s = 2 and 3 s.

Figure 7.14 shows the outputs provided by the tracker. As can be seen, the tracks of T_2 and T_3 are initiated with latencies of 1.2 and 0.7 s, respectively. Indeed, in agreement with the images in Figure 7.13, T_2 and T_3 produce a weak response in the interval [0, 1] s, because they are partly obscured by the internal walls of the room. Once the three tracks are formed, the system is capable of following all targets even when they are in close proximity, and false alarms arising from the ghost targets are effectively filtered out. In the time interval [3.6, 4.5] s, the system begins to lose track of target T_1 (see the top left panel of Figure 7.14), because it is obscured by the largest target, T_2.

It must be stressed that the estimated position along z is characterized by an almost constant offset with respect to the true position. The justification of this phenomenon resides in the fact that the TWR only allows imaging of the upper edge of the object. However, the true position is evaluated by considering the center of the target.

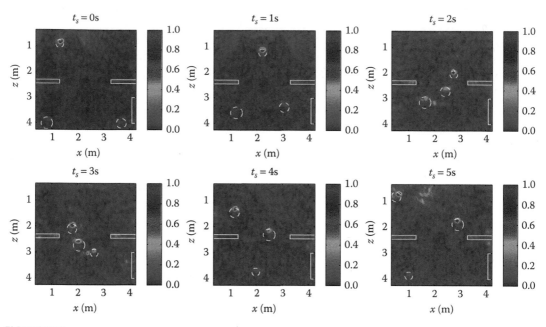

FIGURE 7.13
(Color insert) Tomographic images relevant to data frames collected at times $t_s = 0, 1, 2, 3, 4,$ and 5 s, respectively. (From Gennarelli, G. et al., *IEEE Trans. Geosci. Remote Sens.*, 53, 6482–6494, © 2015 IEEE. With Permission.)

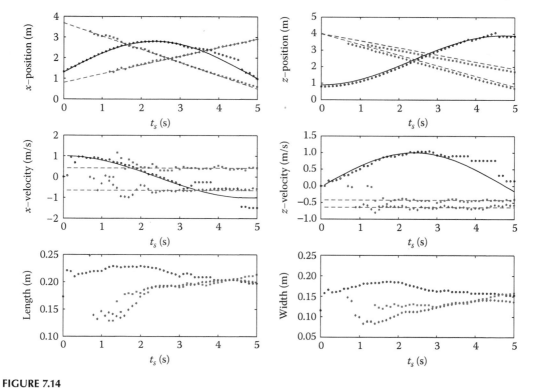

FIGURE 7.14
Tracker outputs (symbols). Solid lines represent the ground truth, whereas dotted lines represent the target size, position and velocity estimates. Different colors are used to indicate the targets: T_1 (black), T_2 (blue), and T_3 (red). (From Gennarelli, G. et al., *IEEE Trans. Geosci. Remote Sens.*, 53, 6482–6494, © 2015 IEEE. With Permission.)

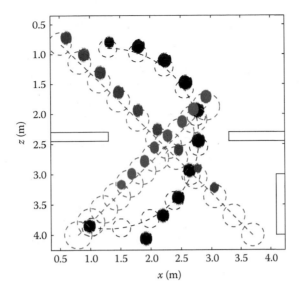

FIGURE 7.15
Tracker outputs (symbols). Solid ellipses represent the ground truth, whereas dotted ellipses represent the target size and position estimates. Different colors are used to indicate the targets: T_1 (black), T_2 (blue), and T_3 (red). (From Gennarelli, G. et al., *IEEE Trans. Geosci. Remote Sens.*, 53, 6482–6494, © 2015 IEEE. With Permission.)

As for estimated object sizes, they can be appreciated in Figure 7.15, which shows the estimated tracks along with the ground truth. As expected, the estimated objects are represented as ellipses; however, their sizes underestimate the true object dimensions because only the upper contours of the object can be imaged by the tomographic approach. Consequently, the ellipse size estimation provides useful information, although it does not accurately represent the target extent.

7.4 Conclusions

This chapter discussed the application of tracking techniques to TWR systems and demonstrated their effectiveness in improving the radar performance for indoor detection of human personnel. Because the major feature of target tracking in these sensing scenarios is suppression of the multipath returns characteristic to this environment, a sizeable portion of this chapter was dedicated to the analysis of multipath propagation and scattering phenomena inside buildings. Knowledge of all the possible multipath return combinations for a given target can be incorporated in the tracker logic and used in separating the direct from multiple scattering paths. Another possible approach, described in Section 7.2.3, is based on the fact that returns generated by the same target (direct and multipath) are strongly correlated with one another.

In terms of radar processing methods, this chapter illustrated both the R/D and the high-resolution imaging approaches. Although there are clear differences between the two in terms of waveforms, number of channels, algorithms, and computational requirements, there are no investigations to date that compare their effectiveness for this specific application. Hopefully, future studies will be able to carefully analyze the trade-offs between the two processing techniques.

The radar tracking algorithms explored in this chapter included two implementations: the MHT and the JPDA. Although a direct comparison between the two methods was not performed here, we can confidently conclude that the large number of false target detections present in the indoor environment requires an advanced treatment of the data-to-track association problem. In this respect, older approaches, such as those based on NN techniques and its variations, may not be adequate in discriminating the correct target tracks from the false alarms. The ETT method was also included in the scenarios analyzed in Section 7.3. Although many of the results in this chapter are based on computer simulations, the experimental results obtained with the L-3 STTW radar system, reported in [14], strongly support our statements regarding the effectiveness of tracking for the indoor sensing application.

One alternative approach to multipath suppression, which could possibly relax the requirements on the tracker implementation, is to incorporate the knowledge of all possible propagation paths inside the image formation algorithm—thus, the target ghosts are effectively eliminated before the detection and tracking stages. This technique was demonstrated in Reference 17. Moreover, a similar approach could be implemented in the case of R/D processing. In both cases, this method can be interpreted as designing a more complex matched filter that takes into account the multipath propagation in the specific environment. Nevertheless, this idea amounts to primarily shifting the computational burden from the tracking algorithm to the matched filter implementation. More recent papers using the sparse reconstruction framework [31–33] demonstrate the techniques for multipath exploitation in the context of TWR sensing of both stationary and moving targets. Other possible approaches to multipath suppression are discussed in [34]. One hopes that future investigations will be able to evaluate the pros and cons of all these methods in designing a high-performance radar system.

Acknowledgment

The authors thank Dr. Giovanni Alli for introducing us to the applications of Householder transform to TWR problems.

References

1. Richards, M., J. Scheer, and W. Holm. 2010. *Principles of Modern Radar*. Raleigh, NC: SciTech Publishing.
2. Bar-Shalom, Y. and X.-R. Li. 1993. *Estimation and Tracking: Principles, Techniques, and Software*. Norwood, MA: Artech House.
3. Blackman, S. and R. Popoli. 1999. *Design and Analysis of Modern Tracking Systems*. Norwood, MA: Artech House.
4. Skolnik, M. 2001. *Introduction to Radar Systems*. New York: McGraw-Hill.
5. Amin, M. G. and F. Ahmad. 2013. Change detection analysis of humans moving behind walls. *IEEE Transactions on Aerospace and Electronic Systems*, 49:1410–1425.
6. Ahmad, F. and M. G. Amin. 2013. Through-the-wall human motion indication using sparsity-driven change detection. *IEEE Transactions on Geoscience and Remote Sensing*, 51:881–890.

7. Gennarelli, G., I. Catapano, and F. Soldovieri. 2013. RF/microwave imaging of sparse targets in urban areas. *IEEE Antennas and Wireless Propagation Letters*, 12:643–646.
8. Martone, A. F., K. Ranney, and C. Le. 2014. Non-coherent approach for through-the-wall moving target indication. *IEEE Transactions on Aerospace and Electronic Systems*, 50:193–206.
9. Rovnakova, J. and D. Kocur. 2009. Compensation of wall effect for through wall tracking of moving targets. *Radioengineering*, 18(2):189–195.
10. He, Y., T. Savelyev, and A. Yarovoy. 2011. Two-stage algorithm for extended target tracking by multistatic UWB radar. *IEEE CIE International Conference on Radar*, 1:795–799.
11. Chen, X., H. Leung, and M. Tian. 2014. Multitarget detection and tracking for through-the-wall radars. *IEEE Transactions on Aerospace and Electronic Systems*, 50:1403–1415.
12. Lisuzzo, A. et al. 2011. Sensing through-the-wall technologies. NATO SET-100 Final Report, RTO-TR-SET-100.
13. Amin, M. G. 2011. *Through-the-Wall Radar Imaging*. Boca Raton, FL: CRC Press.
14. Dogaru, T., C. Ly, K. Ranney, J. Silvious, D. Washington, and K. Tom. 2014. Evaluation of the Army's handheld sensing through the wall radar systems. ARL Technical Report, Adelphi, MD: ARL-TR-6849.
15. Chang, S. et al. 2009. An MHT algorithm for UWB radar-based multiple human target tracking. *Proceedings of the 2009 IEEE International Conference on Ultra-Wideband*, Piscataway, NJ.
16. Dogaru, T., C. Le, and L. Nguyen. 2010. Synthetic aperture radar images of a simple room based on computer models. ARL Technical Report, Adelphi, MD: ARL-TR-5193.
17. Setlur, P., L. Nuzzo, and G. Alli. 2013. Multipath exploitation in through-wall radar imaging via point spread functions. *IEEE Transactions on Image Processing*, 22:4571–4586.
18. Golub, G. and C. Van Loan. 1996. *Matrix Computations*. Baltimore, MD: John Hopkins University Press.
19. Papoulis, A. 1991. *Probability, Random Variables and Stochastic Processes*. New York: McGraw-Hill.
20. Gennarelli, G., G. Vivone, P. Braca, F. Soldovieri, and M. G. Amin. 2015. Multiple extended target tracking for through-wall radars. *IEEE Transactions on Geoscience and Remote Sensing*, 53(12):6482–6494.
21. Bar-Shalom, Y., P. Willett, and X. Tian. 2011. *Tracking and Data Fusion: A Handbook of Algorithms*. Storrs, CT: YBS Publishing.
22. Granström, K., C. Lundquist, and U. Orguner. 2012. Extended target tracking using a Gaussian mixture PHD filter. *IEEE Transactions on Aerospace and Electronic Systems*, 48(4):3268–3286.
23. Granström, K., A. Natale, P. Braca, G. Ludeno, and F. Serafino. 2014. PHD extended target tracking using an incoherent X-band radar: Preliminary real-world experimental results. *Proceedings of the International Conference on Information Fusion*, Salamanca, Spain.
24. Pierri, R., A. Liseno, R. Solimene, and F. Soldovieri. 2006. Beyond physical optics SVD shape reconstruction of metallic cylinders. *IEEE Transactions on Antennas Propagation*, 54(2):655–665.
25. Bertero, M. and P. Boccacci. 1998. *Introduction to Inverse Problems in Imaging*. Boca Raton, FL: CRC Press.
26. Solimene, R., I. Catapano, G. Gennarelli, A. Cuccaro, A. Dell'Aversano, and F. Soldovieri. 2014. SAR imaging algorithms and some unconventional applications. *IEEE Signal Processing Magazine*, 31(4):90–98.
27. Rohling, H. 1983. Radar CFAR thresholding in clutter and multiple target situations. *IEEE Transactions on Aerospace and Electronic Systems*, 19(4):608–621.
28. Errasti-Alcala, B. and P. Braca. 2014. Track before detect algorithm for tracking extended targets applied to real-world data of X-band marine radar. *Proceedings of the International Conference on Information Fusion*, Salamanca, Spain.
29. Haralick, R. M. and L. G. Shapiro. 1992. *Computer and Robot Vision*. Boston, MA: Addison-Wesley.
30. Giannopoulos, A. 2005. *GPRMAX 2D/3D v2.0 User's Manual*.
31. Leigsnering, M., F. Ahmad, M. G. Amin, and A. M. Zoubir. 2014. Multipath exploitation in through-the-wall radar imaging using sparse reconstruction. *IEEE Transactions on Aerospace and Electronic Systems*, 50(2):920–939.

32. Leigsnering, M., F. Ahmad, M. G. Amin, and A. M. Zoubir. 2015. Compressive sensing-based multipath exploitation for stationary and moving indoor target localization. *IEEE Journal of Selected Topics in Signal Processing*, 9(8):1469–1483.
33. Leigsnering, M., F. Ahmad, M. G. Amin, and A. M. Zoubir. 2016. Parametric dictionary learning for sparsity-based TWRI in multipath environments. *IEEE Transactions on Aerospace and Electronic Systems*, 52(2):532–547.
34. Gennarelli, G., G. Vivone, P. Braca, F. Soldovieri, and M. G. Amin. 2016. Comparative analysis of two approaches for multipath ghost suppression in radar imaging. *IEEE Geoscience and Remote Sensing Letters*, 13(9):1226–1230.

8

Bistatic Radar Configuration for Human Body and Limb Motion Detection and Classification

Hugh Griffiths, Matthew Ritchie, and Francesco Fioranelli

CONTENTS

8.1 Introduction

Radar micro-Doppler signatures contain information that can be utilized to classify various actions of targets, including humans. Micro-Doppler itself is defined as the additional frequency modulations on top of the main Doppler shift of moving targets, and these modulations are related to vibrating or rotating parts of vehicles and aircraft (e.g., the tracks of tanks or the blades of helicopters), as well as to swinging limbs and body parts or objects carried by humans [1–2]. These signatures have been investigated in recent years to identify and classify actions for many applications and contexts, such as security, warfare, search and rescue (including the detection of buried humans after earthquakes or avalanches), and human activities monitoring [3–4]. Demonstrated capabilities that have exploited the analysis of micro-Doppler signatures for human limb motion detection and classification include the recognition of humans versus vehicles or animals such as horses or dogs [5–8] and the discrimination between different activities performed by humans such as walking, running, crawling, and carrying objects [9–14]. The identification of specific individuals performing activities based on their signature was also demonstrated in [15–16], showing the identification of individuals walking or running based on their walking gait. Other work on gait classifications using monostatic radar are reported in Chapters 5 and 12. The use of micro-Doppler signatures has been well investigated and reported in the context of indoor monitoring for ambient assisted living and fall detection. The *IET Radar, Sonar & Navigation* journal dedicated a special issue in February 2015 to the use of radar systems to monitor patients remotely and provide care for elderly people. Radar systems can provide accurate monitoring capabilities without raising the privacy concerns associated with camera systems or being as invasive as wearable devices that patients would have to wear all the time. Recent developments in hardware design and processing algorithms in these fields were discussed. In particular, applications included the analysis of different time–frequency distributions to characterize walking with assistive devices such as canes, various approaches for feature extraction to characterize common indoor movements, monitoring of patterns of life and vitality as well as medical parameters such as blood pressure estimation, and finally detecting significant fall events when caring for the elderly [17–21].

The Doppler shift perceived by a radar system depends directly on the cosine of the aspect angle between the trajectory of the target and the radar line of sight. Thus, the micro-Doppler signatures can be significantly attenuated when this angle approaches 90°; this is depicted in Figure 8.1. As a result of this, the classification performance can also be severely degraded. It is reported in Reference 22 that the performance dropped below 40% for aspect angles close to 90°, whereas in Reference 9 it is shown that for angles up to approximately 30° the signatures are only scaled and can be still used to extract suitable features for classification. This demonstrates the importance of geometry when deploying a system within a residential setting in order to provide high classification performance, regardless of the location of the individual in the scene.

Bistatic and multistatic radar systems have been proposed as a potential solution to the issue of classification performance degraded by unfavorable aspect angles, as multiple radar nodes could be deployed, thereby enhancing coverage such that there is always at least one or more radar nodes that are able to observe clear micro-Doppler signatures from the target. Limited results have been presented on human micro-Doppler detection and classification based on experimental data from bistatic/multistatic radar systems. Developing and operating these systems over long distances is a complex challenge, in particular solving the issue of wirelessly providing a coherent clock and phase synchronization at all spatially distributed nodes. This is essential when aiming to extract the Doppler information at each radar node within a multistatic network. As a result of this synchronization challenge, especially over long ranges, the number of available open academic publications using bistatic/multistatic systems is rather limited. These challenges are somewhat mitigated for indoor scenarios, as the relatively small distances allow for wired connectors for synchronization of additional nodes within a bistatic/multistatic radar sensor network.

Previous simulation analyses have discussed how a single spectrogram can be obtained by combining partial spectrograms from different radar nodes within the network, resulting in a compound final spectrogram from which feature extraction and classification can be performed [23–24]. It should be noted that this study did not use experimental data but created simulated data through the Boulic kinematic model and animated a 12-point stick representation of the human body to represent the overall

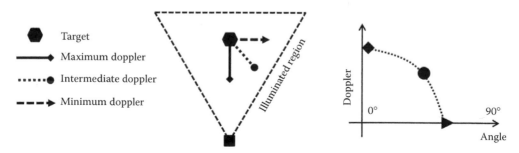

FIGURE 8.1
Diagram showing the effect of target trajectory on the perceived Doppler.

signature as superposition of the responses of individual body parts such as head, upper and lower arms, body, thighs, lower legs, and feet [25–26]. The work in [27] performed experiments with a bistatic radar system to detect the oscillation trajectory of objects such as a pendulum as well as to infer the direction faced by people performing actions such as swinging arms and picking up objects. Other experimental work discussed how the classification accuracy can be improved by combining data from a monostatic radar system and from an acoustic bistatic system, rather than using only the conventional monostatic radar [28]. Analysis presented within [29] showed experimental micro-Doppler signatures of people running and walking in different directions as collected by a multistatic radar system made of three nodes and compared them with simulated results. This work discussed how a multistatic micro-Doppler contains additional information in comparison with monostatic signatures of the same actions; therefore, techniques for automatic target recognition are expected to provide better results when multistatic information is available and correctly exploited.

The radar system used in [29] was also employed to generate the data discussed and analyzed in the following sections. This system is called NetRAD and has been developed over the past decade or more at University College London. Previous research by the authors have presented key results in the context of experimental multistatic human micro-Doppler data, focusing in particular on the classification of unarmed versus potentially armed personnel [30–34] and on the identification of specific individuals based on their walking gait [16,35]. These results are believed to be some of the first examples of experimental multistatic radar data containing human micro-Doppler signatures, and their analysis has provided an opportunity to investigate suitable features and classification methods, as well as approaches to effectively combine information from multiple, spatially distributed nodes.

In review, the key conclusions from these results are that the main benefit of using multistatic radar in comparison with conventional monostatic systems is the possibility of illuminating the target from multiple perspectives to counteract the attenuation on the Doppler signatures at unfavorable aspect angles and to exploit the additional information on targets obtained by looking from multiple viewpoints. Furthermore, multistatic radar can be combined using cognitive radar approaches and *feature diversity*, whereby the different nodes will choose to use different features (or groups of features) based on the target behavior and scenario parameters such as signal-to-noise ratio, polarization, dwell time, and aspect angle, as well as adapt their feature extraction and classification algorithms to respond to changes in those parameters.

The results generated from the NetRAD bistatic/multistatic human micro-Doppler datasets will be discussed in further detail in the remainder of the chapter. This analysis is on the outdoor scenario data that were generated in an open field environment. Data were obtained in an ideal situation with little clutter or obstructions for the detection of the targets. Despite not being generated within an indoor location, these data are directly relevant to this work as the analysis methods and trends shown in classification performance and data fusion algorithms are directly transferrable. The indoor scenario brings additional challenges of multipath, clutter, and added boundaries such as internal walls, but the fundamental micro-Doppler signatures and classification methods are still valid. The compatibility of these analysis and algorithms is clearly demonstrated by their application to the datasets generated indoors by the C-band frequency-modulated continuous wave (FMCW) radar system presented in Section 8.3. The additional information from different, spatially separated radar nodes is expected to be very beneficial to the classification performance of indoor systems, whereby the nodes can be deployed so that at least one of them can see the target of interest from a favorable aspect angle. This is beneficial for more robust classification algorithms to monitor people's activities, for instance, rejecting false alarms for fall detection and providing detailed identification of activities while monitoring daily activities and people's well-being.

The remainder of this chapter is organized as follows: Section 8.2 presents the detailed analysis of multistatic experimental radar data in the context of human micro-Doppler for unarmed versus armed personnel classification and identification of specific individuals based on their walking gait. The experimental setups, the different types of features and classifiers, and the different approaches to use multistatic information will be discussed. Section 8.3 describes some introductory results obtained within an indoor scenario, using an C-band FMCW radar, at the University of Glasgow, Scotland, to analyze the signatures of common movements such as sitting and standing, picking up objects from the floor, walking, and waving with hands. Finally, Section 8.4 draws conclusions and outlines some future research trends that are considered to be significant in this context.

8.2 Analysis of Multistatic NetRAD Data

In this section, the detailed analysis of multistatic radar experimental data is presented. The radar system used for all the experiments discussed in this section was the NetRAD system developed at University College London over the past decade or more [36,37]. NetRAD is a coherent pulsed radar with three separate but identical radar nodes all operating at 2.4 GHz (S-band). The radiofrequency parameters chosen for these experiments were linear up-chirp modulation with 0.6 μs duration and 45 MHz bandwidth, 5 kHz pulse repetition frequency (PRF), and 5 s duration for each recording. The relatively high PRF, for ground-based targets, with respect to the velocity of the body and body parts allows the characterization of the full range of movement from the micro-Doppler signature while avoiding any Doppler ambiguities. The chosen duration for each recording captures several periods of human walking gait, which was shown to be approximately 0.6 s within the results obtained. The radar transmit power is +23 dBm, and the antennas used are vertically polarized mesh reflectors with 24 dBi gain and approximately 10° × 10° beamwidth. Figure 8.2 shows one of the radar nodes deployed during an experiment with the transmitter and receiver antennas on telescopic tripods and the operator. Another node can be seen in the background toward the top-left corner of the figure. The third smaller antenna to the left of the figure is a WiFi antenna operating at a different band to transmit commands and data between the different nodes over the network. All the experiments took place in an open football field at the University College London sports ground in Shenley, to the north of London. This large space was necessary as the pulsed radar has a minimum range of a few tens of meters to be able to detect targets, and the radar was originally designed to collect sea clutter data at longer distances. Nevertheless, it is believed that these data are significant being one of the first examples of experimental multistatic human micro-Doppler data, and that the analysis and results can be easily adapted and applied to multistatic systems designed for indoor detection and classification of humans. The frequency of the system and requirement on beamwidth of the antennas limited the design to be relatively large and bulky in comparison with the one that would be deployed indoors. For an indoor system, the design requirements would be to increase the beamwidth in order to cover the majority of the scene, increase the frequency

FIGURE 8.2
Multistatic transceiver node of the NetRAD system deployed during an experiment.

to reduce the antennas size requirements, and potentially increase the bandwidth to enable finer range resolution due to limited space. As the majority of these points would result in an enhancement of the data provided, it is estimated that the results shown within this section would be improved upon when applied to an indoor scene with a customized designed radar system. The possibility to provide coverage in the whole area under test, the finer range resolution, and above all the additional information available from different nodes that illuminate the target from different aspect angles can be very beneficial to the general problem of improved detection and classification of human activities. This can provide enhanced monitoring capabilities for several applications, including the case of fall detection and health care for elderly and vulnerable people, which is the subject of other chapters in this book.

The data collected by NetRAD are sequential range profiles from a series of N pulses, each consisting of M range bins. The general processing starts with the identification of the range bins containing the target signature, followed by the application of a time–frequency transform to characterize the micro-Doppler signature, that is, the changes in the Doppler pattern as a function of time. Following a common approach in the literature, the short-time Fourier transform (STFT) was used in this study for its simplicity of implementation. This consists of applying a Fourier transform to the time-domain signal, with a Hamming window of a certain duration (0.3 s in this case), which slides over the original signal, with a certain overlap percentage from one window to the next (95% in this case). The overlap of the sliding window was set to this high level in order to produce a smooth representation of the Doppler signatures observed. The analysis within [32] demonstrated that the effect of reducing this overlap window was only a modest reduction on the classification performance while maintaining a window length of 0.3 s. When the window length was increased to 0.7 s, a reduction in overlap had a much more significant impact on the classification performance (a reduction of ~10% classification performance when using 25% overlap).

The result of the STFT is generally referred to as spectrogram, a representation of the Doppler pattern related to the velocity of the movement of the different body parts as a function of time. Example spectrograms of a moving person are shown in Figure 8.3.

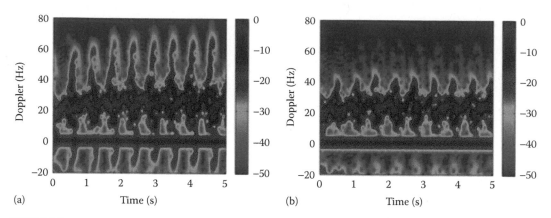

FIGURE 8.3
Example spectrogram of an individual walking with swinging their arms (a) and without swinging their arms (b).

The sinusoidal representation within the spectrogram is typical human micro-Doppler signature where the contributions from swinging arms and legs produce this compound pattern. The micro-Doppler clearly varies between Figure 8.3a and b as the person was walking with and without swinging his arms, respectively.

Feature extraction algorithms are then applied to the generated spectrograms in order to quantify feature samples. This is an essential step within the classification process as performance is more sensitive to samples that are more separable in the defined feature space, rather than to the type of classifier that is used to perform this task. Finally, classifiers based on machine learning techniques are applied to the extracted feature. These are generally based on supervised learning approaches by which the classifiers are trained with a certain percentage of the available data, and then tested using the remaining data to assess their performance, typically repeating this process several times in an iterative Monte Carlo approach that randomly varies the samples used for training and testing to avoid biases in the process.

The classifiers used for this analysis are now briefly discussed from an application perspective and providing the reader with suitable references where more details on the theoretical formulations are given [38,39]. The naïve Bayes (NB) classifier type is based on the assumption of Gaussian distributions and statistical independence for the feature samples belonging to each class. The mean μ and variance σ^2 of these distributions are estimated from the data used to train the classifier. Then at the testing phase, the posterior probability of each sample under test belonging to each class is evaluated, and each sample is assigned to the class presenting the highest probability. Equation 8.1 shows the estimation of the mean and variance of each class, where x_i indicates the training samples for the ith class:

$$\mu_i = \frac{1}{N}\sum_n^N x_{i,n} \quad \sigma_i^2 = \frac{1}{N}\sum_n^N (x_{i,n} - \mu_i)^2 \qquad (8.1)$$

Classifiers based on the concept of discriminant analysis also assume that samples from each class are modeled by a multivariate Gaussian distribution as shown in Equation 8.2, and that the mean μ_k and covariance matrix Σ_k can be estimated at the initial training phase of the classifier. The diagonal–linear (DL) variant of this approach will use a single covariance matrix for all the classes considered and estimate only different mean values for each class. The algorithm divides the feature space into different areas at the training phase. Then a classification cost C is computed for each sample under test with respect to the predicted classification for each possible class. This is shown in Equation 8.3, where $\hat{H}$ is the classification posterior probability whose product with the cost function is minimized.

$$P_{(x|k)} = \frac{1}{\sqrt{2\pi|\Sigma_k|}} \exp\left(-\frac{1}{2}(x-\mu_k)^T \Sigma_k^{-1}(x-\mu_k)\right) \tag{8.2}$$

$$\hat{y} = \underset{y=1,\ldots,K}{\operatorname{argmin}} \sum_{k=1}^{K} \hat{H}(k|x)C(y|k) \tag{8.3}$$

The nearest neighbor classifier with K neighbors computes a distance metric (e.g., Euclidean distance) between the training samples and the test samples, as indicated in Equation 8.4, where x_s is the vector containing training samples of the ith class and x_t is the vector containing samples under test. The N smallest distances are chosen for each sample under test. Then this sample is assigned to the class that generated the highest number of these distances, generally at least $(N-1)/2+1$, where N is an odd integer, that is, the sample under test is assigned to the class whose training samples are closer in terms of the distance metric chosen.

$$d_i = \sqrt{\sum |x_{i,s} - x_t|^2} \tag{8.4}$$

The Binary Tree classifier implements a decision tree to classify samples under test by binary splits from the root node down to a leaf node, that is, a node that assigns the samples to a specific class without any further split. During the training phase, the tree is constructed by considering all the possible binary splits on all the available feature samples, and then choosing the best splits based on an optimization criterion. This is then recursively repeated on the child nodes, until when the resulting child node is *pure*, that is, with samples belonging just to a single class. In this implementation, the Gini Diversity Index (GDI) is used as an optimization criterion. This is defined in Equation 8.5, where i denotes the ith class, n denotes the node, and $p(i)$ is the fraction of classes observed belonging to the ith class that reaches that node. When a node is pure and contains only samples from a specific class (i.e., all the samples belonging to the specific class are assigned to that node), its index will be equal to 0; otherwise, it will be a positive number.

$$GDI_n = 1 - \sum_i p^2(i) \tag{8.5}$$

The analysis of multistatic data highlighted some trends and results, which are common to all the different experiments described in the remainder of this section. A first trend is the fact that the classification performance appears to be better when separate classifiers are implemented to process the feature samples extracted from each radar node, rather than having a single classifier that processes all the multistatic data. In the proposed approach, each separate classifier generates a partial decision on the samples under test, and a final decision is reached by a majority voting procedure by which the final decision needs the vote of at least two out of three NetRAD nodes. This behavior appears to be consistent in the various datasets considered. This is expected to be related to the impact of the different aspect angle and spatial diversity on the data collected at the different radar nodes, whereby the micro-Doppler signatures and the extracted feature samples can be very different at different nodes even for the same recording.

This approach was then improved by taking into account the level of confidence of each separate classifier when making a decision. This is a parameter in the range of 0%–100% related to the confidence of the classifier assigning a feature vector under test to the different possible classes, with 100% meaning that the classifier is completely sure that the sample under test belongs to that specific class. The way these confidence parameters are calculated depends on the specific classifier, but it can be considered in general as a measure of the distance of the feature sample under test from the boundaries of the available classes in the feature space used by the classifier. Using this approach, when two nodes have agreed on a decision, their confidence level is compared to a predefined threshold. If their confidence is above the threshold, their decision will be the final decision. If their confidence is below the threshold, and at the same time, the confidence of the third node is higher than the confidence of both other nodes, then the final decision will be that from the third node. This approach addresses the issue of having two nodes reaching an incorrect decision (and therefore presenting a relatively low confidence level), for instance, because they are both at aspect angles, which are not favorable for good detection of the micro-Doppler target. Different threshold values in the range of 55%–75% have been empirically tested in the different examples, and a good value has been found to be 65%. In the case of a classification problem with more than two classes, the level of confidence was also taken into account when the three nodes provided three different answers.

Another aspect that emerged from the analysis of the multistatic data is the fact that the optimal combination of features, that is, the group of features providing the best classification performance, can change at different radar nodes depending on situational parameters such as the dwell time, the signal-to-noise ratio, and the aspect angle to the target, as they can change from node to node. The exploitation of this *feature diversity* is an advantage provided by multistatic radar systems, in which the additional degrees of freedom given by different feature combinations at different nodes can help increase the overall performance of the system [40]. This can be also combined with emerging cognitive approaches for radar systems, for which each node can potentially operate with different radar parameters (dwell time, polarization, and frequency band) and adapt its behavior and feature extraction and classification algorithm based on the environmental situation and on what the targets of interest are doing.

The first experiment considered was performed to evaluate the effect of the aspect angle of the target on the classification performance. The geometry of the setup is shown in Figure 8.4. The netted radar was directed at a target location where a person would walk on the spot performing two different types of movements. The individual was walking on the spot in order to remove and range gate migration and change in bistatic angle over the recording from the potential variables. The two different actions performed were to walk on the spot with free arms moving up and down, and to walk on the spot holding a simulated rifle-like object. The angles that the individual faced, with respect to the look direction on the central radar node 1, were 0°–300°. Three different individuals were involved in these experiments, and more than 180 different samples were generated with a duration of 5 s each.

The four features extracted from the spectrograms were the bandwidth, the period, the Doppler offset, and the radar cross section (RCS) ratio of the limb to body contributions. These features were empirically extracted rather than using other automated processing methods such as singular value decomposition (SVD). The analysis showed that there was a significant variation of classifier performance with aspect angle, as shown in Figure 8.5.

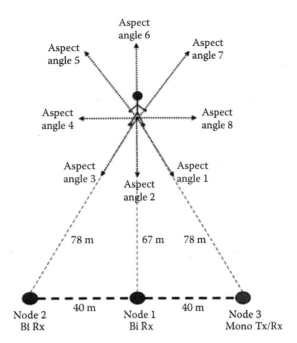

FIGURE 8.4
Sketch of setup for experiment investigating the effect of the aspect angle on classification performance.

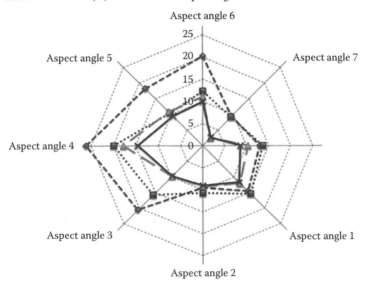

FIGURE 8.5
Plot showing the classification error as a function of the aspect angle and data fusion method.

This figure shows the classification error in percentage as a function of the eight different aspect angles presented in Figure 8.4. The four different lines present the classification error as a function of the different information fusion methods previously discussed. The most effective method of data fusion was found to be the voting threshold technique, which used an independent classifier at each node, and took into account their confidence level within the decision-making processes. The lowest classification error in these results was found to be when the individual was facing angle 2, that is, subject facing the radar node in the middle of the baseline as depicted in Figure 8.4. This is the aspect angle related to the maximum Doppler observed by radar node 1 (i.e., the aspect angle radial to radar node 1). The largest error in classification was shown to be when the individual was facing angle 4, that is, the angle perpendicular to the baseline of the deployed radar network. The range in error across all angles was shown to be up to 13% from the lowest error to the largest, demonstrating the impact that target trajectory can have on the classification performance.

The next experiment considered focused on evaluating the classification performance of NetRAD data when the individual was walking forward, in comparison with walking on the spot. Three different angles while either walking free handed or carrying a simulated rifle, seen in the geometry diagram in Figure 8.6. The measurements taken included three people, three aspect angles, five repetitions, two classes, and three radars, which provided 270 datasets for analysis.

Two different methods for feature extraction were applied, the first using empirically extracted feature samples and the second using techniques based on SVD to generate feature vectors. The empirically extracted parameters are the same as those used in the previous experiment (bandwidth, mean period, offset, and RCS ratio of limbs and body). The SVD-based processing considers the whole spectrogram under test as a matrix of values and applies the SVD decomposition on this matrix to generate the $\mathbf{U}$ and $\mathbf{V}$ matrices. The information initially included in the spectrogram is now contained in these two matrices, and several feature extraction approaches have been proposed in the literature depending on the classification problem to tackle [16,35,40–43]. In this case of armed versus unarmed personnel classification, it was found that the standard deviation of the first vector of the $\mathbf{V}$ matrix was a suitable feature, allowing a good separation of the feature samples belonging to the two different classes considered. Summarizing this approach, the $\mathbf{V}$ matrix is

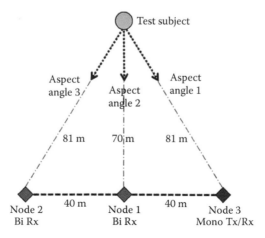

FIGURE 8.6
Experimental geometry for experiment investigating classifier performance with person walking forward.

extracted via SVD from the spectrogram, and then the standard deviation of the first vector in the V matrix is calculated to obtain a feature sample [16,32].

It should be noted that this process is different from principal component analysis, which is often used together with feature extraction techniques to reduce the dimensionality of the available feature space and select only the most informative and relevant features for the classifier [11,44–46]. The use of this feature based on SVD was then compared to the empirically extracted set of four features. A comparison of the results using these features as a function of the aspect angle, as well as presenting results when applying different approaches of combining information from the multistatic radar node, can be seen in Table 8.1. These approaches are using data from the monostatic node only, data from all nodes at a centralized classifier, and data from all nodes at different, separated classifiers followed by a binary voting step. The results show a very good classification performance from the SVD method despite only using a single feature in comparison with the four features used in the empirically extracted examples. The additional benefit of the SVD over the empirically extracted features is that it is an automated process that does not require manual processing of each of the desired parameters, which provides a quick method of assessing different actions in an automated way.

The final experimental results discussed in this section are focused on the recognition of a specific individual rather than an action. This analysis shows that the potential of micro-Doppler classification is not only being able to differentiate various actions but also capable of biometric recognition, an active area of cutting edge research. The experimental setup used for this analysis was directly comparable to that shown in Figure 8.6, where a single individual was located in front of a three-node radar network. The difference in this case was that only a single angle was used, angle 2, when completing the repetitions. Three different people were selected for this experiment, and the action was simply walking at a normal, along angle $2°$ toward the radar system. The key body parameters were 1.70 m and 69 kg for an average body type of subject 1, 1.77 m and 65 kg for a slim body type of subject 2, and 1.87 m and 90 kg for an average body type of subject 3. Each recording was 5 s in duration, but this was subdivided into 1 s sections as this increased the number of samples available and demonstrated shorter timeframes for the decision-making. The feature extracted as part of this analysis was also based on SVD transformed data, but, in contrast to the previously selected feature, it was the sum of the intensity values of the left-hand matrix U from the SVD performed directly on the segmented spectrogram signature [16].

TABLE 8.1

Table Showing Classifier Performance as a Function of Angle and Features Used

Classification Accuracy (%)	Monostatic Only Data	All Node Data at Centralized Classifier	Separate Classifiers and Binary Voting
Empirically Extracted Features			
Aspect angle 1	85.5	83.8	90.9
Aspect angle 2	85.9	74.8	90.1
Aspect angle 3	81.4	66.1	87
SVD Features			
Aspect angle 1	95.8	88.8	97.2
Aspect angle 2	82.5	86.0	90.5
Aspect angle 3	81.1	81.9	91.6

TABLE 8.2

Table Showing Performance of Classifiers as a Function of Different Approaches to Combine Multistatic Data

Classifier Type	Classification Accuracy (%)			
	Monostatic Only Data	All Node Data at Centralized Classifier	Separate Classifiers and Binary Voting	Separate Classifiers and Voting with Threshold
LDA	98.9	72.2	98.8	99.4
QDA	97.6	72.2	98.9	98.9
DLDA	98.9	72.2	98.8	99.4
DQDA	97.6	72.2	98.9	98.9
NB	95.4	72.1	97.9	97.5
NN3	98.7	71.1	99.4	99.4
NN5	98.7	70.9	99.1	99.4
CT	98.8	71.3	99.6	99.6

A number of different classification methods were applied to this single SVD feature to evaluate the importance of classifier on the overall performance. The list of classifiers used is as follows: linear discriminant analysis (LDA), quadratic discriminant analysis (QDA), diagonal LDA (DLDA), diagonal QDA (DQDA), NB with kernel function estimators, nearest neighbors with three samples (NN3) and five samples (NN5), and finally a classification tree (CT) method. The large range of classifiers were applied to both the single monostatic data features and all the data using the different fusion methods as described earlier in this section. The results from the classifiers can be seen in Table 8.2. These results show a minimal variation between the classifier used with only a range of 2.1% from the lowest to the highest success rate when using the best fusion method of threshold voting. The improvement in accuracy when using the threshold voting approach rather than the simpler binary voting is limited to ~1% in this case but is seen to be more consistent in the armed versus unarmed classification problem presented in Figure 8.5 [31]. The classification success rates shown are all very high as the SVD feature selected from the data allowed the different classes (people) to be easily separable within the feature space. If this were not the case, then a greater variation of classification success would potentially be demonstrated. Nevertheless, these results support the previously discussed issue that the correct choice of features is more important than the selection of type of classifier.

In brief summary, this section has presented an analysis that demonstrates that bistatic/multistatic radar systems are more effective than a single monostatic radar when aiming to classifier different actions, recognizing different peoples' limb motions and more resilient to changes in target aspect angle.

8.3 Discussion on Bistatic Indoor Data

This section discusses the collection and analysis of bistatic data for indoor classification of human actions. The data analyzed here were generated by an FMCW off-the-shelf radar sensor operating at 5.8 GHz (C-band). The transmitted power was approximately +19 dBm, the bandwidth and duration of the linear chirp 400 MHz and 1 ms, respectively, and the resulting PRF was 1 kHz, which is sufficient to include the full human micro-Doppler signature within the unambiguous Doppler region. Ten recordings of 10 s each were collected for each movement considered and for each subject, for a total number of datasets equal to 240 (four movements, three subjects, 10 recordings, monostatic and bistatic data). The four movements considered were walking back and forth, sitting and standing on an office chair, bending down to pick up an object from the floor and coming up to stand, and waving with one hand (as this is a type of movement used by a few commercial sensors to lock to a specific user for further, more detailed interaction to follow). The three subjects taking part in the experiment had the following key body parameters: 1.71 m—average body type for person A, 1.89 m—average body type for person B, and 1.84 m—slim body type for person C.

Figure 8.7 shows the sketch geometry for the experimental setup. The measurements were performed in a meeting room at the School of Engineering at the University of Glasgow. Typical pieces of furniture for an office were located in the room, namely, a large meeting table, a couple of smaller tables, and chairs, which can be seen in blue at the bottom of the figure. The radar system was placed near the corner of the room, with the subjects performing different movements located at approximately 4 m from the radar. In the monostatic configuration (red in the figure), the transmitter and receiver antennas were colocated and separated by approximately 30 cm, whereas in the bistatic configuration (green in the figure), the receiver antenna was moved at approximately 1.5 m from the transmitter and aimed at the target, originating a bistatic angle equal to approximately 20°. The antennas were WiFi directional Yagis with 17 dBi gain and 24° × 24° beamwidth, and they were used in vertical polarization. It should be noted that the collection of the monostatic and bistatic data was not simultaneous, as the radar sensor has only one receiver channel, but care was taken to perform the movements as consistently as possible for both cases.

The data were initially processed by applying the STFT to obtain spectrograms. A Hamming window with 200 samples corresponding to 0.2 s was used with 95% overlap. A notch filter was applied during the processing to reduce the contribution of the static clutter around 0 Hz and highlight the actual signatures. Figure 8.8 shows examples of spectrograms for the four considered types of activities performed by the same subject,

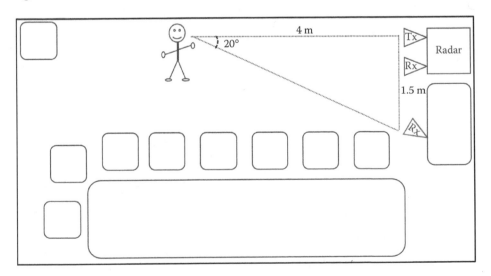

FIGURE 8.7
Sketch of the experimental setup for indoor measurements.

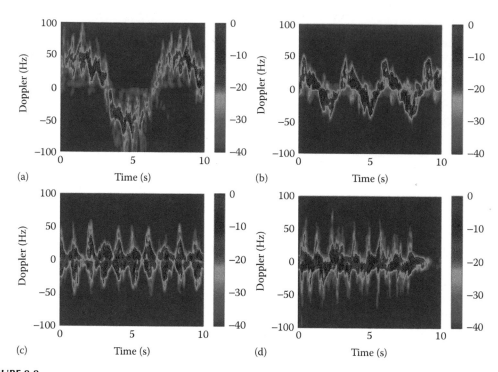

FIGURE 8.8
Example of spectrograms for different movements performed by the same subject. Monostatic data: (a) walking, (b) standing and sitting, (c) picking up an object, and (d) waving one hand.

person A in this case. The most evident difference is between the walking signature, which has a dominant positive and negative main Doppler shift caused by the overall movement of the whole body, and the other three signatures, which appear to be centered around 0 Hz as these were static movements, not involving complete movements of the body. It is therefore expected that the classification of the walking movement will be a fairly straightforward task because of the distinctive main Doppler shift, whereas more detailed analysis will be needed to distinguish between the other types of movements. However, because the differences in the spectrograms can be empirically seen, it is likely that good automatic classification can be achieved, as long as relevant features are developed and extracted from the signatures.

Figure 8.9 shows additional examples of spectrograms for the actions of walking and picking up an object performed by person B for both monostatic and bistatic data. It is interesting to observe the differences in the signatures for the same type of movement, even at the limited bistatic angle of 20° of this setup. The bistatic signatures appear to have slightly lower Doppler values, for example, the peaks for the picking up movement reach approximately 75 Hz for the monostatic case but only around 50 Hz for the bistatic case. This is compatible with the considered measurement setup, as the movements were performed facing the monostatic node; hence, actions such as bending down toward the floor or standing up from the chair are expected to generate the highest Doppler shift in that direction.

Features based on the mean and standard deviation of the centroid and bandwidth of the micro-Doppler signatures were shown to be effective in the analysis of the NetRAD data described in Section 8.2, with the aim of unarmed versus armed classification and personnel

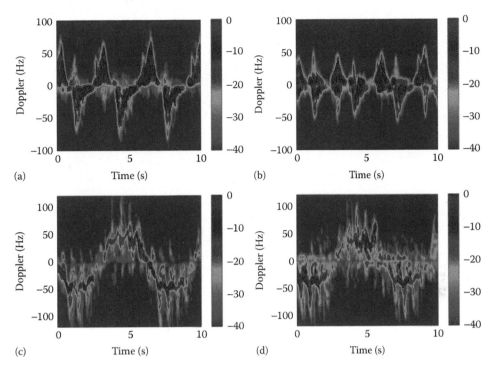

FIGURE 8.9
Example of spectrograms for different movements performed by the same subject to compare monostatic and bistatic data: picking up an object—monostatic (a) and bistatic (b); walking—monostatic (c) and bistatic (d).

recognition. These features have been also extracted for the indoor data just described here. A single feature sample has been extracted from 5 s of data, that is, half of the spectrogram generated for each 10 s measurement. Figure 8.10 presents an example of feature space plots for the collected data. The two features considered here are the standard deviation and mean value of the bandwidth of the signature around the centroid, and these two features were selected empirically as they appeared to provide the best separation between the samples

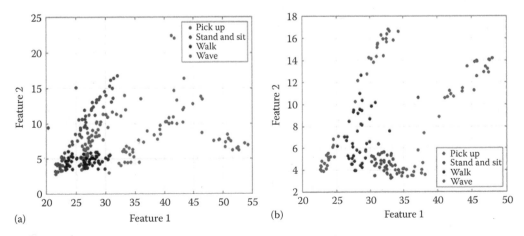

FIGURE 8.10
Feature space plot for samples related to different actions: (a) monostatic data; (b) bistatic data.

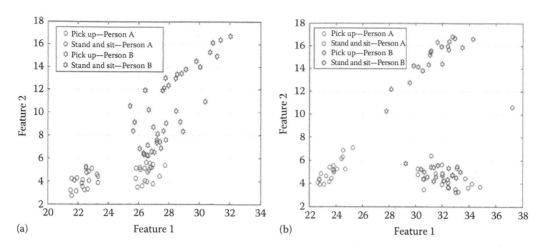

FIGURE 8.11
Feature space plot for samples related to two actions (picking up an object and standing and sitting) performed by two subjects: (a) monostatic data; (b) bistatic data.

belonging to different classes of actions. Figure 8.11 considers only two actions, namely, bending and picking up an object and standing and sitting on a chair, and differentiates not only between the samples for different actions (in different colors) but also between the two subjects (with different shapes of the markers). Both figures show feature samples extracted from monostatic and bistatic data separately, and in general good separation between the samples belonging to different actions can be seen, enabling good classification performance.

An NB classifier was applied to the features extracted to evaluate the success rate of the monostatic configuration in comparison with the bistatic result. For the monostatic classification case, the features from the four different actions using empirically extracted features provided a classification success rate of 71% using 60% training. The training set was greater in this example, but the challenge is a four-class problem resulting in a greater potential for misclassifying each result. In contrast, the bistatic result from the same experiments was found to have a success rate of 88% with the same level of training. This increase is significant and clearly shows some of the advantages of using a bistatic solution.

8.4 Conclusions

This chapter has described the analysis of bistatic and multistatic radar data for the detection and classification of human activities, in particular in the context of identification of unarmed versus potentially armed people and recognition of specific individuals based on their walking gait. These data were collected in a series of experiments performed using the multistatic radar system NetRAD, developed at University College London over the past 10 years. Although the data were collected in outdoor scenarios with target distances on the order of tens of meters because of the minimum range of the pulsed radar, the key methodologies in characterizing the micro-Doppler signatures, extracting suitable features, and applying classification algorithms are still applicable and suitable for indoor systems. Preliminary results on a set of indoor data collected at the University of Glasgow with an FMCW radar system were also shown for identification of typical indoor activities such as walking, waving, bending to pick up an object from the floor, and standing and sitting on a chair.

The analysis of the results showed that the exploitation of multistatic information can improve the overall classification performance in comparison with conventional monostatic systems. In particular, it was shown that separate classifiers to process the feature samples at each radar node independently provide better classification accuracy than a single classifier that processes the feature samples from all nodes. This is thought to be related to the different situational parameters affecting the data at each node, such as the different aspect angle with respect to the target trajectory, thus generating different values of feature samples even from data recorded for the same action, which is therefore best to process separately. The partial decisions produced by each separate classifier were then combined either by a simple majority voting or by setting thresholds to take into account the level of confidence of each classifier to reach the final decision, which is beneficial to multiclass problems in which the separate classifiers may reach all different decisions one from the others.

Bistatic and multistatic radar systems are believed to provide significant performance improvement for indoor monitoring applications related to eldercare, such as checking the level of daily activities in nonhospital environments and fall detection and localization for prompt rescue. The main advantage and benefit of a bistatic/multistatic configuration against a conventional monostatic radar sensor is the additional information related to the multi-perspective views on the targets and on the scene under test. This means that different radar sensors that are spatially separated can view the target from different aspect angles, thus generating different micro-Doppler signatures that can contain more information than the signatures from a single aspect angle. In indoor scenarios, these multiple views on the person of interest (e.g., an elderly person being cared for) can mitigate the presence of clutter (if, for instance, pieces of furniture are partially obscuring the target from one specific radar sensor) and can provide improved characterization of micro-Doppler signatures, in which at least one node can view the target from a favorable aspect angle and information from multiple nodes can be effectively combined. This is expected to provide more robust classification schemes for indoor activity recognition and rejection of false alarms for fall detection, for instance, recognizing fast bending toward the floor or fast sitting actions that may be mistaken for falling, as well as for more detailed identification of specific actions, which can be relevant for the well-being of the person cared for, for instance, verifying food intake activities while the person is sitting at the table.

Multistatic systems are also well suited to be used with emerging cognitive radar approaches. Each radar node can potentially work with different radar parameters (dwell time, polarization, and frequency band) and adapt its behavior to the changing environmental conditions and actions performed by targets of interests. These cognitive changes can also include feature extraction and algorithms approaches, varying, for instance, the number and type of features used for a certain classification problem while the target is moving in different trajectories. Some of the results mentioned in this chapter and in the relevant literature can contribute to the development of these emerging systems and algorithms, for instance, the effect of the aspect angle parameter on the overall classification performance given a set of features.

Acknowledgment

The authors are grateful for the support of the IET A. F. Harvey Prize awarded to Hugh Griffiths (2013). They also thank S. Alhuwaimel, F. Wei, A. Amiri, B. Beaudouin, X. Savalle, H. Sherwani, J. Weatherwax, Dr. V. Jeauneau, Dr. A. Gning, J. Patel, and A. Angelov for their help with the various experiments described in this chapter.

References

1. V. C. Chen, Doppler signatures of radar backscattering from objects with micro-motions, *IET Signal Processing*, 2, 291–300, 2008.
2. R. G. Raj, V. C. Chen, and R. Lipps, Analysis of radar human gait signatures, *IET Signal Processing*, 4, 234–244, 2010.
3. V. C. Chen, D. Tahmoush, and W. J. Miceli, *Radar Micro-Doppler Signatures: Processing and Applications*, Institution of Engineering and Technology, Stevenage, 2014.
4. D. Tahmoush, Review of micro-Doppler signatures, *IET Radar, Sonar & Navigation*, 9 (9), 1140–1146, 2015.
5. D. Tahmoush and J. Silvious, Remote detection of humans and animals, *2009 IEEE Applied Imagery Pattern Recognition Workshop (AIPRW)*, pp. 1–8, October 14–16, Washington, DC.
6. K. Youngwook, H. Sungjae, and K. Jihoon, Human detection using Doppler radar based on physical characteristics of targets, *IEEE Geoscience and Remote Sensing Letters*, 12, 289–293, 2015.
7. Y. Kim and T. Moon, Human detection and activity classification based on micro-Doppler signatures using deep convolutional neural networks, *IEEE Geoscience and Remote Sensing Letters*, 13 (1), 8–12, 2016.
8. X. Shi, F. Zhou, L. Liu, B. Zhao, and Z. Zhang, Textural feature extraction based on time–frequency spectrograms of humans and vehicles, *IET Radar, Sonar & Navigation*, 9 (9), 1251–1259, 2015.
9. K. Youngwook and L. Hao, Human activity classification based on micro-Doppler signatures using a support vector machine, *IEEE Transactions on Geoscience and Remote Sensing*, 47, 1328–1337, 2009.
10. B. Cagliyan and S. Z. Gurbuz, Micro-Doppler-based human activity classification using the mote-scale BumbleBee radar, *IEEE Geoscience and Remote Sensing Letters*, 12, 2135–2139, 2015.
11. J. Zabalza, C. Clemente, G. Di Caterina, J. Ren, J. Soraghan, and S. Marshall, Robust PCA micro-Doppler classification using SVM on embedded systems, *IEEE Transactions on Aerospace and Electronic Systems*, 50 (3), 2304–2310, 2014.
12. S. Z. Gürbüz, B. Erol, B. Çağlıyan, and B. Tekeli, Operational assessment and adaptive selection of micro-Doppler features, *IET Radar, Sonar & Navigation*, 9 (9), 1196–1204, 2015.
13. R. M. Narayanan and M. Zenaldin, Radar micro-Doppler signatures of various human activities, *IET Radar, Sonar & Navigation*, 9 (9), 1205–1215, 2015.
14. S. Björklund, H. Petersson, and G. Hendeby, Features for micro-Doppler based activity classification, *IET Radar, Sonar & Navigation*, 9 (9), 1181–1187, 2015.
15. R. Ricci and A. Balleri, Recognition of humans based on radar micro-Doppler shape spectrum features, *IET Radar, Sonar & Navigation*, 9 (9), 1216–1223, 2015.
16. F. Fioranelli, M. Ritchie, and H. Griffiths, Performance analysis of centroid and SVD features for personnel recognition using multistatic micro-Doppler, *IEEE Geoscience and Remote Sensing Letters*, 13 (5), 725–729, 2016.

17. Q. Wu, Y. D. Zhang, W. Tao, and M. G. Amin, Radar-based fall detection based on Doppler time–frequency signatures for assisted living, *IET Radar, Sonar & Navigation*, 9 (2), 164–172, 2015.
18. B. Jokanovic, M. G. Amin, Y. D. Zhang, and F. Ahmad, Multi-window time–frequency signature reconstruction from undersampled continuous-wave radar measurements for fall detection, *IET Radar, Sonar & Navigation*, 9 (2), 173–183, 2015.
19. M. G. Amin, F. Ahmad, Y. D. Zhang, and B. Boashash, Human gait recognition with cane assistive device using quadratic time–frequency distributions, *IET Radar, Sonar & Navigation*, 9 (9), 1224–1230, 2015.
20. B. Erol, M. Amin, Z. Zhou, and J. Zhang, Range information for reducing fall false alarms in assisted living, *2016 IEEE Radar Conference*, pp. 1–6, 2016, Philadelphia, PA.
21. B. Jokanovic, M. Amin, and F. Ahmad, Radar fall motion detection using deep learning, *2016 IEEE Radar Conference*, pp. 1–6, 2016, Philadelphia, PA.
22. D. Tahmoush and J. Silvious, Radar micro-Doppler for long range front-view gait recognition, *IEEE 3rd International Conference on Biometrics: Theory, Applications, and Systems BTAS '09*, pp. 1–6, September, Washington, DC.
23. C. Karabacak, S. Z. Gürbüz, M. B. Guldogan, and A. C. Gürbüz, Multi-aspect angle classification of human radar signatures, *Proceedings of the SPIE 8734, Active and Passive Signatures IV*, 873408, May 23, 2013.
24. B. Tekeli, S. Z. Gurbuz, M. Yuksel, A. C. Gurbuz, and M. B. Guldogan, Classification of human micro-Doppler in a radar network, *2013 IEEE Radar Conference*, pp. 1–6, May 2013, Ottawa, Canada.
25. R. Boulic, M. N. Thalmann, and D. Thalmann, A global walking model with real-time kinematic personification, *Visual Computing*, 6, 344–358, 1990.
26. P. Van Dorp and F. C. A. Groen, Human walking estimation with radar, *IEE Proceedings on Radar, Sonar and Navigation*, 150 (5), 356–365, 2003.
27. D. P. Fairchild and R. M. Narayanan, Multistatic micro-Doppler radar for determining target orientation and activity classification, *IEEE Transactions on Aerospace and Electronic Systems*, 52(1), 512–521, 2016.
28. M. Perassoli, A. Balleri, and K. Woodbridge, Measurements and analysis of multistatic and multimodal micro-Doppler signatures for automatic target classification, *2014 IEEE Radar Conference*, pp. 0324–0328, May 19–23, Cincinnati, OH.
29. G. E. Smith, K. Woodbridge, C. J. Baker, and H. Griffiths, Multistatic micro-Doppler radar signatures of personnel targets, *IET Signal Processing*, 4, 224–233, 2010.
30. F. Fioranelli, M. Ritchie, and H. Griffiths, Multistatic human micro-Doppler classification of armed/unarmed personnel, *IET Radar, Sonar & Navigation*, 9 (7), 857–865, 2015.
31. F. Fioranelli, M. Ritchie, and H. Griffiths, Aspect angle dependence and multistatic data fusion for micro-Doppler classification of armed/unarmed personnel, *IET Radar, Sonar & Navigation*, 9 (9), 1231–1239, 2015.
32. F. Fioranelli, M. Ritchie, and H. Griffiths, Centroid features for classification of armed/unarmed multiple personnel using multistatic human micro-Doppler, *IET Radar, Sonar & Navigation*, 10, 1751–8784, 2016.
33. F. Fioranelli, M. Ritchie, and H. Griffiths, Classification of unarmed/armed personnel using the NetRAD multistatic radar for micro-Doppler and singular value decomposition features, *IEEE Geoscience and Remote Sensing Letters*, 12 (9), 1933–1937, 2015.
34. F. Fioranelli, M. Ritchie, and H. Griffiths, Analysis of polarimetric multistatic human micro-Doppler classification of armed/unarmed personnel, *IEEE International Radar Conference*, pp. 0432–0437, May 2015, Arlington, VA.
35. F. Fioranelli, M. Ritchie, and H. Griffiths, Personnel recognition based on multistatic micro-Doppler and singular value decomposition features, *Electronics Letters*, 51 (25), 2143–2145, 2015.
36. T. E. Derham, S. Doughty, K. Woodbridge, and C. J. Baker, Design and evaluation of a low-cost multistatic netted radar system, *IET Radar, Sonar & Navigation*, 1, 362–368, 2007.

37. W. A. Al-Ashwal, Measurement and modelling of bistatic sea clutter, PhD thesis, Department of Electronic and Electrical Engineering, University College London, London, 2011, http://discovery.ucl.ac.uk/1334081/1/1334081.pdf.

38. T. Hastie, R. Tibshirani, and J. Friedman, *The Elements of Statistical Learning: Data Mining, Inference, and Prediction*, 2nd ed., Springer, New York, 2009.

39. C. D. Manning, P. Raghavan, and M. Schütze, *Introduction to Information Retrieval*, Cambridge University Press, New York, 2008.

40. F. Fioranelli, M. Ritchie, S. Gürbüz, and H. Griffiths, Feature diversity for optimized human micro-Doppler classification using multistatic radar, *IEEE Transactions on Aerospace and Electronic Systems*, 53(2), pp 640–645, 2017.

41. M. Ritchie, F. Fioranelli, H. Borrion, and H. Griffiths, Multistatic micro-Doppler radar feature extraction for classification of unloaded/loaded micro-drones, *IET Radar, Sonar & Navigation*, 11, 116–124, 2017.

42. F. Fioranelli, M. Ritchie, H. Borrion, and H. Griffiths, Classification of loaded/unloaded micro-Drones using multistatic radar, *Electronics Letters*, 51 (22), 1813–1815, 2015.

43. J. J. M. de Wit, R. I. A. Harmanny, and P. Molchanov, Radar micro-Doppler feature extraction using the singular value decomposition, *2014 International Radar Conference*, SEE, October 13–17, 2014, Lille, France.

44. B. Jokanovic, M. Amin, F. Ahmad, and B. Boashash, Radar fall detection using principal component analysis, *Proceedings of the SPIE Radar Sensor Technology XX Conference 2016*, vol. 9829, pp. 982916–982919.

45. C. Clemente, A. W. Miller, and J. J. Soraghan, Robust principal component analysis for micro-doppler based automatic target recognition, *3rd IMA Conference on Mathematics in Defence*, Malvern, October 24, 2013.

46. A. G. Stove, A Doppler-based target classifier using linear discriminants and principal components, *2006 IET Seminar on High Resolution Imaging and Target Classification*, London, pp. 107–125, 2006.

9

Microwave and Millimeter-Wave Radars for Vital Sign Monitoring

Changzhan Gu, Tien-Yu Huang, Changzhi Li, and Jenshan Lin

CONTENTS

9.1 Background of Doppler Radar

Doppler radar has been widely used for remote sensing applications for many years, including velocity measurement [1], navigation and position sensing [2], weather sensing [3], automobile speed sensing [4,5], and vital sign detection [6,7]. In principle, a Doppler radar transmits a continuous wave (CW) toward a moving target and measures the relative velocity by detecting the frequency shift in the received signal as illustrated in Figure 9.1. The frequency shift of the received signal is known as the Doppler effect. Based on the effect, Doppler radar can be used to detect periodic movements such as mechanical vibration and physiological movement.

The radar-based technique features noninvasiveness, noncontact detection, continuous operation, low cost, and low power consumption, which have drawn increasing interest in biomedical and physiological research communities. The development of Doppler radar for vital sign applications has grown rapidly in the areas of health-care monitoring and life-sign detection [8,9] such as observation of sleep apnea [10,11] and search-and-rescue in earthquake rubble [12]. The first demonstration of measuring respiratory movements of human and animals using microwave radar was published in the 1970s [13], whereas later, an X-band life detector for detecting both heartbeat and respiration movements of human targets was demonstrated [14]. Doppler radar has also been applied to monitoring the physiological movement and behavior of animals, including rodents and fish [15–19].

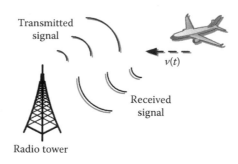

FIGURE 9.1
Target velocity measurement based on the Doppler effect.

The operating frequencies of Doppler radar discussed in this chapter range from microwave to millimeter wave. In the electromagnetic spectrum, microwave falls in the frequency range of 1–30 GHz, whereas millimeter wave lies in the frequency range of 30–300 GHz. Both microwave and millimeter wave feature large bandwidth, little radiofrequency (RF) interference and fading problem, and small antenna size, leading to successful demonstration of the effectiveness and robustness for remote sensing. For vital sign detection of human subjects, both 2.4 and 5.8 GHz Doppler radars are proved to successfully detect both respiration and heartbeat signals [20,21], and radars using shorter wavelengths have been proposed at frequencies such as 24 [22], 60 [23], and even 228 GHz [24]. The optimal carrier frequency was found to be approximately 27 GHz for a normal male subject of 1.75 m height [25]. However, measuring a small animal's vital signs with radar is more challenging compared to measuring a human's vital signs due to smaller physiological movements. A 60 GHz radar has been demonstrated to measure heartbeat and respiration signals of a laboratory rat due to its capability to detect even smaller vibration movements.

This chapter starts with the working principle of CW Doppler radar, including radar architecture, the choice of radar carrier frequency, and the design consideration about DC offset, phase noise, and in-phase and quadrature (I/Q) imbalance. In Section 9.3, the digital intermediate-frequency (IF) Doppler radar, baseband signal processing methods dealing with signal distortion and nonlinear phase modulation effect, and random body movement cancellation are also reviewed. Finally, a 60 GHz radar for the detection of respiration and heartbeat of human and laboratory animal is presented in Section 9.4.

9.2 Working Principle of CW Doppler Radar

The Doppler effect modulates the phase of the transmitted radar signal linearly with respect to the displacement of the target. Therefore, even in the case of single-tone radar and linear target (radial) movement, the received baseband signal has a nonlinear relationship with the displacement of the target. Consequently, the received baseband signal needs to be processed in order to extract the actual movement of the target.

Figure 9.2 shows the fundamental mechanism of using a single-tone Doppler radar for noncontact vital sign detection. The signal source generates a single-tone unmodulated RF

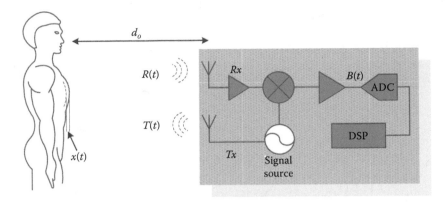

FIGURE 9.2
Setup of Doppler radar for noncontact vital sign detection.

signal $T(t)$ that is transmitted via the Tx antenna to the target at a nominal distance d_o away from the radar. The transmitted signal could be represented as

$$T(t) = A_T \cdot \cos(2\pi f t + \theta(t)) \tag{9.1}$$

where:
 f is the carrier frequency
 A_T is the magnitude of the transmitted signal
 $\theta(t)$ is a random variable that is the phase noise from the signal source

The transmitted signal hits the subject's chest wall, which modulates the RF signal in phase and reflects part of the signal back to the radar. The signal at the input of radar receiver antenna can be represented as

$$R(t) \approx \partial \cdot A_T \cdot \cos\left[2\pi f t - 4\pi d_o/\lambda - 4\pi x(t)/\lambda + \theta(t - 2d_o/c)\right] \tag{9.2}$$

where:
 $\partial \cdot A_T$ is the received signal amplitude
 ∂ is the coefficient including the impact of path loss and reflection loss
 λ is the carrier wavelength
 c is the speed of light in free space
 $x(t)$ is the chest wall movement, including respiration and heartbeat
 $\theta(t - 2d_o/c)$ is the phase noise with a delay of $2d_o/c$

The same signal source is used for both the transmitter and the receiver so that the radar detection is coherent as shown in Figure 9.2. Part of RF signal from the signal source is transmitted out to the subject, and another part of the signal is used as the local oscillation (LO) signal to down-convert the received signal in Equation 9.2 to baseband $B(t)$:

$$B(t) = \cos\left[\theta_o + 4\pi x(t)/\lambda + \Delta\theta(t)\right] \tag{9.3}$$

where:
 $\theta_o = 4\pi d_o/\lambda + \sigma$ is the summation of phase shift due to the nominal detection distance d_o
 σ is the phase shift at the reflection surface
 $\Delta\theta(t) = \theta(t) - \theta(t - 2d_o/c)$ is the residual phase noise

It should be noted that the baseband signal amplitude in Equation 9.3 is normalized and approximated to be constant, which does not affect the analysis. The target movement is assumed to be much smaller than the distance to the radar, such that the amplitude modulation is negligible. The amplitude modulation may always be ignored in short-range vital sign detection because human's respiration and heartbeat motions are very weak. As shown in Figure 9.2, both Tx and LO signals share the same signal source so that they have the same phase noise characteristics. Because the term $2d_o/c$ is usually small for noncontact vital sign detection (short range), the phase noise of the signal backscattered from the subject is strongly correlated with that of the LO signal. Due to the range correlation effect, the residual phase $\Delta\theta(t)$ in Equation 9.3 is negligible. The range correlation effect describes how the baseband noise spectrum would be affected by the residual phase noise $\Delta\theta(t)$. It is the foundation of radar vital sign detection because it explains why the weak vital sign signals are not overwhelmed by the noise. The baseband noise spectral density can be expressed as follows [26]:

$$S_{\Delta\theta}(f_o) = S_\theta(f_o)\left\{4\left[\sin\left(2\pi\frac{d_o f_o}{c}\right)\right]^2\right\} \tag{9.4}$$

where:
  d_o is the target distance
  f_o is the offset frequency
  $S_\theta(f)$ is the RF phase noise density

Small-angle approximation is valid because $d_o f_o/c$ will be very small on the order of 10^{-9} for radar vital sign monitoring. Therefore, Equation 9.4 can be approximated as

$$S_{\Delta\theta}(f_o) \approx S_\theta(f_o)\left\{16\left[\pi\frac{d_o f_o}{c}\right]^2\right\} \tag{9.5}$$

It is seen that, due to range correlation, the baseband noise spectrum will increase in proportion to the square of the target distance and the square of the offset frequency. In short-range applications, the residual phase noise can be significantly reduced. For example, for a 0.5 m range, the residual phase noise is decreased by over 150 dB compared to that without the range correlation effect at an offset frequency of 1 Hz, which is close to vital sign frequencies. Amplitude noise is less affected by range correlation. Therefore, the target movement is always assumed to be much smaller than the distance, so that the amplitude modulation is negligible. The phase noise cancellation, owing to the range correlation effect, is expected for radars operating in short ranges of less than a few meters. At long range, it would become less effective because of the increased time delay due to an increasing distance.

The range correlation effect makes it possible to overcome the phase noise of the signal generated by the transceiver to design high-sensitivity Doppler radar for detecting the weak vital signs. Owing to the range correlation effect, a free-running voltage-controlled oscillator (VCO) is usually sufficient for a Doppler radar. Frequency synthesis such as phase-locked loop (PLL) or direct digital synthesis (DDS) is not necessary dramatically reducing the radar system complexity and allowing higher level of chip integration. However, although phase noise can be largely canceled in short-range detection, the frequency drift in a free-running VCO may still affect the detection accuracy. This is because, as shown in Equation 9.3, nonlinear phase modulation in radar sensing is inversely proportional to the wavelength. As the frequency drifts, the wavelength drifts, which leads to the variation in phase modulation. Accurate recovery of the phase information would require the complete time-varying frequency drift information, which is obviously difficult to be obtained from a free-running VCO. This effect can be serious for higher frequencies (e.g., millimeter wave) and long-term monitoring. Therefore, frequency synthesizers such as PLL and DDS may be a better choice due to higher frequency stability.

Although Figure 9.2 describes a homodyne architecture, other transceiver architectures are also available for Doppler radar, and each of them has pros and cons. The selection of radar architecture is driven by the specific application and should be pondered over both hardware complexity and system specifications. An architecture based on digital IF could completely solve the *I/Q* imbalance issue but at the cost of requiring higher speed analog-to-digital converters (ADCs). Homodyne direct conversion allows high level of electronic device integration, but it is prone to strong DC offset and high flicker noise level at the mixer output.

The choice of radar carrier frequency is very important to the sensing performance in noncontact vital sign detection. It is seen from Equation 9.3 that the phase modulation is inversely proportional to the carrier wavelength, and therefore, a higher frequency should lead to better detection sensitivity. However, theoretical analysis and experiments have revealed that, for CW Doppler radar, there exists a strict relationship between the optimal carrier frequency and the amplitude of target movement to be detected [22]; for a given displacement, there is a specific carrier frequency that results in the best detection. In practice, however, not all carrier frequencies can be chosen due to the cost and availability of RF components. Among all possible frequencies, 2.4 and 5.8 GHz are probably the most popular ones due to the vast availability of low-cost RF/microwave components and the integrated circuits. It has been proved that both 2.4 and 5.8 GHz Doppler radars are able to successfully detect human vital signs involving respiration and heartbeat [20,21] to optimize the detection of the heartbeat signal with small displacement of about only 1 m.

Figure 9.3 illustrates a quadrature radar receiver with *I/Q* channels. The received RF signal is amplified, filtered, and down-converted directly to baseband *I/Q* signals. Another stage of baseband amplifier may be employed to further boost the signal. The *I/Q* signals before ADC can be expressed as

$$B_I(t) = A_I \cdot \cos\left[\theta_o + 4\pi x(t)/\lambda + \Delta\theta(t)\right] + DC_I \qquad (9.6)$$

$$B_Q(t) = A_Q \cdot \sin\left[\theta_o + 4\pi x(t)/\lambda + \Delta\theta(t)\right] + DC_Q \qquad (9.7)$$

where:

A_I and A_Q represent the amplitude of *I/Q* channels

DC_I and DC_Q are the DC offsets, which are inevitable in a homodyne receiver

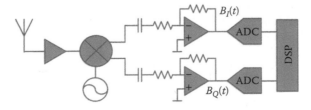

FIGURE 9.3

Block diagram of Doppler radar with quadrature *I/Q* architecture.

The DC offsets occur at the down-conversion mixer output, which may saturate the amplifiers in the following stages or limit their performance of dynamic range. The sources of DC offset mainly include (1) direct coupling from *Tx* to *Rx*, (2) reflections from stationary objects around the subject, and (3) circuit imperfections such as LO self-mixing and interferers. If the *I/Q* signals can be calibrated to remove the DC offsets DC_I and DC_Q and normalize the amplitude, they will become

$$B_I(t) = \cos\left[\theta_o + 4\pi x(t)/\lambda + \Delta\theta(t)\right] \tag{9.8}$$

$$B_Q(t) = \sin\left[\theta_o + 4\pi x(t)/\lambda + \Delta\theta(t)\right] \tag{9.9}$$

As already stated, Doppler radar vital sign detection using Equation 9.3 is based on nonlinear phase modulation because the cosine transfer function in Equation 9.3 is nonlinear. Several techniques have been developed to recover the actual movement information from nonlinear phase modulation; the most straightforward is that of small angle approximation which can be applied when the displacement caused by respiration and heartbeat is very small compared with the carrier wavelength. As shown in Equation 9.3, if θ_o is an odd multiple of $\pi/2$, the radar baseband output can approximately be linearly expressed as

$$B(t) \approx 4\pi x(t)/\lambda + \Delta\theta(t) \tag{9.10}$$

This is called the optimum detection point because the baseband output is linearly proportional to the time-varying motion displacement $x(t)$. However, there are situations where θ_o is an even multiple of $\pi/2$, as θ_o is a function of the distance to the target; in this case, a *null point* occurs because the baseband output is no longer proportional to $x(t)$. The *null point* issue not only leads to weak baseband output but also entails inaccurate measurement results [27]. In order to overcome the null point issue, several effective approaches have been proposed. In [28] a frequency tuning technique was developed to adjust the *null point* distribution along the path away from the radar; by utilizing the double-sideband transmission technique, the null point problem of one sideband can be overcome by the other sideband, as shown in Figure 9.4.

However, the frequency tuning technique is not convenient because it needs hardware tuning to achieve good detection accuracy. An alternative approach to solve the null point issue is to employ the quadrature receiver, in which the *I/Q* channels can always ensure that there is one channel that is not at the null detection point without any need to tune the hardware.

In quadrature Doppler radar, the *I/Q* channels at the receiver baseband output are illustrated in Equations 9.8 and 9.9. According to small-angle approximation, when θ_o is an odd multiple of $\pi/2$, the *I* channel signal is at the optimum point, whereas the *Q* channel is at

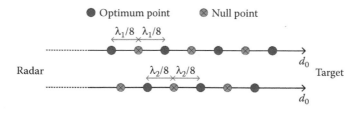

FIGURE 9.4
Frequency tuning technique using double-sideband transmission to overcome the null point issue. λ_1 is the wavelength of one sideband and λ_2 is that of the other.

the null point. Similarly, when θ_o is an even multiple of $\pi/2$, the I channel becomes the null point, whereas the Q channel is at the optimum point. Therefore, good detection accuracy can be ensured because there is always one channel that is not at the null point. The worst case is when θ_o is an integral multiple of $\pi/4$; in this case, none of the I/Q channels is at the optimum detection point.

Although phase estimation based on small-angle approximation is a linear demodulation technique that only applies for small displacement, an alternative demodulation technique that can precisely recover the full phase information from the nonlinear sinusoidal signal is arctangent demodulation. As illustrated in Equations 9.8 and 9.9, the I/Q signals will fit with the unit circle in the I/Q plane as shown in Figure 9.5:

The subject's movement $x(t)$ can be recovered using arctangent demodulation:

$$\theta_o + 4\pi x(t)/\lambda + \Delta\theta(t) = \arctan\left(\frac{B_Q(t)}{B_I(t)}\right)$$

$$= \arctan\left(\frac{\sin\left[\theta_o + 4\pi x(t)/\lambda + \Delta\theta(t)\right]}{\cos\left[\theta_o + 4\pi x(t)/\lambda + \Delta\theta(t)\right]}\right)$$

(9.11)

In the coherent Doppler radar system, the residual phase $\Delta\theta(t)$ is negligible due to the range correlation effect [26], whereas the constant phase shift θ_o is dependent on the nominal distance to the subject, so it is a DC value that can be easily removed. Therefore, with a known carrier wavelength λ, $x(t)$ can be easily recovered using arctangent demodulation.

For effective arctangent demodulation, accurate DC offset calibration is required. The calibration is challenging as the DC offset values are unpredictable. Because the calibrated I/Q signals should sit on the unit circle in the I/Q plane, a possible method is to manually adjust the DC offset values for the I/Q channels until the trajectory fits with the unit circle. The downside is that the manual circle-fitting method is not efficient, is sensitive to environmental changes, and sometimes is inaccurate. Furthermore, the estimation of calibration parameters in the manual method is a subjective approach, which may vary from one person to another and thus leads to inevitable error. Therefore, manual fitting is only possible as a postprocessing analysis tool rather than an automatic real-time mechanism. In order to have accurate and robust DC calibration, researchers proposed an

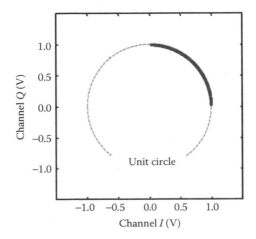

FIGURE 9.5
Calibrated I/Q signals sitting on the unit circle as an arch.

algorithm based on compressed sensing using the ℓ_1 minimization [29], which has been experimentally proven to be effective in arctangent demodulation and robust against large interferences.

Phase discontinuity is another challenge in arctangent demodulation. An arctangent function mathematically only allows a native codomain range of $(-\pi/2, +\pi/2)$. If the target's movement is large and comparable to the carrier wavelength, the demodulation may exceed this range so that a phase discontinuity will occur. Figure 9.6a shows the demodulation of a 0.2 Hz 20 mm sinusoidal movement in a 2.4 GHz Doppler radar using the regular arctangent demodulation showing the occurrence of the phase discontinuity. Theoretically, this discontinuity can be compensated in digital signal processing by shifting an integer multiple of π (i.e., phase unwrapping), but in practical applications, it is hard for a hardware or software to automatically make a judicious choice on exactly which point to shift. It becomes extremely difficult when both the displacement and the environmental clutter are large because it would cause severe discontinuity issues and the discontinuous data points may need to be compensated by different integer multiples of π.

In order to tackle the phase discontinuity issue, an extended differentiate and cross-multiply (DACM) algorithm is proposed for automatic phase reconstruction without ambiguities [30]. In the digital (i.e., sampled) domain, the phase information $\varphi[n]$ can be recovered using DACM as follows:

$$\varphi[n] = \sum_{k=2}^{n} \frac{I[k] \cdot \{Q[k] - Q[k-1]\} - Q[k] \cdot \{I[k] - I[k-1]\}}{I[k]^2 + Q[k]^2} \tag{9.12}$$

where $\varphi[n]$, $I[n]$, and $Q[n]$ are the digital representations of the signal as in $\varphi(n\delta t) = \arctan\left[Q(n\delta t)/I(n\delta t)\right]$, where δt is the sampling interval. Compared to phase unwrapping, phase demodulation using DACM algorithm allows the phase information to be directly retrieved from the calibrated I/Q signals without phase ambiguities.

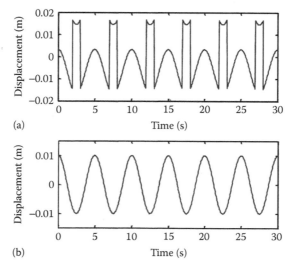

FIGURE 9.6
Demodulated displacement of a 0.2 Hz 20 mm sinusoidal movement in 2.4 GHz Doppler radar using (a) regular arctangent demodulation and (b) extended DACM algorithm.

Figure 9.6b shows the demodulated result of the same 0.2 Hz 20 mm sinusoidal movement using the extended DACM algorithm, which naturally solves the phase discontinuity problem. Phase demodulation using DACM algorithm results in robust and accurate phase information recovery from nonlinear phase modulation, regardless of the magnitude of vibration motions.

Although many of the aforementioned working principles were demonstrated in the lower 2.4 or 5 GHz band, they can also be applied as general theories to the higher millimeter-wave bands.

9.3 Advancements in Microwave Doppler Radar

9.3.1 Digital IF Doppler Radar

The conventional I/Q receiver, as shown in Figure 9.3, are custom-built analog circuitry that suffers from I/Q unbalances, and therefore, a vital sign detection radar requires a careful design of the RF segment. An alternative implementation based on a digital IF architecture was reported in [31] employing general-purpose RF and communication instruments that are widely available in RF/microwave laboratories. The proposed system can be easily assembled by connecting the instruments, and it can be quickly disassembled for other microwave laboratory applications. Furthermore, it allows many RF/microwave laboratories to quickly engage in radar research without spending a long-time building custom designed radar hardware. The system is composed of the following instruments: Agilent E4407B spectrum analyzer, Agilent E8267C vector signal generator, and Agilent 89600S vector signal analyzer, Agilent (now Keysight), Santa Clara, CA.

The Doppler radar with digital IF technique mitigates I/Q mismatch that occurs in the analog quadrature down-converter architecture and eliminates the complicated DC offset calibration required for arctangent demodulation. Furthermore, the carrier frequency can be tuned across a wide range of frequencies from ultra high frequency (UHF) to K band. The wide tuning range allows electromagnetic waves to penetrate through different depths of obstacles, which is useful for studying radar-based earthquake rescue. It also allows selection of the optimal carrier frequencies for different people with different physiological movement amplitudes. Figure 9.7 shows the simplified block diagram of the instrument-based digital IF Doppler radar system. The vector signal generator E8267C generates a single-tone signal and transmits it via the T_antenna to the target. The frequency could be tuned up to 20 GHz, and the Tx power could also be adjusted independently according to different applications. The reflected microwave signal that is modulated by the vital signs is received by the Rx antenna and sent into the spectrum analyzer, whose internal down-conversion module down-converts the microwave signal to the 70 MHz IF output. The IF signal is then fed to the vector signal analyzer 89600S for further processing. The digitization of the 70 MHz IF is carried out at the ADC module E1439. The digital demodulation in IF leads to very high accuracy because it ensures accurate matching of the I/Q channels. The baseband digital output is filtered, windowed, and autocorrelated in MATLAB®, whereas further data analysis is also performed in MATLAB.

Experiments show that, compared to a pulse sensor attached to the subject's index finger, the proposed heterodyne digital IF system is able to achieve high accuracy of >80% with 0 dBm transmit power at a distance of 2.4 m.

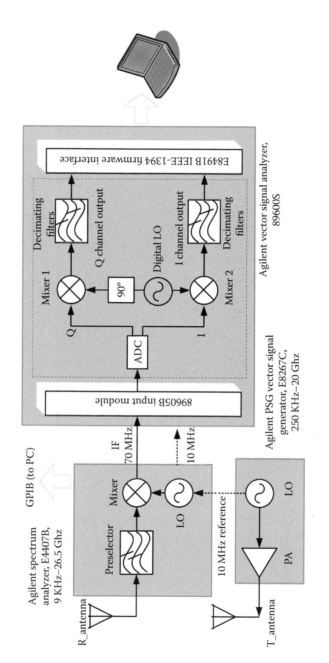

FIGURE 9.7
Simplified block diagram of the instrument-based digital IF Doppler radar.

9.3.2 Signal Distortion Compensation

Signal distortion may happen in homodyne radar receiver. To avoid the null point problem and leverage on the range correlation effect to reduce the effect of phase noise, homodyne radar receiver with quadrature I/Q architecture is widely used for noncontact vital sign detection. Ideally, the I/Q signals as in Equations 9.8 and 9.9 are expected to form an ideal arch sitting on the unit circle. Figure 9.8 illustrates the ideal I/Q signals in the time domain and in the I/Q plane. In the practical radar circuitry, due to circuit imperfections and the clutter reflections from the stationary objects around the target, a DC offset is present at the mixer output of the radar homodyne receiver. The straightforward and simplest way to remove the DC offset is to use AC coupling at the mixer output. However, the shortcoming is that AC coupling essentially has high-pass characteristics, which results in signal distortion when the target motion is very slow or when there is stationary moment [32]. The distorted I/Q signals no longer form an ideal arch but a ribbonlike trajectory, as shown in Figure 9.9.

This trajectory can be explained considering a sinusoidal motion $x(t) = m \cdot \sin(\omega\tau)$, where m is the amplitude and ω is the frequency. The signal resulting from this sinusoidal modulation of the phase through Equation 9.4 can be expanded as summation of Bessel functions:

$$B(t) = \sum_{n=-\infty}^{n=\infty} J_n\left(\frac{4\pi m}{\lambda}\right) \cdot \cos(n\omega t + \varphi) \tag{9.13}$$

where:
  φ is the total residual phase
  λ is the wavelength
  $J_n(x)$ is the nth-order Bessel function of the first kind

As Equation 9.13 indicates, a single sinusoidal movement would generate harmonics at the radar baseband output. Owing to the slope of the high-pass filter, these harmonics would be subject to different degrees of attenuation, leading to a loss of the inherent spectrum characteristics and resulting in the ribbonlike trajectory in the I/Q plane as shown

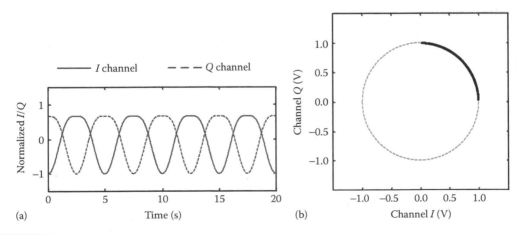

(a)

(b)

FIGURE 9.8
The ideal I/Q signals at the quadrature Doppler radar receiver output: (a) time domain and (b) an ideal arch on the unit circle.

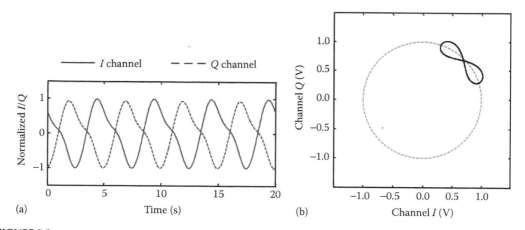

FIGURE 9.9

The distorted *I/Q* signals at the quadrature Doppler radar receiver output: (a) time domain and (b) a ribbonlike trajectory on the unit circle.

in Figure 9.9. As discussed in Section 9.2, proper DC calibration to the baseband *I/Q* signals needs to be carried out in order to utilize the arctangent demodulation technique to accurately recover the phase information. In radar nonlinear phase modulation, DC information is the constant term from the expansion of Equation 9.13. During DC calibration, the DC information needs to be kept as it is an essential part of the radar signal, whereas the DC offset values need to be removed. However, the distorted ribbon shape challenges the calibration algorithms to find the DC offset values while keeping the useful DC information [32]. Furthermore, the demodulation accuracy may also be affected because the length of the *I/Q* trajectory, as illustrated in Figure 9.8b, is strictly related to the vibration amplitude [32].

There are several techniques proposed to tackle the signal distortion problem. A straight-forward method to avoid signal distortion due to AC coupling is to utilize DC coupling for the radar receiver. The DC-coupled baseband has all-pass characteristics, unlike AC coupling that is high pass, whereas all the harmonics of nonlinear phase modulation undergo the same degree of attenuation, which means that no distortion will occur. Hardware modifications need to be made to employ DC offset in order to avoid saturation of the stages following the mixer. A 2.4 GHz Doppler radar with DC-coupled architecture was proposed in [33], the block diagram of which is shown in Figure 9.10. The radar system includes two stages of adaptive DC tuning: RF coarse tuning and baseband fine-tuning. To deal with the DC offset caused by the direct coupling from the transmitter to the receiver, the RF coarse tuning was implemented using an extra signal path made of an attenuator and a phase shifter at the RF front end of the Doppler radar system. It feeds a portion of the transmitter signal back to the receiver to cancel most of the DC offset at the mixer output. To further calibrate the remaining DC offset and to fully utilize the ADC's dynamic range, the baseband fine-tuning circuitry is added to adaptively adjust the amplifier bias to the desired level. The two DC tuning stages allow high-gain amplification as well as the maximum dynamic range. The proposed radar system with DC tuning architecture can accurately measure the slow movement even when it stops for a little while.

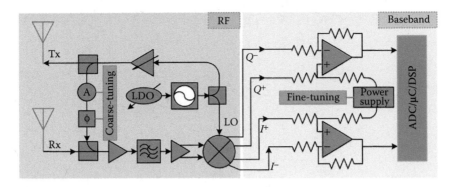

FIGURE 9.10
The block diagram of a 2.4 GHz Doppler radar architecture with DC tuning technique, including RF coarse tuning and baseband fine-tuning.

While the DC-coupled radar can measure slow movements without signal distortion at the cost of increased hardware complexity, several other approaches have been proposed to use the simpler AC-coupled architecture for accurate displacement measurement without adding to the hardware complexity. A recent advancement is digital post-distortion (DPoD) technique that digitally compensates the signal distortions introduced by the AC coupling [34]. Without any change to the radar hardware, the simple quadrature direct conversion receiver can be used to detect the complete pattern of slow periodic movements. The post-distortion technique applies signal compensation in the digital baseband to recover the signal information that is lost in the AC-coupled receiver. As shown in Figure 9.11, the distortion compensation is performed in the digital domain by applying an algorithm whose system response is the inverse function of that of the high-pass filter at the AC-coupled baseband. In other words, the DPoD technique tends to *linearize* the AC-coupled baseband so that the resulting transfer function has virtually an *all-pass* characteristic. The proposed DPoD technique has been experimentally verified to be effective using a 2.4 GHz radar to measure a healthy adult at rest. The regular respiratory motion for an adult at rest includes a short period of stationary moment, which means that the respiration tends to rest for a moment at the end of each expiration phase. The

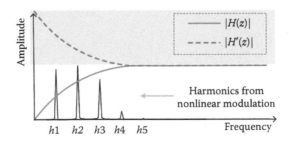

FIGURE 9.11
DPoD applies a system response $H'(z)$, which is the inverse function of the AC-coupled baseband response $H(z)$.

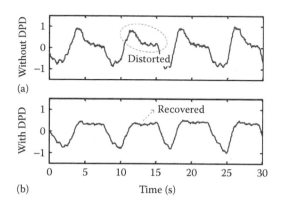

(a)

(b)

FIGURE 9.12
Respiration motion measured by AC-coupled Doppler radar (a) without and (b) with the DPoD technique.

experimental results, as illustrated in Figure 9.12, show that although the information of the stationary moment is lost without using DPoD, and therefore, the respiration signal is distorted, DPoD allows the undistorted recovery of the complete respiratory pattern.

9.3.3 Random Body Movement Cancelation

One challenge of noncontact vital sign detection is the filtering of random body movement, which hinders the ubiquitous use of the radar technology in many realistic scenarios. If the target (e.g., human) is resting still, the only movement modulating the phase of the transmitted signal would be that of the chest wall vibrations due to physiological activities. However, if the target is not stationary, the radar signal would be modulated also by the macroscopic movement of the body and mixed together with heartbeat and respiration. Body movement creates a challenge for extracting the smaller signal due to vital signs. It hinders the extraction of the expected vital sign signals, which is usually much stronger in magnitude. If the body movement has a regular motion pattern, it may be removed in the baseband digital processing; nevertheless, the body often moves randomly in practical situations, which challenges the signal processing to predict the precise motion pattern.

To deal with this issue, several approaches have been proposed such as the multiradar system described in [35]. Figure 9.13 illustrates the setup of using two radars measuring

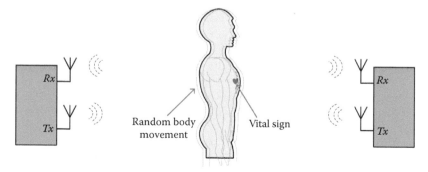

FIGURE 9.13
Random body movement cancelation using two radars for noncontact vital sign detection.

from the back and the front of the body simultaneously to cancel the random body movement but keeping the target vital sign signals. It utilizes two identical radars with the same linearly polarized patch antennas. To reduce the mutual interference, antennas on one radar are rotated 90 so that they are orthogonal to the antennas on the other radar. The heartbeat and respiration have the same movement direction from the radar perspective as they are both periodic movements with small amplitude relative to the radar wavelength. The nature of the chest cavity expansion would create the same Doppler shift for each radar. The random body movement, however, is the large displacement so that when the body leans to one of the radars, it moves away from the other. In other words, if the distance between the body and one radar increases, the distance between the body and the other radar decreases. By combining the signals measured by the two radars, most of the clutter from the random body movement can be canceled. A similar approach using two injection locking radar arrays was proposed in [36] for canceling the body movement artifacts.

The approaches to cancel the random body movement using multiple radars or arrays have obvious shortcomings. For instance, they inevitably add up to hardware complexity, cost, and power consumption, and they are not user-friendly solutions, as they are applicable only to some specific situations where there is enough space to set up the multisensor system. In [37], a different approach for the random body movement compensation is proposed using a radar–camera hybrid system as shown in Figure 9.14. In this case, an ordinary smartphone camera was used to capture the body movement demonstrating that this information can be used to cancel the body movement in radar-measured signals. Considering the ubiquity of today's smartphones, this approach sounds very attractive. The working principle is that the external phase information, which is captured by the camera and is opposite to the phase information of random body movement, is added in the radar-received signals. The phase compensation can be applied either at the RF front end using a phase shifter or in the baseband digital domain. Because the large body movement signal may saturate the baseband amplifier, the phase compensation using phase shifter at the RF front end helps to avoid the saturation of the radar receiver, thus maintaining maximum sensitivity. If the body movement does not saturate the radar receiver, the phase compensation can also be implemented in the baseband digital domain so as to be more precisely to remove the artifacts.

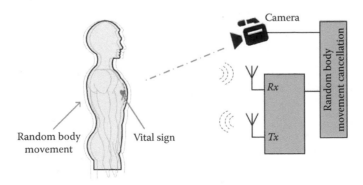

FIGURE 9.14
A radar–camera hybrid system for noncontact vital sign detection with random body movement cancelation.

9.4 Design of a 60 GHz (Millimeter-Wave) Radar for the Detection of Respiration and Heart Rates

Recent progresses on microwave radar for noncontact vital sign detection have been made with growing interests in health-care and biomedical sciences. When the operating frequency in radar increases from microwave to millimeter range, the sensitivity to small movement can be increased, and the size of antenna and circuit components can be shrunk down because the antenna size is on the order of wavelength. Small antennas can be integrated on chip. The antenna gain, which is inversely proportional to the wavelength, will be increased as the frequency increases if the antenna area is fixed [38,39]. In addition, the radar cross section also increases with the frequency in far-field condition. If the antenna effective area and distance between the radar and the target are fixed and air absorption is negligible, the received power at millimeter wave will be higher than that at microwave even when a smaller antenna is used. Furthermore, although lower frequency bands have been occupied by growing communication applications and cannot be easily occupied by other services such as vital sign monitoring, millimeter waves offer wider bandwidth and less RF interference from other wireless technologies. A single-channel, 94 GHz complementary metal–oxide–semiconductor (CMOS) Doppler radar transceiver [40] was proposed and can be potentially used to detect small movements, although this specific transceiver is subject to the null point problem. Another heterodyne radar system at 228 GHz [41] was demonstrated to successfully detect the vital sign signal at 50 m distance.

For the same displacement of $x(t)$ in Equation 9.4, a shorter wavelength λ creates a larger phase modulation angle θ, which will increase the sensitivity to small displacement as shown in Figure 9.15. When the displacement of $x(t)$ is small enough compared to λ, the system has an approximately linear transfer function near the optimal point as shown in Figure 9.16. The null point, as discussed in Section 9.2, can be solved by the quadrature I/Q architecture without frequency tuning. Arctangent demodulation, however, requires recalibration with changes in target distance to address the DC offset problem. Alternatively, complex signal demodulation (CSD) combines the I and Q baseband outputs [42] and solves the null detection point and DC offset problem without tuning the carrier frequency [28]. The CSD baseband output can be expressed as Equation 9.14 in which the DC offset is removed by simply extracting from the averaged time-domain sliding window because it only affects the DC term of the received baseband signal. Therefore, the CSD simplifies the demodulation procedure. Nevertheless, for vital sign detection, harmonics arising from

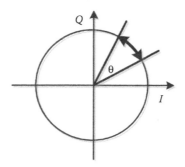

FIGURE 9.15
Doppler phase modulation in the I/Q plane.

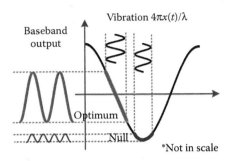

FIGURE 9.16
Illustration of vibration displacement detection at the optimum and null points of a Doppler radar system. Vibration and baseband output waveform are not drawn to scale.

respiration signal might interfere with heartbeat signal and degrade the accuracy of heart rate (HR) detection. For radar using millimeter-wave carrier frequency, the harmonics due to nonlinearity will be more significant:

$$B_I(t) + j \cdot B_Q(t) = \cos(4\pi x(t)/\lambda + \Delta\theta(t)) + j \cdot \sin(4\pi x(t)/\lambda + \Delta\theta(t)) \tag{9.14}$$

9.4.1 60 GHz Radar-on-Chip in CMOS for Vital Sign Detection

Implementation of Doppler radar for vital sign detection can be realized by discrete components on printed circuit boards (PCBs) [43]. In addition to the board-level realization, system-on-chip or system-in-package (SiP) can provide another cost-effective and low-power solution for compact system integration. Single-chip Doppler radar transceivers on CMOS technology have been successfully developed for noncontact vital sign detection [7,44,45]. As the frequency increases to a millimeter-wave range, design challenges, including circuit performance, noise and loss, package and antenna integration, and strong nonlinear effect due to Doppler phase demodulation method, should be carefully considered.

The quadrature I/Q system architecture is proven to be effective in solving the null point problem, and it is thus used in the 60 GHz radar transceiver. For vital sign detection, the flicker noise in CMOS technology is a major concern due to the vital sign signal being close to DC. A simple and power-efficient way to separate I/Q at an IF (6 GHz) is using a ring oscillator, which not only provides a compact topology but also enables driving a IF passive mixer with low flicker noise [46].

Figure 9.17 shows the block diagram of the 60 GHz micro-radar system, including a 90 nm CMOS transceiver chip on an RT/duroid 5870 laminate PCB board. The RF pads are on-chip and are attached to the metal traces on the PCB with flip-chip integration. The G–S–G–S–G transition (see Figure 9.17) achieves good antenna impedance matching while maximizing antenna isolation, which helps to reduce DC offsets.

The sensitivity calculation for the Rx focuses on the received power to achieve the required output signal-to-noise ratio, as shown in Figure 9.18. The noise figure (NF) of F_4 and F_5 are estimated to be high [7] due to flicker noise, which is critical for the detected vital sign signals near DC (~1 Hz), and the total NF is approximately 31.2 dB even with a high-gain (35 dB) first stage. In the baseband, the time-domain window size of Fourier transform determines the resolution bandwidth (RBW). The smaller the RBW, the lower will be the overall noise at the output. According to [7], the relative strength of flicker noise and white noise is controlled by the sampling frequency (f_s) and the equivalent noise bandwidth of

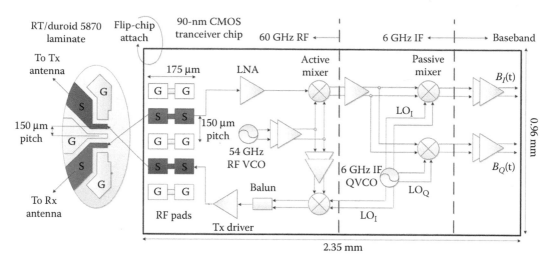

FIGURE 9.17
Block diagram of 60 GHz CMOS micro-radar system.

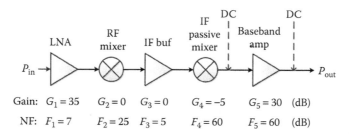

FIGURE 9.18
Sensitivity estimation for the radar receiver.

the receiver (B). If a low f_s is used (e.g., 20–50 Hz), the noise level will be dominated by the folded white noise, and the Rx sensitivity is estimated as

$$\text{Sensitivity}_{Rx} = kT \cdot \text{RBW(dBm)} + \text{NF}_{wh}(\text{dB}) + B/f_s(\text{dB}) + \text{SNR}_{re}(\text{dB}) \qquad (9.15)$$

where:
   kT is thermal noise floor per hertz at the input
   NF_{wh} is the white noise
   SNR_{re} is the required signal-to-noise ratio at the output

The white noise level before aliasing can be represented by $kT \cdot \text{RBW} + \text{NF}_{wh}$, where NF_{wh} is approximately 7.1 dB in the first two stages of Figure 9.18. The corresponding sensitivity is −117 dBm at $\text{SNR}_{re} = 20$ dB.

The 60 GHz radar chip microphotograph in 90 nm CMOS is shown in Figure 9.19. The 60 GHz RF core on the left is 0.73 mm², and the down-converted baseband differential I/Q outputs are on the right. DC biases are supplied from the DC pads on the top and bottom sides, which will not interfere with the antennas.

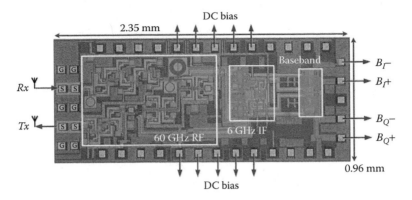

FIGURE 9.19
Microphotograph of the 60 GHz micro-radar in 90 nm CMOS technology. (From Kao, T. et al., *IEEE Trans. Microw. Theory Techn.*, 61, 1649–1659, 2013. With Permission.)

Patch antennas are designed for both *Tx* and *Rx* at the operating frequency (55 GHz). The simulated *S*-parameter after flip-chip packaging and the simulation setup are shown in Figure 9.20. Port 1 is the *Tx* antenna and port 2 is the *Rx* antenna. S_{11} shows the input matching of the antenna, and S_{12} represents the isolation between two ports. The isolation between port 1 and port 2 reaches −34 dB at 55 GHz. The simulated gain of the single patch antenna is approximately 5 dBi as shown in Figure 9.21.

The final system configuration of the 60 GHz micro-radar SiP is shown in Figure 9.22, including the CMOS transceiver chip, two PCB patch antennas, and DC biasing through the blue wires. Bypass capacitors (22 μF) are used on the PCB to suppress the power supply noise. The weight of the system shown in the photo is less than 10 g (0.3 ounce). The total power consumption of the micro-radar SiP is 377 mW at 1.2 V, and the transmit power is approximately 0 dBm.

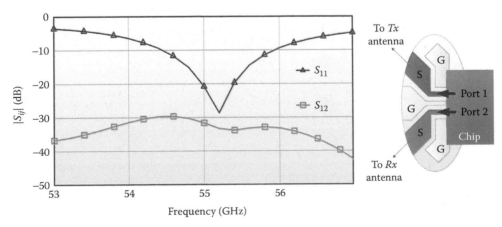

FIGURE 9.20
Simulated patch antenna *S*-parameter after flip-chip packaging.

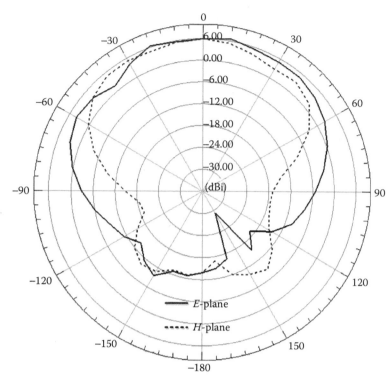

FIGURE 9.21
The simulated gain of the single patch antenna after flip-chip packaging.

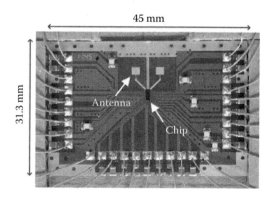

FIGURE 9.22
The final system configuration of the 60 GHz micro-radar SiP. (From Kao, T. et al., *IEEE Trans. Microw. Theory Techn.*, 61, 1649–1659, 2013. With Permission.)

9.4.2 Human Vital Sign Measurement

Human vital sign can be modeled as a two-tone sinusoidal vibration [48], including respiration and heartbeat signal, which can be expressed as

$$x(t) = x_r(t) + x_h(t) = m_r \sin(2\pi f_r t) + m_h \sin(2\pi f_h t) \tag{9.16}$$

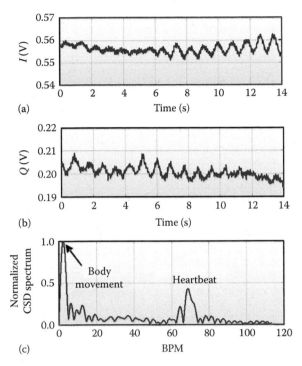

FIGURE 9.23
Heartbeat detection when the target holds the breath at 0.3 m away. The baseband output signals in (a) the *I* channel, (b) the *Q* channel, and (c) the normalized CSD spectrum.

where:
f_r and f_h are the respiration rate (RR) and the HR, respectively
m_r is respiration displacement
m_h is heartbeat displacement

In general, the human RR is 12–16 breaths/minute (BR/min) and HR is 60–100 beats/minute (BPM). In the experiment, the 60 GHz CMOS radar was placed in front of a human target sitting still on a chair, and the radar was pointed to his chest wall 0.3 m away. In order to detect the heartbeat signal, the target was asked to hold their breath to avoid interference from respiration harmonics. As shown in Figure 9.23, the heartbeat signals from *I* and *Q* channels were directly sampled by the oscilloscope. In the period $t = 0$–7 s, the *Q* channel was around the optimal point, and the *I* channel was near the null point. After $t = 9$ s, the *Q* channel was near the null point due to body movement, and the *I* channel started to take over the detection. The normalized CSD spectrum shows a strong body movement, and the detected HR was at 69 BPM, which agrees with the result of human counting.

9.4.3 Laboratory Rat Vital Sign Measurement

Compared to humans, rats have significantly higher RR (~85 BR/min) and HR (330–480 BPM), which are associated with smaller chest wall movements. A 60 GHz radar can be used to detect a rat's chest wall movements due to its high sensitivity to small vibrations.

As described in Equation 9.17, the detected baseband signal with the CSD method can be further expanded with Bessel function:

$$B(t) = I(t) + jQ(t)$$

$$= \sum_{p=-\infty}^{\infty} \sum_{q=(x-f_hp)/f_r}^{\infty} J_p\left(\frac{4\pi m_h}{\lambda}\right) J_q\left(\frac{4\pi m_r}{\lambda}\right) \cdot e^{j2\pi xt} \cdot e^{j\phi} \tag{9.17}$$

where:
  p and q are the integers that satisfy $x = f_h p + f_r q$
  x represents the nonlinearity-caused harmonic frequency

The amplitudes of the harmonics in the baseband spectrum are determined by the residual phase ϕ, the wavelength λ, and the displacement of vibration m. The term $e^{j\phi}$ can be eliminated because a quadrature architecture is used and the effect of the residual phase on the amplitude can be neglected. The large respiratory movement in humans (m_r = 1–6 mm) makes heartbeat detection using millimeter-wave radar more difficult because $J_1(4\pi m_r/\lambda)$ has several zero-crossing points [47], which results in an unrecognizable respiration peak in the detected spectrum. Therefore, the small respiratory movement of a laboratory rat makes heartbeat extraction much simpler by avoiding the zero-crossing problem. This is due to the fact that the chest displacement during respiration of the rat is estimated to be less than 2 mm, which is comparable to the 5 mm wavelength at 60 GHz; therefore, the baseband spectrum suffers from strong nonlinear Doppler-phase modulation effects.

The displacement of vibration movements can be extracted from the ratio of the harmonics measured in the baseband spectrum [49]. The amplitude of the harmonics that contain the frequency components from respiration and heartbeat can be expressed as

$$H_x = \left| \sum_{p=-\infty}^{\infty} \sum_{q=(x-f_hp)/f_r}^{\infty} J_p\left(\frac{4\pi m_h}{\lambda}\right) J_q\left(\frac{4\pi m_r}{\lambda}\right) \right| \tag{9.18}$$

Harmonics can be grouped into two categories (see Figure 9.25b): the respiration harmonics (squares) and the frequency mixing products of RR and HR (dots). Similar to the concept of the intermodulation or frequency mixing, the HR mixes with the RR and its harmonics resulting in these mixing products. The mixing products that are symmetrically centered around the HR peak are used to determine the correct HR peak. After taking the ratio of the chosen harmonics, the desired displacement can be extracted [50].

In order to calculate the displacements of respiration and heartbeat, the harmonics should be properly chosen. First, the H_{RR}/H_{2RR} ratio is used for calculating the respiration displacement because they are the two strongest peaks on the spectrum with the highest SNR and therefore are less affected by noise. As shown in Equation 9.19, the $J_0(4\pi m_h/\lambda)$ term can be canceled out, leaving $J_q(4\pi m_r/\lambda)$ only for calculating m_r. Once the absolute value of the ratio is obtained, m_r can be determined by fitting the measured harmonic ratio to the theoretical value from the Bessel function. Similarly, the heartbeat displacement is extracted by utilizing the strong mixing products between the HR and respiration harmonics because the harmonics of the HR are too small. Two mixing products, H_{HR-RR} and H_{HR-2RR}, are chosen and paired up with the respiration harmonics for calculating m_h, as shown in Equation 9.20.

$$\frac{H_{RR}}{H_{2RR}} = \left| \frac{J_0\left(\dfrac{4\pi m_h}{\lambda}\right) J_1\left(\dfrac{4\pi m_r}{\lambda}\right)}{J_0\left(\dfrac{4\pi m_h}{\lambda}\right) J_2\left(\dfrac{4\pi m_r}{\lambda}\right)} \right| = \left| \frac{J_1\left(\dfrac{4\pi m_r}{\lambda}\right)}{J_2\left(\dfrac{4\pi m_r}{\lambda}\right)} \right| \tag{9.19}$$

$$\frac{H_{RR}}{H_{HR-RR}} = \frac{H_{2RR}}{H_{HR-2RR}} = \left| \frac{J_0\left(\dfrac{4\pi m_h}{\lambda}\right)}{J_1\left(\dfrac{4\pi m_h}{\lambda}\right)} \right| \tag{9.20}$$

The experimental setup that uses a 60 GHz CMOS radar for noncontact monitoring of an anesthetized rat is depicted in Figure 9.24. To demonstrate the effectiveness of this technique using the 60 GHz radar, two experiments were performed using anesthetized rats specifically instrumented to record their vital data. Anesthesia of a healthy adult rat was induced with urethane (1.3 mg/kg), and an arterial catheter was placed in the femoral artery for blood pressure (BP) and HR monitoring. After instrumentation, the anesthetized animal was placed in a whole-body plethysmograph (chamber volume 4 L; Buxco Electronic Inc., Wilmington, NC, USA) for monitoring the respiratory movement. BP, HR, RR, and body movement were simultaneously recorded with Spike2 software, CED, Cambridge, UK, (200 Hz sampling rate per channel) for reference. The radar was placed at 0.3 m from the chamber to detect the rat's vital signs. The baseband signal was then sent, through a National Instruments data acquisition board, to a laptop equipped with LabVIEW, National Instruments, Austin, TX, for analysis.

Figure 9.25 shows the results of the first experiment, including the measured 15 s time-domain waveform and the corresponding baseband spectrum. Harmonics and intermodulation products shown in the baseband spectrum were labeled in Figure 9.25b. The frequencies and amplitudes of the detected harmonics are listed in Table 9.1. The measured rate and the calculated displacement for the observation time ($t = 0$–15 s) are listed in Table 9.2. Variations of both rates and displacements within the measurement time (47 s) can be observed from Figure 9.26, in which each data point represents a 15 s observation time window.

The second experiment was performed on three different rats and on different dates. The correlation diagram of the instrument-recorded data and the radar-recorded data for the three test subjects are shown in Figure 9.27. The variability in HR and RR among rats

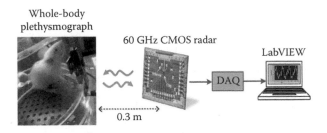

FIGURE 9.24
Measurement of a laboratory rat in a whole-body plethysmography using the 60 GHz CMOS radar. (From Huang, T. et al., Non-invasive measurement of laboratory rat's cardiorespiratory movement using a 60-GHz radar and nonlinear Doppler phase modulation, *IEEE International Microwave Workshop Series on RF and Wireless Technologies for Biomedical and Healthcare Applications*, pp. 83–84, September 2015. With Permission.)

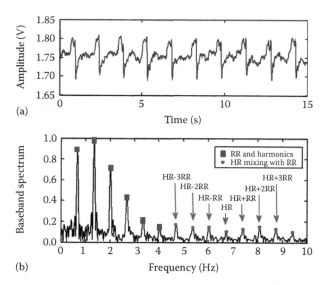

FIGURE 9.25
Radar measurement of a rat's vital signs: (a) time-domain waveform and (b) normalized baseband spectrum.

TABLE 9.1

Frequency of the Harmonics Satisfying $x = f_h p + f_r q$ and Its Amplitude H_x with Bessel Function Expression

Harmonic	x (Hz)	p	q	$|H_x|$
H_{RR}	0.66	0	1	$J_0(a_h)J_1(a_r)$[a]
H_{2RR}	1.32	0	2	$J_0(a_h)J_2(a_r)$
H_{3RR}	1.98	0	3	$J_0(a_h)J_3(a_r)$
H_{4RR}	2.64	0	4	$J_0(a_h)J_4(a_r)$
H_{5RR}	3.3	0	5	$J_0(a_h)J_5(a_r)$
H_{6RR}	3.96	0	6	$J_0(a_h)J_6(a_r)$
H_{HR-3RR}	4.75	1	−3	$J_1(a_h)J_{-3}(a_r)$[b]
H_{HR-2RR}	5.41	1	−2	$J_1(a_h)J_{-2}(a_r)$
H_{HR-RR}	6.07	1	−1	$J_1(a_h)J_{-1}(a_r)$
H_{HR}	6.73	1	0	$J_1(a_h)J_0(a_r)$
H_{HR+RR}	7.39	1	1	$J_1(a_h)J_1(a_r)$
H_{HR+2RR}	8.05	1	2	$J_1(a_h)J_2(a_r)$
H_{HR+3RR}	8.71	1	3	$J_1(a_h)J_3(a_r)$

[a] $a_h = 4\pi m_h/\lambda$, $a_r = 4\pi m_r/\lambda$.
[b] $J_{-n}(a) = -J_n(a)$ for odd n or $J_n(a)$ for even n.

TABLE 9.2

Measurement of a Healthy Rat under Anesthesia at $t = 0–15$ s

	Respiration	Heartbeat
Measured rate (Hz)	0.66	6.73
Calculated displacement (mm)	1.19	0.13

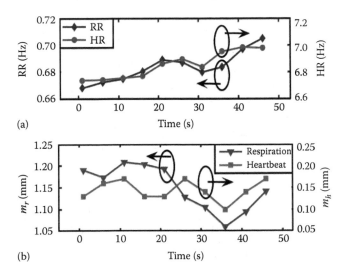

FIGURE 9.26
Radar measurement results: (a) RR and HR versus time and (b) displacements of both respiration and heartbeat versus time.

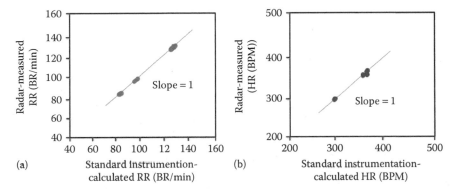

FIGURE 9.27
Correlation diagram of instrument and radar recorded data: (a) RR and (b) HR (slope = 1 for reference).

may be due to how the urethane was delivered to the test subjects. The average errors of RR and HR are less than 0.1% and 0.33% in all the three test subjects, respectively.

These results show the advantages of using millimeter-wave radar to measure cardiorespiratory displacement with the aid of harmonics because the measured cardiorespiratory variations can provide additional objective parameters in animal study.

9.5 Conclusion

Doppler radar enables remote sensing of vital signs in a stress-free and long-term manner [51]. It provides an alternative method to obtain the cardiorespiratory information without the need for body contact or implanted devices in the traditional way.

The advancement in detection and analysis techniques introduced in this chapter has a lot of potential for future development. Although the discussed approaches compensate the body movement when the radar system is stationary, researchers have also proposed approaches to tackle the movement of the radar platform itself, because it can be useful if radar sensors could be integrated in mobile platforms such as smartphone or tablet and used while these devices are moving. A complete compensated single transceiver radar system has been proposed in Reference 52 for vital sign detection in the presence of platform movement. In this case, a custom designed tag near the subject produces a reference signal for the radar platform to compensate the movement itself. In addition, researchers are using biomedical radar in chronobiology and pharmacology for behavioral study or physiological effect analysis of laboratory animals, such as monitoring sleep stages [19] and locomotor activities [53] in animals. The choice of operating frequency for monitoring a small laboratory rat's vital signs has also been discussed in Reference 54 by comparing 4 and 14 GHz radars. The efforts in the development of vital sign radar range from microwave to millimeter wave and system level to chip level, which creates many possibilities for future radar techniques. Although wireless technology gradually becomes part of our life, sensors as connections among living things and the world would greatly change our life.

References

1. S. Kim and C. Nguyen, On the development of a multifunction millimeter-wave sensor for displacement sensing and low-velocity measurement, *IEEE Trans. Microw. Theory Techn.*, 52, 2503–2512, 2004.

2. A. Stezer, C. G. Diskus, K. Lubke, and H. W. Thim, Microwave position sensor with sub millimeter accuracy, *IEEE Trans. Microw. Theory Techn.*, 47 (12), 2621–2624, 1999.

3. R. J. Doviak, R. M. Rabin, and A. J. Koscielny, Doppler weather radar for profiling and mapping winds in the prestorm environnment, *IEEE Trans. Geosci. Remote Sens.*, GE- 21 (1), 25–33, 1983.

4. P. Heide, V. Mdgori, and R. Schwarte, Coded 24 GHz Doppler radar sensors: A new approach to high-precision vehicle position and ground-speed sensing in railway and automobile applications, *IEEE MTT-S Int. Microw. Symp. Digest.*, 2, 965–968, 1995.

5. H. H. Meinel, Commercial applications of millimeter waves history, present status, and future trends, *IEEE Trans. Microw. Theory Techn.*, 43 (7), 1639–1653, 1995.

6. A. D. Droitcour, V. M. Lubecke, J. Lin, and O. Boric-Lubecke, A microwave radio for Doppler radar sensing of vital signs, *IEEE MTT-S International Microwave Symposium Digest*, pp. 176–178, May 2001.

7. C. Li, X. Yu, C. Lee, L. Ran, and J. Lin, High-sensitivity software configurable 5.8 GHz radar sensor receiver chip in 0.13 µm CMOS for non-contact vital sign detection, *IEEE Trans. Microw. Theory Techn.*, 58 (5), 1410–1419, 2010.

8. H.-R. Chuang, Y. F. Chen, and K.-M. Chen, Automatic clutter-canceler for microwave life-detection systems, *IEEE Trans. Instrum. Meas.*, 40 (4), 747–750, 1991.

9. J. C. Lin, Microwave sensing of physiological movement and volume change—A review, *Bioelectromagnetics*, 13, 557–565, 1992.

10. P. K. Capp, P. L. Pearl, and D. Lewin, Pediatric sleep disorders, *Primary Care*, 32, 549–562, 2005.

11. H. Forster, O. Ipsiroglu, R. Kerbl, and E. Paditz, Sudden infant death and pediatric sleep disorders, *Wiener Klin. Wochenschrift*, 115, 847–849, 2003.

12. K.-M. Chen, Y. Huang, J. Zhang, and A. Norman, Microwave life-detection systems for searching human subjects under earthquake rubble or behind barrier, *IEEE Trans. Biomed. Eng.*, 27 (1), 105–114, 2000.

13. J. C. Lin, Noninvasive microwave measurement of respiration, *Proc. IEEE*, 63 (10), 1530, 1975.
14. K.-M. Chen, D. Misra, H. Wang, H.-R. Chuang, and E. Postow, An X-band microwave life-detection system, *IEEE Trans. Biomed. Eng.*, 33 (7), 697–701, 1986.
15. C. J. Gordon and J. S. Ali, Measurement of ventilatory frequency in unrestrained rodents using microwave radiation, *Respir. Physiol.*, 56, 73–79, 1984.
16. D. Kropveld and R. A. F. M. Chamuleau, Doppler radar device as a useful tool to quantify the liveliness of the experimental animal, *Med. Biol. Eng. Comput.*, 31, 340–342, 1993.
17. N. Hafner, J. C. Drazen, and V. M. Lubecke, Fish heart rate monitoring by body-contact Doppler radar, *IEEE Sens. J.*, 13 (1), 408–414, 2013.
18. T. Huang, J. Lin, and L. Hayward, Non-invasive measurement of laboratory rat's cardiorespiratory movement using a 60-GHz radar and nonlinear Doppler phase modulation, *IEEE International Microwave Workshop Series on RF and Wireless Technologies for Biomedical and Healthcare Applications*, pp. 83–84, September 2015.
19. T. Zeng, C. Mottband, D. Molliconeb, and L. D. Sanforda, Automated determination of wakefulness and sleep in rats based on non-invasively acquired measures of movement and respiratory activity, *J. Neurosci. Methods*, 24, 276–287, 2012.
20. C. Li, V. Lubecke, O. Boric-Lubecke, and J. Lin, A review on recent advances in Doppler radar sensors for noncontact healthcare monitoring, *IEEE Trans. Microw. Theory Techn.*, 61, 2046–2060, 2013.
21. C. Gu and C. Li, From tumor targeting to speech monitoring: Accurate respiratory monitoring using medical continuous-wave radar sensors, *IEEE Microw. Mag.*, 15, 66–76, 2014.
22. C. Gu, T. Inoue, and C. Li, Analysis and experiment on the modulation sensitivity of Doppler radar vibration measurement, *IEEE Microw. Wirel. Compon. Lett.*, 23, 566–568, 2013.
23. S. Bakhtiari, T. W. Elmer, N. M. Cox, N. Gopalsami, A. C. Raptis, S. Liao, I. Mikhelson, and A. Sahakian. Compact millimeter-wave sensor for remote monitoring of vital signs, *IEEE Trans. Instrum. Meas.*, 61, 830–841, 2012.
24. A. J. Gatesman, A. Danylov, T. M. Goyette, J. C. Dickinson, R. H. Giles, W. Goodhue, J. Waldman, W. E. Nixon, W. Hoen. Terahertz behavior of optical components and common materials—Art, no. 62120E. *Terahertz for Military and Security Applications IV*, 2006, 6212, E2120.
25. C. Li and J. Lin, Optimal carrier frequency of non-contact vital sign detectors, *IEEE Radio and Wireless Symposium*, pp. 281–284, January 2007.
26. A. D. Droitcour, O. Boric-Lubecke, V. M. Lubecke, J. Lin, G. T. A. Kovac, Range correlation and I/Q performance benefits in single-chip silicon Doppler radars for noncontact cardiopulmonary monitoring, *IEEE Trans. Microw. Theory Techn.*, 52, 838–848, 2004.
27. B. K. Park, O. Boric-Lubecke, and V. M. Lubecke, Arctangent demodulation with DC offset compensation in quadrature Doppler radar receiver systems, *IEEE Trans. Microw. Theory Techn.*, 55, 1073–1079, 2007.
28. Y. Xiao, J. Lin, O. Boric-Lubecke, and M. Lubecke, Frequency-tuning technique for remote detection of heartbeat and respiration using low-power double-sideband transmission in the Ka-band, *IEEE Trans. Microw. Theory Techn.*, 54 (5), 2023–2032, 2006.
29. W. Xu, C. Gu, C. Li, and M. Sarrafzadeh, Robust Doppler radar demodulation via compressed sensing, *IET Electron. Lett.*, 48, 1428–1430, 2012.
30. J. Wang, X. Wang, L. Chen, J. Huangfu, C. Li, and L. Ran, Non-contact distance and amplitude independent vibration measurement based on an extended DACM algorithm, *IEEE Trans. Instrum. Meas.*, 63, 145–153, 2014.
31. C. Gu, C. Li, J. Lin, J. Long, J. Huangfu, and L. Ran, Instrument-based noncontact Doppler radar vital sign detection system using heterodyne digital quadrature demodulation architecture, *IEEE Trans. Instrum. Meas.*, 59, 1580–1588, 2010.
32. C. Gu and C. Li, Frequency-selective distortion in continuous-wave radar displacement sensor, *IET Electron. Lett.*, 48, 1495–1497, 2012.
33. C. Gu, R. Li, H. Zhang, A. Fung, C. Torres, S. Jiang, and C. Li, Accurate respiration measurement using DC-coupled continuous-wave radar sensor for motion-adaptive cancer radiotherapy, *IEEE Trans. Biomed. Eng.*, 59, 3117–3123, 2012.

34. C. Gu, Z. Peng, and C. Li, High-precision motion detection using low-complexity Doppler radar with digital post-distortion technique, *IEEE Trans. Microw. Theory Techn.*, 64, 961–971, 2016.

35. C. Li and J. Lin, Random body movement cancellation in Doppler radar vital sign detection, *IEEE Trans. Microw. Theory Techn.*, 56, 3143–3152, 2008.

36. F.-K. Wang, T.-S. Horng, K.-C. Peng, J.-K. Jau, J.-Y. Li, and C.-C. Chen, Single-antenna Doppler radars using self and mutual injection locking for vital sign detection with random body movement cancellation, *IEEE Trans. Microw. Theory Techn.*, 59, 3577–3587, 2011.

37. C. Gu, G. Wang, T. Inoue, and C. Li, A hybrid radar-camera sensing system with phase compensation for random body movement cancellation in Doppler vital sign detection, *IEEE Trans. Microw. Theory Techn.*, 61, 4678–4688, 2013.

38. C. Balanis, *Antenna Theory Analysis and Design*. Hoboken, NJ: John Wiley & Sons, 2005, ch. 2, pp. 110–112.

39. E. Knott, *Radar Handbook*. New York: McGraw-Hill, 2008, ch. 14, p. 10.

40. E. Laskin, M. Khanpour, S. T. Nicolson, A. Tomkins, P. Garcia, A. Cathelin, D. Belot, and S. P. Oinigescu, Nanoscale M S transceiver design in the 0-170 GHz range, *IEEE Trans. Microw. Theory Techn.*, 57, 3477–3490, 2009.

41. D. T. Petkie, C. Benton, and E. Bryan, Millimeter wave radar for remote measurement of vital signs, *IEEE Radar Conference*, pp. 1–3, May 2009.

42. C. Li and J. Lin, Complex signal demodulation and random body movement cancellation techniques for non-contact vital sign detection, *IEEE MTT-S International Microwave Symposium*, pp. 567–570, June 2008.

43. C. Li and J. Lin, Non-contact measurement of periodic movements by a 22-40 GHz radar sensor using nonlinear phase modulation, *IEEE MTT-S International Microwave Symposium Digest*, pp. 579–582, June 2007.

44. A. D. Droitcour, O. Boric-Lubecke, V. M. Lubecke, and J. Lin, 0.25um CMOS and BiCMOS single chip direct conversion Doppler radars for remote sensing of vital signs, *IEEE International Solid State Circuits Conference, Digest of Technical Papers*, pp. 348–349, February 2002.

45. C. Li, Y. Xiao, and J. Lin, A 5 GHz double-sideband radar sensor chip in 0.18 µm CMOS for non-contact vital sign detection, *IEEE Microw. Wirel. Compon. Lett.*, 18 (7), 495–496, 2008.

46. B. Razavi, *Design of Analog CMOS Integrated Circuits*. New York: McGraw-Hill, 2001, ch. 7, p. 215.

47. T. Kao, Y. Yan, T. Shen, A. Chen, and J. Lin, Design and analysis of a 60-GHz CMOS Doppler micro-radar system-in-package for vital-sign and vibration detection, *IEEE Trans. Microw. Theory Techn.*, 61 (4), 1649–1659, 2013.

48. Y. Yan, L. Cattafesta, C. Li, and J. Lin, Analysis of detection methods and realization of a real-time monitoring RF vibrometer, *IEEE Trans. Microw. Theory Techn.*, 59 (12), 3556–3566, 2011.

49. Y. Yan, L. Cattafesta, C. Li, and J. Lin, Analysis of detection methods of RF vibrometer for complex motion measurement, *IEEE Trans. Microw. Theory Techn.*, 59 (12), 3556–3566, 2011.

50. T.-Y. Huang, L. Hayward, and J. Lin, Non-invasive measurement and analysis of laboratory rat's cardiorespiratory movement, *IEEE Trans. Microw. Theory Techn.*, 65, 574–581.

51. C. Li, J. Lin, and Y. Xiao, Robust overnight monitoring of human vital signs by a non-contact respiration and heartbeat detector, *IEEE Engineering in Medicine and Biology Society*, pp. 2235–2238, September 2006.

52. A. Rahman, E. Yavari, A. Singh, V.-M. Lubecke, and O. Boric-Lubecke, A low-IF tag-based motion compensation technique for mobile Doppler radar life signs monitoring, *IEEE Trans. Microw. Theory Techn.*, 63, 3034–3041, 2015.

53. V. Pasquali, E. Scannapieco, and P. Renzi, Validation of a microwave radar system for the monitoring of locomotor activity in mice, *J. Circadian Rhythms*, 4, 1–7, 2006.

54. L. Anishchenko and E. Gaysina, Comparison of 4 GHz and 14 GHz SFCW radars in measuring of small laboratory animals vital signs, *IEEE International Conference on Microwaves, Communications, Antennas and Electronic Systems*, pp. 1–3, 2015.

10

Multiple-Input Multiple-Output Radar for Monitoring of Bed-Ridden Patients

Chi Xu and Jeffrey Krolik

CONTENTS

10.1 Introduction

Human motion monitoring has become a highly active topic in various research areas because of its potential for understanding people's activities, intents, and even health statuses. The ability to continuously and consistently monitor human motion is an important function in numerous applications, including surveillance, control, and analysis [1]. A typical surveillance application is abnormality detection in public areas. Human–computer interface is a well-known example of control applications. In terms of analysis applications, patient diagnostics and athletic performance analysis are receiving more and more attention from both researchers and practitioners [2–4].

Among all the applications of human motion monitoring mentioned previously, this chapter specifically addresses the issue of patient pressure ulcer prevention, which is one of the patient monitoring applications. However, the systems and methods discussed here can be applied to other applications as well, such as fall detection [5–7] and monitoring of human activities in supermarkets or malls for commercial interests. In the literature, there are three main types of sensors that have been used for human motion monitoring, including body-fixed sensors [2,3], optical cameras [8,9], and radiofrequency (RF) sensors [7,10,11]. In this chapter, we propose RF-based systems and methods for monitoring motions of multiple human targets in wide-area indoor environments, with the application of monitoring bed-ridden patients for pressure ulcer prevention in long-term care nursing homes.

10.1.1 Pressure Ulcer Prevention

"Pressure ulcers are localized areas of soft-tissue injury resulting from compression between a bony prominence and an external surface and can lead to septic infection and premature death" [12]. By definition, pressure ulcers occur most commonly in elderly people due to their limited mobility [13], and nursing home residence is a major risk [14]. Evidence shows that at least one of every nine patients in nursing homes is experiencing a pressure ulcer [15]. The cost to treat pressure ulcers is estimated to be from $21,000 to $152,000 per case [16], which makes low-cost prevention a priority in most U.S. nursing homes. Also, the number of pressure ulcers can increase in the future, because one in every five adults will be more than 65 years by 2030 [13]. Therefore, pressure ulcer prevention is an important issue for nursing homes nowadays and in the future, and effective prevention can reduce the number of pressure ulcers and thus avoid expensive treatment costs.

The existing pressure ulcer prevention standard is repositioning bed-ridden patients every 2 hours by the nursing staff, if the patients do not turn over themselves [17]. Despite several assessment methods for evaluating the risk of developing pressure ulcers [18], automatic detection of patient repositioning is helpful in long-term care facilities for obtaining reliable repositioning records. Current patient monitoring methods quantify the activity level and classify the motion of a bed-ridden patient using either patient-worn accelerometers [2] or bed-installed pressure sensors [4], both of which require deployment on an individual basis, which is challenging in nursing home settings with a large number of patients. In addition, a wearable device can cause pressure ulcers to the patient by the device itself because it is attached to the human body for 24/7 monitoring. Therefore, methods that can monitor multiple targets remotely (sometimes called *nonintrusive* [4] or *device free* [19]) are highly demanded in nursing homes to help patient pressure ulcer prevention. Ruling out wearable sensors, existing human motion monitoring methods include two categories: video based (optical) and RF based (microwave).

10.1.2 RF-Based Methods

Video-based methods typically extract global or local representations of video frames using spatial–temporal features [9]. Data-dependent training is usually necessary in these methods, which have been successfully used for classifying various human motions including walking, jogging, running, boxing, hand waving, sitting down, turning around, jumping in place, and jumping jack. Higher level activities such as checking watch, scratching head, answering phone, and hugging can also be recognized from camera measurements [20]. However, video-based methods may become less reliable under less-than-ideal illuminations or in situations where the targets are substantially occluded. In contrast to video-based methods, RF-based methods are more robust to less-than-ideal illuminations and occlusions, which often occur in wide-area indoor environments. RF-based human motion monitoring methods classify different motions based on their time-varying micro-Doppler signatures. When a point target is moving with a constant velocity, the backscattered signal will contain a frequency shift proportional to the target radial velocity, which is known as the *Doppler effect* (Chapter 3). If the target's movement is nonrigid, that is, body parts of the target are moving with different velocities, then the backscattered signal will contain a Doppler spread around the main frequency shift. The spread is the so-called micro-Doppler signature [21], which contains unique characteristics of the target motion. The signature changes as a function of time when the target moves continuously, and time–frequency analysis such as the spectrogram can be used to extract the unique features for classifying the motion [22].

In the literature, most RF-based human motion classification researches only include the results for monitoring one single target [11,23–25]. To achieve this, a continuous-wave radar is used to measure the micro-Doppler signature from a moving person. In order to monitor the motions of multiple people simultaneously in a wide-area, multiple transmitters and/or multiple receivers are needed, and the probing signals should be wideband. On the one hand, if the transmitters and receivers are coherent, that is, all sensors are time synchronized and phase calibrated, array processing can be applied to resolve the motion characteristics at different locations [26,27]. In the coherent system, sensors need to be placed to form an array with half-wavelength spacing. This setup is often referred as *monostatic*, as shown in Figure 10.1. As a result, a larger array aperture (which gives better angular resolution) needs more coherent sensors/channels, which increases the complexity of the system and thus not practical. On the other hand, if the transmitters and receivers are not coherent, signals sampled in each channel should be processed noncoherently to estimate the location and velocity of the targets [28,29]. In the noncoherent system, sensors have independent clocks and different phases, which makes such a system low cost and thus more practical. This setup is often referred as *multistatic*, as shown in Figure 10.2. In this chapter, both systems are discussed and evaluated with real-data experiments for monitoring patient repositioning activities.

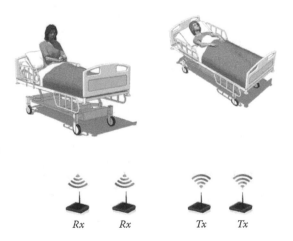

FIGURE 10.1
Coherent monostatic RF sensing setup for wide-area indoor human motion monitoring.

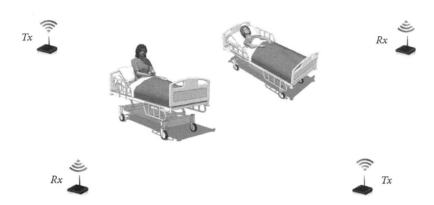

FIGURE 10.2
Noncoherent multistatic RF sensing setup for wide-area indoor human motion monitoring.

10.2 RF Monitoring Using a Coherent Monostatic Multiple-Input Multiple-Output Array

This section discusses the coherent RF system for monitoring multiple human motions in a wide area. In Reference 26, a sensor array was used to separate the Doppler characteristics of multiple targets for detecting and tracking multiple human movers. Array processing separates the targets in signal direction based on the phases across the array spatial samples. In particular, an incoming planewave signal with some angle θ with respect to the broadside of the array provides phase shifts across the array sensors. The phase shifts form a complex exponential array wavefront for a uniform linear array (ULA), and the spatial frequency is a linear function of sin θ. Therefore, the direction of arrival (DoA) of the signal can be used to separate the targets at different locations in a wide area in addition to the time delay. These arrays are called phased arrays, and estimating the spatial frequency is referred as beamforming because it steers the look direction of the array. However, a major challenge for single-input multiple-output (SIMO) systems, for example, phased-array receivers, in indoor environments is the difficulty in discriminating direct-path target returns from multipath ghost returns. A multiple-input multiple-output (MIMO) system, however, employs multiple transmit elements, each transmitting an orthogonal waveform. This allows beamforming of both the transmit and receive arrays by the receiver to discern both the direction of departure (DoD) and the DoA of a signal. The key advantage of MIMO compared to SIMO is the ability to discriminate both the DoD and the DoA via so-called noncausal beamforming [30], which allows for design of the transmit beampattern after signal reception [31]. The spatial diversity provided by MIMO leads to more reliable target detection, more accurate parameter estimation, and decreased minimum detectable velocity [32]. Furthermore, MIMO was shown to increase the degree of freedom (DOF) by a factor of approximately K compared to SIMO, where K is the number of transmit channels [33].

In Reference 34, the authors used an ultra-wideband MIMO probe to mitigate ghost motion returns due to indoor multipath scattering. However, conventional nonadaptive methods for ranging and beamforming degrade dramatically when the signal bandwidth is limited. Although some range resolution improvement can be achieved by trading sidelobe level for mainlobe width, high sidelobe levels result in motion artifacts that are easily confused with

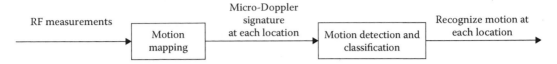

FIGURE 10.3
Diagram of wide-area indoor human motion monitoring. The micro-Doppler signature of each target at a specific location is resolved, and then motion detection and classification are applied to each target separately.

weaker targets in indoor environments. Typically, indoor or urban RF localization and imaging methods require ultra-wideband signals to achieve the time-delay resolution required for separation of returns in complex multipath environments [35–38]. Due to restrictions on spectrum use, however, commercially viable systems must operate using smaller bandwidth, for example, for use in the industrial, scientific, and medical (ISM) bands.

In Reference 39, space-delay adaptive processing (SDAP) is proposed for linear MIMO *Tx–Rx* arrays with the aim of mitigating ghost motion returns due to indoor multipath scattering and bandwidth limitation. In particular, the Doppler characteristics (micro-Doppler signatures) of multiple targets at different locations are resolved by jointly processing the fast-time samples (i.e., those within a pulse) and array spatial samples (i.e., those at each antenna). Although adaptive delay and adaptive array processing can be applied sequentially for motion mapping, this separable processing is suboptimal in multipath environments because of the complex coupling between the delay and the direction of a signal. Therefore, SDAP is proposed here to suppress nonseparable returns and resolve micro-Doppler signatures from multiple targets. This idea is related to space–time adaptive processing [40], which is used on moving platforms to suppress nonseparable clutter return in Doppler and angle. Given accurate motion detection/localization, motion classification methods based on micro-Doppler analysis are applied separately to each location where motion has been detected. The proposed method includes two steps: (1) resolve the micro-Doppler signature at each location in a wide area (motion mapping) and (2) detect/localize and classify human motion at each location using the signature obtained in the first step. The processing flow is shown in Figure 10.3.

10.2.1 Signal Model

Consider a MIMO array transmitting a sequence of M linear frequency-modulated (LFM) pulses in each coherent processing interval (CPI). The array has K transmitters and N receivers for a total of NK *Tx–Rx* pairs. For general MIMO systems, the signals transmitted from different transmit elements are orthogonal to each other to enable transmit channel separation after signal reception by the receiver. In this chapter, we use the slow-time MIMO waveform as the probing signals, in which the transmitted signal from the kth transmitter is given by

$$s(t,k) = \alpha \sum_{m=0}^{M-1} u(t - mT_r)e^{j2\pi\frac{k}{KT_r}t} \tag{10.1}$$

where:
$u(t) = e^{j2\pi\left(f_c - \frac{f_b}{2}\right)t} e^{j\pi\frac{f_b}{T_r}t^2}$ is the LFM waveform
f_c and f_b are the LFM center frequency and bandwidth, respectively
α is the transmitter gain
T_r is the pulse repetition interval

The signal $s(t,k)$ from each transmit element is modulated with $e^{j2\pi(k/KT_r)t}$ for $k = 0,1,\ldots,K-1$, and thus orthogonal (modulated with sinusoids at different frequencies) and can be separated by the receiver. Assuming L samples in each received pulse, the total return during a CPI after MIMO separation can be built into an $L \times M \times N \times K$ four-dimensional (4D) data hypercube. To model the 4D signal, let $\tau(m,n,k)$ denote the time delay from the kth transmitter to a far-field point target and back to the nth receiver at the mth pulse. The returned signal after dechirp processing is given by

$$z(l,m,n,k) \propto e^{-j2\pi\left(f_c - \frac{f_b}{2}\right)\tau(m,n,k)} e^{-j2\pi\frac{f_b}{T_r}\tau(m,n,k)T_s l} \tag{10.2}$$

where:
$l = 0,1,\ldots,L-1$ denotes the fast-time samples
$m = 0,1,\ldots,M-1$ denotes the pulses
$n = 0,1,\ldots,N-1$ denotes the receivers
$k = 0,1,\ldots,K-1$ denotes the transmitters

Here, we have omitted the leading constant for convenience. In Equation 10.2, T_s is the sampling interval. The number of fast-time samples in each pulse is therefore $L = T_r/T_s$, and the CPI is $T_c = MT_r$. It is important to choose small enough T_c to ensure the scene is approximately stationary during the CPI. The time delay $\tau(m,n,k)$ consists of a group delay $\bar{\tau}(m)$ between the target and the phase center of the Tx–Rx array and a relative phase delay between the transmit and receive elements. The phase delay can be further decomposed into receive array delay $\tilde{\tau}_r(n)$ and transmit array delay $\tilde{\tau}_t(k)$:

$$\tau(m,n,k) = \bar{\tau}(m) + \tilde{\tau}_r(n) + \tilde{\tau}_t(k)$$
$$= \bar{\tau}(m) + \frac{nd_r\sin\theta_r}{c} + \frac{kd_t\sin\theta_t}{c} \tag{10.3}$$

where:
d_r and d_t are the inter-element spacings of the receive and transmit arrays (both are ULA's), respectively
θ_r and θ_t are the DoA and DoD of the signal, respectively
c is the speed of light

In Equation 10.3, we use the 0th receiver and 0th transmitter as the phase center. Assuming that the point target is in the far field from the array and the waveform is sufficiently narrowband ($f_c \gg f_b$), the signal (Equation 10.2) can be approximated as

$$z(l,m,n,k) \propto e^{-j2\pi\frac{f_b}{T_r}\tau_d T_s l} e^{-j2\pi f_c \bar{\tau}(m)} e^{-j2\pi f_c \frac{nd_r\sin\theta_r}{c}} e^{-j2\pi f_c \frac{kd_t\sin\theta_t}{c}} \tag{10.4}$$

where we assume the target time delay τ_d is approximately the same for each m, n, k for target ranging. Therefore, the return from a far-field point target is a 4D complex exponential. More complicated targets and environments can be modeled by an ensemble of individual point targets.

The main idea of SDAP is to process fast-time samples and array spatial samples jointly to suppress the interference that is nonseparable in signal delay and direction. The signal snapshot for SDAP is defined as a vectorized version of $z(l,m,n,k)$:

$$z(m) = \begin{bmatrix} z(0,m,0,0) \\ z(0,m,0,1) \\ \vdots \\ z(0,m,0,K-1) \\ z(0,m,1,0) \\ z(0,m,1,1) \\ \vdots \\ z(0,m,N-1,K-1) \\ z(1,m,0,0) \\ z(1,m,0,1) \\ \vdots \\ z(L-1,m,N-1,K-1) \end{bmatrix} \tag{10.5}$$

The snapshot (Equation 10.5) is an $LNK \times 1$ vector consisting of the fast-time samples at all receivers from all transmitters at the mth pulse, which is proportional to the Kronecker product of the fast-time vector $a_d(\tau_d)$ (the vector consisting of pulse samples at each receiver from each transmitter), the receive array vector $a_r(\theta_r)$ (wavefront across receivers), and the transmit array vector $a_t(\theta_t)$ (wavefront across transmitters):

$$z(m) \propto e^{-j2\pi f_c \tau(m)} a_d(\tau_d) \otimes a_r(\theta_r) \otimes a_t(\theta_t) \tag{10.6}$$

where:

$$a_d(\tau_d) = \begin{bmatrix} 1, e^{-j2\pi \frac{f_b}{T_r}\tau_d T_s 1}, \ldots, e^{-j2\pi \frac{f_b}{T_r}\tau_d T_s (L-1)} \end{bmatrix}^T \tag{10.7}$$

and

$$a_r(\theta_r) = \begin{bmatrix} 1, e^{-j2\pi f_c \frac{d_r \sin\theta_r}{c} 1}, \ldots, e^{-j2\pi f_c \frac{d_r \sin\theta_r}{c}(N-1)} \end{bmatrix}^T \tag{10.8}$$

$$a_t(\theta_t) = \begin{bmatrix} 1, e^{-j2\pi f_c \frac{d_t \sin\theta_t}{c} 1}, \ldots, e^{-j2\pi f_c \frac{d_t \sin\theta_t}{c}(K-1)} \end{bmatrix}^T \tag{10.9}$$

The SDAP snapshot (Equation 10.6) is for a particular mode of returns from a point target with respect to the Tx–Rx array. In indoor environments, the total return $x(m)$ includes direct-path signals $z(m)$, multipath returns $p(m)$, clutter return $c(m)$, and noise $n(m)$:

$$x(m) = \sum_{i=1}^{N_s} z(m)^{(i)} + \sum_{i=1}^{N_p} p(m)^{(i)} + c(m) + n(m) \tag{10.10}$$

where N_s and N_p are the number of direct-path signals and multipath returns, respectively. The direct-path signal $z(m)$ is the line-of-sight (LOS) return from a moving target and thus has identical DoD and DoA, which can be modeled as Equation 10.6 with $\theta_r = \theta_t$. In indoor environments, the transmitted signal can propagate to moving targets and bounce off the walls, floors, and ceilings for multiple times before being received. Different from direct-path signals, the non-LOS multipath return $p(m)$ can have arbitrary DoD and DoA. Therefore, multipath returns with different DoDs and DoAs can be discriminated from direct-path targets and thus mitigated using MIMO. For fixed transmit and receive systems, the clutter return $c(m)$, that is, backscatter from nonmoving objects, is at zero Doppler and can be removed by high-pass filtering over pulses. With the clutter removed, the goal is to resolve the micro-Doppler signature as a function of range (delay) and angle (direction) for LOS targets at each location and mitigate ghost returns due to indoor multipath scattering and filter sidelobes.

10.2.2 Motion Mapping Using Space-Delay Adaptive Processing

As discussed previously, the clutter returns are at zero Doppler and can be removed using a high-pass filter. Let $h(m)$ denote the impulse response of a high-pass filter. The *clutter-free* SDAP snapshot $y(m)$ is given by

$$y(m) = h(m) * x(m) \tag{10.11}$$

where * denotes convolution. We can choose a cutoff frequency for the high-pass filter by assuming some amount of Doppler spread of the nonmoving indoor environment.

The goal of motion mapping is designing space-delay filter weights $w(\tau, \theta)$ to resolve the micro-Doppler signature at a specific location in a wide area while suppressing Doppler contributions from moving targets at other locations through multipath/sidelobes. The micro-Doppler signature is resolved at each range τ and angle θ for wide-area motion monitoring. The processing flow is shown in Figure 10.4. Unlike conventional range or array processing, the proposed SDAP method jointly performs ranging and beamforming by using a filter on the signal vector (Equation 10.6). As a result, SDAP can resolve micro-Doppler signatures at each location while suppressing sidelobes of nonseparable returns from other targets/multipath. To design the weights that both filter the data to resolve the return from a hypothesized motion location and put nulls at locations of other moving targets, adaptive processing is applied, which uses the covariance matrix of the SDAP snapshot. The well-known Wiener filtering problem tells us that the optimal filter weights are obtained by minimizing the expected power of the filter output for interference and noise, and are given by

$$w_{\text{opt}}(\tau, \theta) = R_u(\tau, \theta)^{-1} d(\tau, \theta) \tag{10.12}$$

where:

$R_u(\tau, \theta)$ is the covariance matrix of interference and noise (nontarget components)
$d(\tau, \theta) = a_d(\tau) \otimes a_r(\theta) \otimes a_t(\theta)$ is the target signal vector for a specific location

The target signal vector $d(\tau, \theta)$ is used for matched-filtering the data snapshot to resolve the return from a specific location. To suppress the contributions from other moving targets, the covariance matrix $R_u(\tau, \theta)$ is required in Wiener filtering. In this application, interference includes returns from other targets as well as multipath returns from the target at the

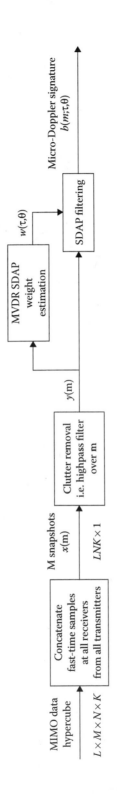

FIGURE 10.4
Diagram of SDAP. LOS returns from moving targets at each location are resolved by SDAP, and the obtained micro-Doppler signature is a function of range and angle for motion detection and classification at each location.

hypothesized location. However, because the measured data snapshot given by Equation 10.10 contains direct-path and multipath returns from moving targets at all locations, the *target-free* covariance matrix $R_u(\tau,\theta)$ is unavailable. Therefore, the data covariance matrix R_y is used and the minimum variance distortionless response (MVDR) weights are obtained by solving the constrained optimization problem:

$$\min_w w^H R_y w, \text{ subject to } w^H d(\tau,\theta) = 1 \tag{10.13}$$

with solution given by

$$w_{\text{mvdr}}(\tau,\theta) = \frac{R_y^{-1} d(\tau,\theta)}{d(\tau,\theta)^H R_y^{-1} d(\tau,\theta)} \tag{10.14}$$

The MVDR-based SDAP weights (Equation 10.14) minimize the expected power of all returns from a wide area but ensure that the return from the hypothesized target location is fixed. Consequently, the filter response of Equation 10.14 has a fixed gain at (τ,θ) and puts nulls at other ranges and angles. Although MVDR filters have been used for high-resolution frequency estimation and adaptive array processing, this section develops an MVDR-based joint spatial–temporal filter to deal with the complicated and nonseparable returns for indoor motion mapping. To obtain the solution given by Equation 10.14, the covariance matrix R_y is required. Due to the high-dimensionality of R_y, which is an $LNK \times LNK$ matrix, the MVDR-based solution is impractical as an accurate estimate of R_y and its inverse is needed. Therefore, we propose a low-rank SDAP solution, which is based on the idea of beamspace beamforming.

Here, we propose a general low-rank SDAP weight vector given by

$$w_{\text{lr}}(\tau,\theta) = U(\tau,\theta) r(\tau,\theta) \tag{10.15}$$

where:
  $U(\tau,\theta)$ is an $LNK \times D$ matrix with orthonormal columns
  $r(\tau,\theta)$ is a $D \times 1$ vector

In Equation 10.15, we assume that the SDAP weight vector can be sufficiently approximated by a linear combination of D orthonormal basis vectors. In other words, the SDAP weights lie in some D-dimensional subspace instead of being an arbitrary vector. The orthonormal matrix $U(\tau,\theta)$ can be chosen in a similar manner to those used in [39]. Also, the data covariance matrix R_y is estimated using the snapshots over pulses in every CPI [39]:

$$\hat{R}_y = \frac{1}{M} \sum_{m=0}^{M-1} y(m) y(m)^H \tag{10.16}$$

Using the $U(\tau,\theta)$ we choose and the estimated data covariance matrix, the optimal $r(\tau,\theta)$ can then be obtained. Finally, the micro-Doppler signature at a specific location (τ,θ) is given by

$$b(m;\tau,\theta) = w_{\text{lr}}(\tau,\theta)^H y(m) \tag{10.17}$$

10.2.3 Motion Detection/Localization and Classification

Given accurate resolved micro-Doppler signature $b(m; \tau, \theta)$ at each location in a wide area, motion detection/localization and classification are achieved based on the analysis of $b(m; \tau, \theta)$. The processing flow is shown in Figure 10.5. In this section, a motion event is detected first, and then micro-Doppler analysis is used for motion classification.

In wide-area indoor environments, a moving target at a specific location can generate motion returns at further ranges that are actually ghost returns due to multipath scattering. Without prior knowledge of the environment and target locations, ghost returns can be misinterpreted as motion generated by another moving target and create false alarms in motion detection. These multipath returns may have different DoDs and DoAs, so the use of MIMO mitigates these returns and enhances the performance of motion detection when multiple targets are present. In addition, filter sidelobes are suppressed using SDAP, which further improves the accuracy of motion detection for multitarget scenarios.

Denote $b(m; \tau, \theta)$ for $m = 0, 1, \ldots, M-1$ in vector form as

$$\boldsymbol{b}(\tau, \theta) = \left[b(0; \tau, \theta), b(1; \tau, \theta), \ldots, b(M-1; \tau, \theta) \right]^{T} \tag{10.18}$$

When a target moves at location (τ, θ), the micro-Doppler signature $\boldsymbol{b}(\tau, \theta)$ will exhibit motion features and have nonzero power. On the contrary, if no motion occurs, $\boldsymbol{b}(\tau, \theta)$ will only contain noise because the clutter return has been removed. Therefore, the two hypotheses of motion detection are given by

$$H_1 : \boldsymbol{b}(\tau, \theta) = \boldsymbol{s}(\tau, \theta) + \boldsymbol{n}(\tau, \theta) \tag{10.19}$$

$$H_0 : \boldsymbol{b}(\tau, \theta) = \boldsymbol{n}(\tau, \theta) \tag{10.20}$$

where:
$\boldsymbol{n}(\tau, \theta) \sim CN(0, \sigma_n^2 \boldsymbol{I})$ is assumed to be circularly symmetric complex Gaussian noise
$\boldsymbol{s}(\tau, \theta)$ is the micro-Doppler signal due to motion

Due to limited range and angular resolution, the micro-Doppler signal comprises the Doppler signatures of multiple body components of the target at the specific location. Assuming the velocity of each body component is constant during the CPI, the signal $\boldsymbol{s}(\tau, \theta)$

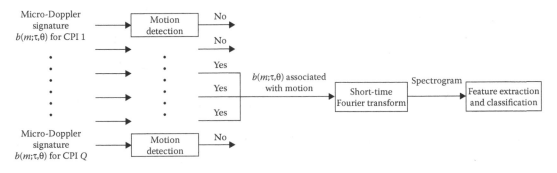

FIGURE 10.5
Diagram of motion detection and classification. The micro-Doppler signatures corresponding to a motion are selected and then used to classify the motion.

can be modeled as a sum of complex exponentials with different amplitudes and Doppler frequencies (except for the zero-Doppler component, which has been filtered out):

$$\alpha(\tau,\theta) = \boldsymbol{F}s(\tau,\theta) \tag{10.21}$$

$$s(\tau,\theta) = \boldsymbol{F}^{H}\alpha(\tau,\theta) \tag{10.22}$$

where:
   $\boldsymbol{F}$ is a unitary discrete Fourier transform (DFT) matrix with the zero-frequency column removed
   $\alpha(\tau,\theta)$ consists of the Fourier coefficients

To detect a general motion event, assume that $\alpha(\tau,\theta) \sim CN(0,\sigma_s^2 \boldsymbol{I})$. In this case, $s(\tau,\theta)$ is also zero-mean complex Gaussian with covariance matrix given by

$$\boldsymbol{R}_s = \boldsymbol{F}^{H}(\sigma_s^2 \boldsymbol{I})\boldsymbol{F} = \sigma_s^2 \boldsymbol{I} \tag{10.23}$$

Therefore, $s(\tau,\theta)$ is white Gaussian. It is well known that the optimal detector of white Gaussian signals in white Gaussian noise is the energy detector given by

$$E(\tau,\theta) = b(\tau,\theta)^{H} b(\tau,\theta) \tag{10.24}$$

The micro-Doppler signature $b(m;\tau,\theta)$ at each location is obtained by jointly processing the fast-time samples and array spatial samples, and motion events are detected based on the power of $b(m;\tau,\theta)$. Once a motion event is detected at a specific location, time–frequency analysis is used to extract the motion features for classification. In this section, the short-time Fourier transform (STFT) is used for the time–frequency analysis. The spectrogram is the power of the STFT output (Chapters 3 through 5]:

$$S(t_s, f_D; \tau,\theta) = \left| \sum_{m} b(m;\tau,\theta)w(m - t_s)e^{-j2\pi f_D m T_r} \right|^2 \tag{10.25}$$

where:
   $w(m)$ is a general sliding window (a Hamming window is used in this section)
   f_D is the Doppler frequency

The spectrogram represents the signal in the time–frequency domain, which details the time-varying Doppler characteristics of the moving target. Motion features are extracted from the spectrogram to classify different types of motion.

In this section, six motion features are extracted from the spectrogram for classifying different types of human motion relevant to patient repositioning activities. First, the spatial extent that the motion covers in the (τ,θ) space is estimated. In addition, five features across the slow-time (Doppler) dimension are included. The five features are illustrated in Figure 10.6. Although identifying appropriate features is important in any classification algorithm, optimizing over the set of features is beyond the scope of this chapter. Furthermore, the most relevant features will be application dependent. The selected features provide acceptable classification rates for recognizing patient turnovers from arm and leg movements, which is relevant in pressure ulcer prevention.

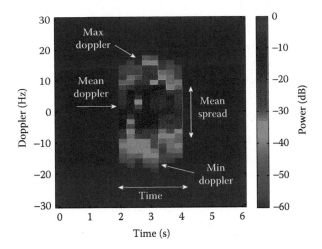

FIGURE 10.6
Motion features extracted from the spectrogram for patient repositioning activity recognition.

To evaluate the ability to classify various patient movements in a wide area using the proposed method, several classifiers are used along with the features discussed previously. In particular, the logistic discriminant classifier, the naive Bayes classifier, the k-nearest neighborhood classifier, and the support vector machine (SVM) classifier are used in this section. We use the pattern recognition toolbox (PRT) implementation, which is a set of MATLAB-compatible codes available online for classification. Interestingly, these classifiers provide similar classification rates assuming accurate motion detection and localization. Among all classifiers, the SVM provides the most accurate classification results. More details are found in Section 10.4.

10.3 RF Monitoring Using Noncoherent Multistatic Sensors

This section discusses the noncoherent RF system for monitoring the motions of multiple human targets using distributed sensors in wide-area indoor environments. Although it has been shown that using coherent arrays with wideband signals that can monitor multiple human targets in a wide area, there are several disadvantages for the coherent system. First of all, the hardware complexity increases dramatically as the number of sensors increases because the coherent system requires time synchronization and phase calibration among all transmitters and receivers. As a result, it is impractical to have a large array of sensors to achieve a desired cross-range resolution [41]. Moreover, array processing methods require precise sensor placements, and their beamforming performances can degrade if there are errors in sensor locations. In contrast, a noncoherent system does not require precise time synchronization or phase calibration among the sensors. Also, noncoherent systems allow for more flexible sensor placements, which can be used to capture target features from different aspects and can be more robust to occlusions in highly cluttered environments, such as indoor environments. Therefore, a noncoherent system is promising to provide better detection, localization, and classification using fewer sensors and much cheaper implementations than that is needed in coherent systems [42].

The noncoherent system discussed here is related to RF sensor networks, which have been used for device-free localization [19]. Ultra-wideband (UWB) sensor networks use multistatic transmitters and receivers to monitor targets in a wide area, and target localization is achieved using the bistatic range measurements from all Tx–Rx pairs. Because the sensors have independent clocks, the range difference between the direct blast path from the transmitter to the receiver and the bistatic target path is estimated, and the bistatic range of the target can then be obtained given the locations of the transmitter and receiver. Estimating the range difference is equivalent to synchronizing the transmitter and receiver to some level using the direct blast signal, which is sometimes referred as *difference-time-of-arrival* estimation [43]. Based on the bistatic range of the target, an ellipse in which the target can locate is obtained for each Tx–Rx pair, and multiple ellipses obtained from multiple pairs are combined to localize the target. Here, we introduce a noncoherent system to localize and classify human motion in a wide-area indoor environment using multistatic/distributed RF sensors. Motion detection/localization is achieved by noncoherently combining bistatic range profiles from all Tx–Rx pairs, and motion classification is achieved by noncoherently combining micro-Doppler signatures from the multistatic sensor pairs. The proposed method includes two steps: (1) detect and localize motions from multiple targets and then (2) classify the motion for each target. The processing flows of the two steps are shown in Figures 10.7 and 10.8, respectively.

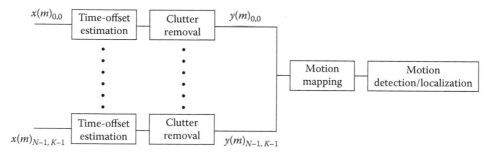

FIGURE 10.7
Diagram of noncoherent motion mapping. Time offset between each Tx–Rx pair is estimated, and clutter return is removed before noncoherently combining range profiles from multiple Tx–Rx pairs for motion mapping.

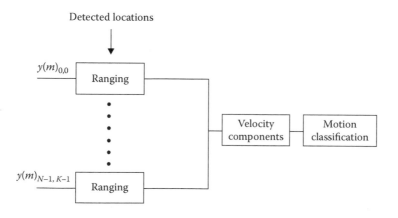

FIGURE 10.8
Diagram of noncoherent motion classification. Velocity components are resolved for each target location by noncoherently combining the micro-Doppler signatures from multiple Tx–Rx pairs for motion classification.

10.3.1 Signal Model

Consider a noncoherent MIMO system with K transmitters and N receivers that are distributed in a wide-area indoor environment. Again, we use the slow-time MIMO waveform as the probing signal, in which the transmitted signal from the kth transmitter is given by

$$s(t)_k = \alpha_k \sum_{m=0}^{M-1} u(t - \eta_k - mT_r) e^{j2\pi \frac{k}{KT_r}t} \tag{10.26}$$

where η_k is the sensor time for the kth transmitter. For coherent processing, the transmitter gain α_k and the sensor time η_k should be the same (or known) for all transmitters. For the noncoherent case, these parameters can be different and are unknown to all transmitters and receivers. Assuming L samples in each received pulse, the returned signal from the kth transmitter to the nth receiver can be organized into an $L \times M$ data matrix $z(l,m)_{n,k}$. Because the signals for different Tx–Rx pairs need to be processed separately and then combined noncoherently, we omit the transmitter index k and the receiver index n for notation simplicity. To model $z(l,m)$, consider the return from a point target after dechirp processing:

$$z(l,m) \propto e^{-j2\pi\left(f_c - \frac{f_b}{2}\right)(\tau(m)+\Delta)} e^{-j2\pi \frac{f_b}{T_r} \tilde{\tau}_d T_s l} \tag{10.27}$$

where:

$$\tilde{\tau}_d = \tau_d + \Delta \approx \tau(m) + \Delta \tag{10.28}$$

where:
τ_d is the true bistatic delay
Δ is the time offset between the transmitter and the receiver

The phase of $z(l,m)$ over pulses, that is, $e^{-j2\pi\left(f_c - \frac{f_b}{2}\right)(\tau(m)+\Delta)}$, is changing due to the varying bistatic delay $\tau(m)$ of the target; thus, it can be used to separate a moving target from a nonmoving clutter and to classify the target's motion. In order to estimate the true bistatic delay of the target, we need to estimate the time offset Δ for each Tx–Rx pair. This is discussed in Section 10.3.2. Assume that Δ is known and define the signal vector:

$$z(m) = \begin{bmatrix} z(0,m) \\ z(1,m) \\ \vdots \\ z(L-1,m) \end{bmatrix} \propto a_d(\tilde{\tau}_d) \tag{10.29}$$

where:

$$a_d(\tilde{\tau}_d) = \begin{bmatrix} 1, & e^{-j2\pi \frac{f_b}{T_r} \tilde{\tau}_d T_s 1}, \ldots, & e^{-j2\pi \frac{f_b}{T_r} \tilde{\tau}_d T_s (L-1)} \end{bmatrix}^T \tag{10.30}$$

A more explicit relationship between the target location $d_s = [d_x, d_y]^T$ on a two-dimensional plane and the returned signal can be obtained by defining

$$h(d_s) = a_d(\tau_d + \Delta) \tag{10.31}$$

and the true bistatic delay is given by

$$\tau_d = \frac{\|d_t - d_s\| + \|d_s - d_r\|}{c} \tag{10.32}$$

where d_t and d_r are the locations of the transmitter and receiver, respectively. In indoor environments, the total return $x(m)$ is the sum of the direct blast return, s, and returns from reflections off the ensemble of point targets $z(m)$, including nonmoving objects (walls, floors, ceilings) and motion returns. Also, there are ghost motion returns $p(m)$ due to indoor multipath scattering and noise $n(m)$. To model the total return, we grid the space (d_x, d_y) into $N_s = N_x \times N_y$ locations, and $x(m)$ can be written as

$$x(m) = s + \sum_{i=1}^{N_s} z(m)^{(i)} + \sum_{i=1}^{N_p} p(m)^{(i)} + n(m) \tag{10.33}$$

where the multipath returns come from none of the N_s locations, but from paths with multiple bounces off the environment. These multipath returns will have larger time delays than the moving target from which it bounces off, which can be mitigated based on the fact that the extra bounces give bistatic delays that are not consistent with multiple Tx–Rx pairs. Therefore, we expect improved multipath mitigation using more sensor pairs.

10.3.2 Motion Mapping for Motion Detection/Localization

The time offset between each Tx–Rx pair needs to be estimated from the direct blast signal for localizing the moving targets. Assuming that the direct blast signal s is present in the received signal, it can be used to estimate Δ for each Tx–Rx pair, if the transmitter and receiver locations are known. The direct blast signal for each Tx–Rx pair travels the shortest distance from the transmitter to the receiver, and thus has the smallest time delay in the received signal. Therefore, the delay of the direct blast in the received signal can be estimated by finding the first peak in the spectrum of $x(m)_{n,k}$ for the kth transmitter and the nth receiver. Let $t_{n,k}$ denote the estimated delay of the direct blast between the kth transmitter and the nth receiver. Then the time offset between the two sensors is given by

$$\Delta_{n,k} = t_{n,k} - \frac{\|d_t^{(k)} - d_r^{(n)}\|}{c} \tag{10.34}$$

For each Tx–Rx pair, there are returns from the nonmoving clutter in the total received signal $x(m)_{n,k}$, which can be coming from all of the N_s locations on the grid. Similar to the coherent system and method, we apply a high-pass filter over the pulses to remove the clutter return and denote the *clutter-free* signal $y(m)_{n,k}$.

Assuming the time on each sensor is estimated and the clutter return is removed, motion locations for multiple targets can be obtained by estimating the bistatic ranges of the targets for each Tx–Rx pair. The response of a hypothesized target location d_s obtained from the received signal for the kth transmitter and n-th receiver is given by

$$\lambda(m; d_s)_{n,k} = \frac{1}{L} h(d_s)_{n,k}^H y(m)_{n,k} \tag{10.35}$$

Note that the response given by Equation 10.35 is ambiguous because there are different locations that give the same received signal for a single Tx–Rx pair. This is because that the two locations have the same bistatic range to the transmitter and receiver. The equirange contour for a single *Tx–Rx* pair gives an ellipse, with foci being the transmitter and receiver, which is well known for UWB localization. The ambiguities on the ellipse for each single *Tx–Rx* pair can be eliminated by combining the results from different *Tx–Rx* pairs. In this section, this is achieved by calculating the geometric mean of the signal powers obtained from all *Tx–Rx* pairs for each location:

$$E(d_s) = \left(\prod_{n=0}^{N-1} \prod_{k=0}^{K-1} \sum_{m=0}^{M-1} |\lambda(m; d_s)_{n,k}|^2 \right)^{1/(NK)} \tag{10.36}$$

Motion detection and localization are achieved by thresholding $E(d_s)$ and finding peaks on the (d_x, d_y) grid. False alarms may occur because there are locations where a subset of ellipses intersects, which create ghost returns. These ghost return locations can be rejected if the range measurements are perfect, because there are less than *NK* ellipses going through the locations. In other words, these locations are not consistent with all *NK* *Tx–Rx* pairs. However, there are errors in range estimates for the targets, making every location not consistent with all sensor pairs. In this case, people lower the criterion for detecting a target to a least square solution to the bistatic ranges [44]. In this section, we leverage the fact that because less than *NK* ellipses are intersecting at ghost return locations, the powers at these locations are smaller than the strongest target, but can be larger than some of the weaker targets. Therefore, instead of detecting all moving targets based on the motion map for one time, we propose an iterative motion mapping approach in order to reduce the number of false alarms for detecting multiple targets, which is shown in Figure 10.9. Similar ideas can be found in, for example, [45]. In particular, let d_s denote the detected/localized location of the strongest target on the motion map, and we form a grid around d_s by

$$H(d_s) = \left[\ldots, h(d_s + d_\delta), \ldots \right] \tag{10.37}$$

where $d_\delta < R$ defines a circle grid with radius R. In Equation 10.37, the matrix $H(d_s)$ consists of signal wavefronts around d_s. The subspace of signal at location d_s can be obtained by

$$U(d_s) = \left[u(d_s)_1, u(d_s)_2, \ldots, u(d_s)_D \right] \tag{10.38}$$

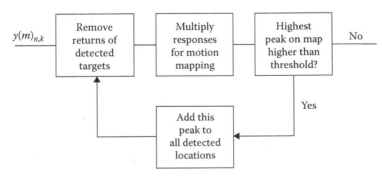

FIGURE 10.9
Diagram of iterative motion mapping for de-ghosting.

which consists of D singular vectors corresponding to the D largest singular values of $H(d_s)$. The signal return from the detected target can be removed by

$$y(m)' = y(m) - \rho U(d_s) U(d_s)^H y(m) \tag{10.39}$$

where ρ determines how much signal return from the first target to suppress for detecting the second target. After removing the signal return from the first (strongest) target, motion detection for the second strongest target can be achieved based on the motion mapping using $y(m)'$. Then the signal return from the second target is removed using the same steps listed previously for detecting the third target. This motion mapping is applied iteratively until nothing but noises exist, that is, the peak on the motion map is not sufficiently large to be discriminated from noises. In addition, there are multipath ghost returns in the received signal. Because the ghost return locations, from either less than NK ellipses intersecting or multipath returns, are less likely to be consistent with multiple Tx–Rx pairs as the true target locations, we expect to suppress and mitigate ghost returns using more sensor pairs.

10.3.3 Motion Classification

After resolving the locations of the targets, motion classification can be achieved using micro-Doppler signatures from the multistatic sensor pairs for each target location. First, velocity components at each target location are resolved, and then motion features are extracted and used in a classifier to classify different types of human motion. Let d_s denote the target location; then the responses at ds for each Tx–Rx pair are given by Equation 10.35. Instead of calculating the signal power of $\lambda(m; d_s)_{n,k}$ for localization as in Equation 10.36, Doppler processing is performed to resolve the velocity components from the target. For a hypothesized velocity $v_s = [v_x, v_y]^T$ at target location d_s, the response for each Tx–Rx pair is given by

$$\gamma(d_s, v_s)_{n,k} = \frac{1}{M} g(v_s)^H \lambda(d_s)_{n,k} \tag{10.40}$$

where:

$$g(v_s) = a_D(f_D) \tag{10.41}$$

where:

$$a_D(f_D) = \left[1, e^{j2\pi f_D T_r 1}, \ldots, e^{j2\pi f_D T_r (M-1)}\right]^T \tag{10.42}$$

$$f_D = \frac{2v_s}{c} f_c \cos(\phi + \beta) \cos(\beta) \tag{10.43}$$

$$\lambda(d_s)_{n,k} = \left[\lambda(0; d_s)_{n,k}, \lambda(1; d_s)_{n,k}, \ldots, \lambda(M-1; d_s)_{n,k}\right]^T \tag{10.44}$$

The bistatic Doppler shift f_D is a function of the target location d_s and the hypothesized velocity v_s, which is given in Equation 10.43, where ϕ and β are the angles shown in Figure 10.10. Similar to the location ambiguities discussed earlier, there are velocity ambiguities for a single Tx–Rx pair as well. In particular, all velocities in two dimensions

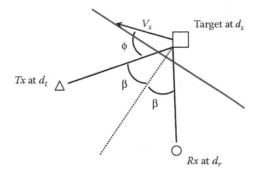

FIGURE 10.10
Bistatic Doppler of a target.

with arrow on the red line in Figure 10.10 will have the same Doppler shift. Although the location ambiguities form an ellipse on the (d_x, d_y) plane, the velocity ambiguities form a line on the (v_x, v_y) plane. Again, these ambiguities can be eliminated by combining the results from multiple *Tx–Rx* pairs:

$$F(d_s, v_s) = \left(\prod_{n=0}^{N-1} \prod_{k=0}^{K-1} |\gamma(d_s, v_s)_{n,k}|^2 \right)^{1(NK)} \tag{10.45}$$

After detecting/localizing multiple targets, the velocity components for each target are resolved using Equation 10.45. Then motion classification can be achieved based on the features from $F(d_s, v_s)$. We extract the features to classify different types of human motion based on our observations of the key differences between repositioning activities and non-repositioning activities. These features are discussed in more details with real data in Section 10.4. Similar to the coherent RF system and method, we use several classifiers to classify the body turning over from lifting an arm and bending a leg. We use the PRT implementations for the classifiers and evaluate the classification rates using cross-validations.

10.4 Experimental Results

10.4.1 Sensor System

To evaluate the proposed method for wide-area indoor motion monitoring, real-data experiments were conducted in the Michael W. Krzyzewski Human Performance Lab and the School of Nursing at Duke University, Durham, North Carolina, which allowed the collection of motion ground truth in realistic environments. In particular, the motion of bed-ridden patients at multiple locations in a wide-area indoor environment was measured and classified. In this chapter, repositioning activities such as turning over were classified from non-repositioning activities such as lifting an arm and bending a leg for patient pressure ulcer prevention.

The MIMO RF system used in the experiment, which is shown in Figure 10.11, was developed in the Sensor Array and Multipath Signal Processing Lab at Duke University. The

FIGURE 10.11
MIMO RF system. There are 4 transmit channels and 16 receive channels.

TABLE 10.1

Experimental Parameters

Center frequency	2.4 GHz
Maximum bandwidth	600 MHz
Transmit power	100 mW
Pulse repetition frequency	200 Hz
CPI	0.5 s

testbed has four transmit channels and 16 receive channels, operating at a center frequency of 2.4 GHz with a maximum of 600 MHz bandwidth. The frequency of the transmitted LFM chirp sweeps from 2.1 to 2.7 GHz in each pulse. Because the system transmits LFM pulses, it is robust to indoor RF interference such as WiFi signals. The transmitting power of the system is 100 mW, so it satisfies Federal Communications Commission (FCC) rules and will not interfere with other users. In the coherent approach, the transmit elements are uniformly spaced by 11.43 cm (approximately twice the Nyquist spacing), and the receive elements are uniformly spaced by 5.715 cm (the Nyquist spacing). The advantage of spacing the transmit array by more than a half-wavelength is a narrower transmit mainlobe. The experimental parameters are summarized in Table 10.1. For a MIMO system with K transmit channels implemented using the double-sideband slow-time MIMO, the highest and lowest Doppler frequencies of the resolved micro-Doppler signature are given by

$$f_D^{\text{high}} = -f_D^{\text{low}} = \frac{1}{4KT_r} = 12.5\,\text{Hz} \tag{10.46}$$

which corresponds to a velocity of about 0.78 m/s. Motion with velocities higher than this value are filtered out during processing. To measure motion with higher velocities, the pulse repetition frequency should be increased, which decreases the maximum unambiguous range. Nevertheless, for measuring patient repositioning-related motion in this study, the maximum Doppler frequency of 12.5 Hz is sufficient.

10.4.2 Coherent Approach

In the experiment, the examining table in Figure 10.12 was used as a surrogate patient bed that was moved to different locations at which RF returns were measured. We also used the infrared motion capture system to obtain the motion ground truth. Measurements at multiple locations were summed to synthesize a multitarget scenario. In particular, the table was moved to four different locations, as shown in Figure 10.13, and three types of

FIGURE 10.12
Motion capture markers to track human motion for ground truth.

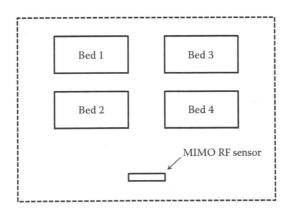

FIGURE 10.13
Geometry of multiple candidate motion locations in a room. Each bed is of size 1 m by 2 m. The lower two beds are about 3 m away from the sensor array, and the upper two beds are about 5 m away from the sensor array.

motion were measured at each location. For each motion, three realizations were measured and each realization was recorded for 30 s. When measuring motion returns, we
made sure that there were no other moving objects outside of the monitoring area. The
background WiFi signals have the potential to cause interference, but the transmitted LFM
pulses were robust to this type of interference.

To compare the different processing methods, the power of micro-Doppler signatures
obtained using each method is plotted as a function of τ and θ. In this section, the full
600 MHz bandwidth measurements are used along with conventional nonadaptive processing. Figure 10.14a shows the motion map generated using the simulated ground truth,
which indicates that there is a true motion happening at about 3 m and −30°. Figure 10.14b
and c shows the motion maps generated using the full-bandwidth (600 MHz) RF measurements with SIMO and MIMO processing, respectively. The SIMO measurements are
obtained by summing the received data over all transmit channels. Both the SIMO and
MIMO maps are normalized for clear comparison. Figure 10.14b shows that in addition to
a return at the true motion location, motion energy is also returned from further ranges
because of the multipath propagation. These returns are actually ghost returns and can
be misinterpreted as motion happening at locations other than the true target location.

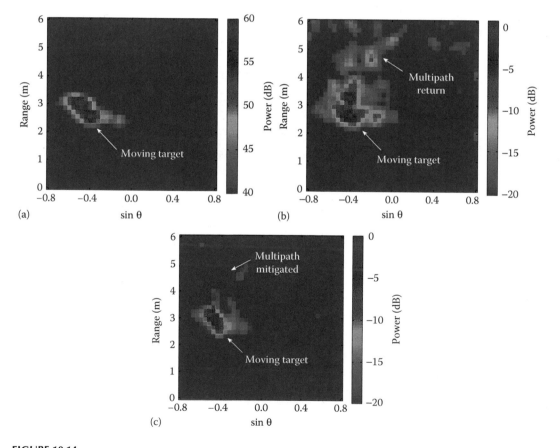

FIGURE 10.14
Motion maps of one moving target obtained using SIMO and MIMO with 600 MHz bandwidth. Note that the
multipath ghost return is mitigated using MIMO. (a) Ground truth obtained using simulated LOS returns,
(b) SIMO processing with 600 MHz bandwidth, and (c) MIMO processing with 600 MHz bandwidth.

The ghost returns are effectively mitigated by MIMO processing as shown in Figure 10.14c, which demonstrates the ability of MIMO to mitigate the multipath in wide-area indoor environments. Statistical comparison between SIMO and MIMO processing will be discussed using signal detection theory.

Due to the spectrum restrictions in the bands likely to be used for such applications, for example, the ISM bands, UWB signals are not practical for industrial and medical applications. Motion monitoring with limited bandwidth is therefore discussed in this section. The limited bandwidth measurements are obtained by processing only a portion of the full-bandwidth measurements corresponding to 150 MHz bandwidth around the center frequency. Note that 150 MHz is not available for the ISM band at 2.4 GHz, but it is available at 5 GHz. Three different processing methods are used along with MIMO: (1) conventional nonadaptive processing, (2) separable adaptive processing, and (3) SDAP. The resulting motion maps are normalized and compared. For all maps, the truth is that there are two motion events happening: one at 3 m and the other at 5 m range. Figure 10.15a shows the motion map obtained using the full-bandwidth measurements, processed by conventional ranging and MIMO beamforming. The large sidelobes

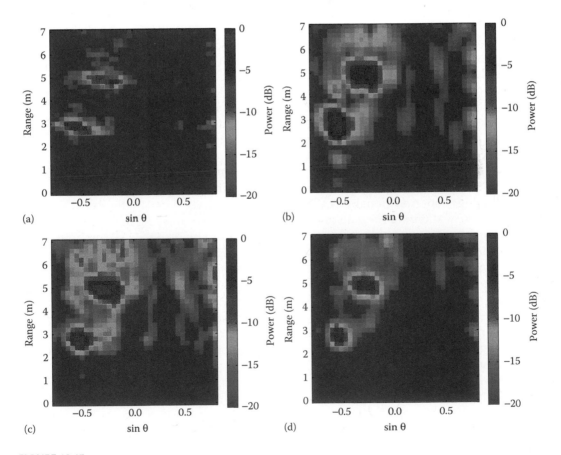

FIGURE 10.15
Motion maps of two moving targets obtained using MIMO with several different processing methods. Note that SDAP achieves similar results to conventional processing with 600 MHz bandwidth but uses only 150 MHz bandwidth. (a) Nonadaptive with 600 MHz bandwidth, (b) nonadaptive with 150 MHz bandwidth, (c) separable adaptive with 150 MHz bandwidth, and (d) SDAP with 150 MHz bandwidth.

are not observable on the map because they are close to the mainlobe and thus overlap with the target. Figure 10.15b–d shows the motion maps obtained using 150 MHz bandwidth RF measurements. Figure 10.15a is regarded as the benchmark for comparison. Figure 10.15b shows that processing artifacts appear due to filter sidelobes. These artifacts happen because conventional nonadaptive processing uses filter weights that have a fixed response pattern. Doppler characteristics from moving targets leak into other locations. Figure 10.15c shows the motion map obtained by applying adaptive ranging and adaptive beamforming separately. This method successfully mitigates the sidelobes. However, this separable filter is suboptimal for suppressing nonseparable returns. In indoor environments, moving targets generate direct path returns as well as multipath returns due to local scattering. Unlike point targets, the real target return typically spreads in delay and direction. As a result, the separable filter squeezes its mainlobe to suppress closely located interference and thus amplifies the background noise at other locations, as can be seen on the right side of Figure 10.15c. In contrast, the SDAP filter operates in a higher dimensional space where the returns from different targets are well separated and can be suppressed using a few DOFs. The motion map obtained using SDAP is shown in Figure 10.15d, which outperforms the nonadaptive and separable adaptive methods.

As discussed in Section 10.2.3, the power of the resolved micro-Doppler signature offers the optimal motion detector and is calculated as the detection statistic at each candidate location. Doppler characteristics from one moving target can leak into other locations due to the limited bandwidth (resolution) and indoor multipath scattering, creating false alarms in motion detection. The receiver operating characteristic (ROC) curves for different methods can be obtained using the detection statistics calculated at multiple bed locations. We generate ROC curves using the measurements corresponding to the CPIs when at least one motion event happens that creates ghost returns at other candidate locations. A better motion monitoring method suppresses those artifacts and provides better detection performance, that is, a ROC curve that is closer to the upper left corner. The ROC curves are shown in Figure 10.16. It can be seen that MIMO outperforms SIMO using either conventional nonadaptive processing or SDAP, validating the gain of using MIMO for indoor multipath mitigation. Additionally, SDAP outperforms conventional methods using either MIMO or SIMO, validating the gain of using joint adaptive processing methods compared to nonadaptive methods. The best performance is achieved by using SDAP combined with MIMO. In addition, no significant improvement is obtained using separable adaptive processing compared to conventional nonadaptive processing, which demonstrates the gain of processing the fast-time samples and array spatial samples jointly to suppress nonseparable returns in realistic complex indoor environments.

After correctly locating and detecting a motion event at a specific location, time–frequency analysis is performed to extract the motion features for classification. Figure 10.17 shows the spectrograms for three different types of human motion: lifting an arm, bending a leg, and turning over. For each motion, the spectrogram is shown from the real RF measurements and also from the simulated measurements using the ground truth motion data. Note that the zero-Doppler ridge is removed in each spectrogram because of the clutter removal. In each spectrogram, there are three realizations of a particular motion. The RF measurements are pretty well matched to the simulated ground truth motion, demonstrating that motion features are adequately captured by MIMO RF sensing. Furthermore, the patterns of turning over generally have larger Doppler spread and longer time duration than those of lifting an arm or bending a leg, and the patterns of lifting an arm have slightly larger Doppler spread than those of bending a leg. This observation leads to the selection of features for motion classification described previously. To generate data for

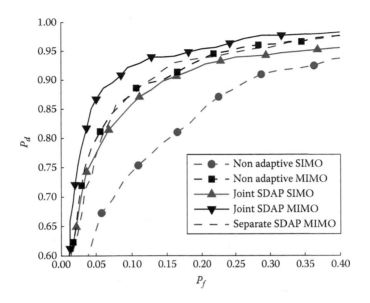

FIGURE 10.16
ROC curves of different methods to resolve micro-Doppler signatures at each location for detecting motion at multiple locations in a wide area.

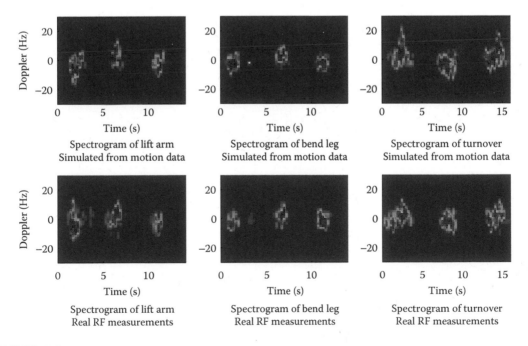

FIGURE 10.17
Spectrograms of lifting an arm, bending a leg, and turning over. The first row includes spectrograms generated from simulated RF returns using the ground truth motion data. The second row includes spectrograms generated from real RF measurements. The dynamic range of all plots is 25 dB.

TABLE 10.2

Averaged 10-Fold Cross-Validation Rates of All Data Using SVM (the Coherent System)

Decision Truth	"Lift Arm" (%)	"Bend Leg" (%)	"Turnover" (%)
Lift arm	90.4	6.7	2.9
Bend leg	9.5	88.2	2.3
Turnover	4.9	3.7	91.4

classification, the motion features of the three types of motion are extracted from the spectrogram to form a feature vector. In total, there are 76 vectors of lifting an arm, 79 vectors of bending a leg, and 64 vectors of turning over, collected from two people and four locations. To evaluate the ability of the proposed method to classify the different types of motion, 10-fold cross-validation is performed on the dataset. In particular, a random partition of the dataset is used for each cross-validation, and 50 independent cross-validations are performed, and the results are averaged. In this chapter, four classifiers are used, including the logistic discriminant classifier, the naive Bayes classifier, the *k*-nearest neighbor classifier, and the SVM classifier. In terms of recognizing turnovers, the four classifiers provide similar results, whereas the SVM performs the best (logistic 88.6%, Bayes 87.5%, nearest neighbor 85.4%, SVM 91.4%). Table 10.2 shows the averaged confusion matrix of SVM with the six motion features extracted to classify the three types of motion.

10.4.3 Noncoherent Approach

Here, we used a subset of the channels provided by the MIMO system and distributed antennas using long cables. The cables delay the received signals in the system because of their nonnegligible lengths, and we evaluated the noncoherent method by not measuring the cable lengths but to estimate them using the direct blast signal. Also, we added random phases to the channels so that they had noncoherent phases. The geometry of the experiments is shown in Figure 10.18. In particular, four transmitters and two receivers

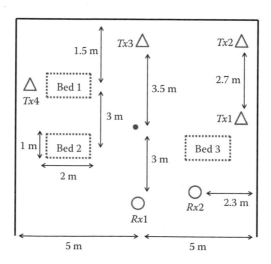

FIGURE 10.18
Experiment geometry.

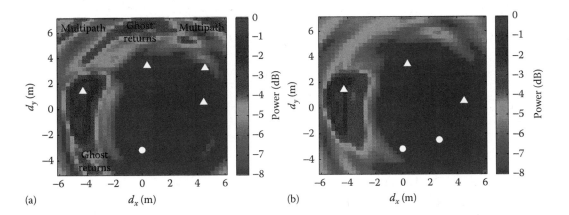

FIGURE 10.19
Motion maps of one moving target obtained from 4Tx/1Rx SIMO and 3Tx/2Rx MIMO. (a) SIMO and (b) MIMO.

were distributed in the room, and motion data were measured at three different beds from three different people. For each person on each bed, three types of human motion were measured, including lifting an arm, bending a leg, and turning over. Each dataset was measured for 30 s.

First, results obtained using four Tx's and one Rx ($4Tx/1Rx$) as well as three Tx's and two Rx's ($3Tx/2Rx$) are shown and compared. These results are obtained by processing only the corresponding channels measured in the $4Tx/2Rx$ experiment. The $4Tx/1Rx$ SIMO is equivalent to one Tx and four Rx's. For fair comparison, the total number of sensors is fixed to five for both cases. Motion maps from $4Tx/1Rx$ (SIMO) and $3Tx/2Rx$ (MIMO) are shown in Figure 10.19. For both maps, the maximum power is normalized to unit, and the powers at different locations are plotted in dB. It can be seen that in both maps, there is one strong motion return at the location of bed 2. Moreover, there are ghost returns at other locations in both maps. The ghost returns form elliptical shapes on the (d_x, d_y) grid, validating that the location ambiguities form an ellipse. In addition, there are ghost returns at further ranges (outer ellipse) than the target, which are caused by multipath scattering. Note that the ghost returns due to both ambiguities and multipath are mitigated using MIMO. The reason is that using $3Tx/2Rx$ gives six sensor pairs, whereas using $4Tx/1Rx$ gives four pairs. It can be seen that using more sensor pairs can effectively suppress ghost returns because the ambiguities and multipath are less likely to be consistent with multiple Tx–Rx pairs than the targets.

Moreover, Figure 10.20 shows the peak value on the motion map over time for different types of human motion. It can be seen that in this problem, motion detection is achieved based on thresholding the maximum power on the motion map over time, and motion localization is achieved based on finding peaks on the motion map at each time when motion has been detected. Figure 10.21 shows the localization errors using motion maps obtained from SIMO and MIMO as a function of signal-to-noise ratio (SNR) for the first (strongest) target. A localization error is defined as associating a motion with a wrong bed. It can be seen that MIMO outperforms SIMO in this case because ghost returns are suppressed using more sensor pairs. At 40 dB SNR, the probability of error using MIMO is about 14%.

In this chapter, we are interested in monitoring the motions of multiple human targets in a wide area. Therefore, results for detecting/localizing the second target are shown

254

Radar for Indoor Monitoring

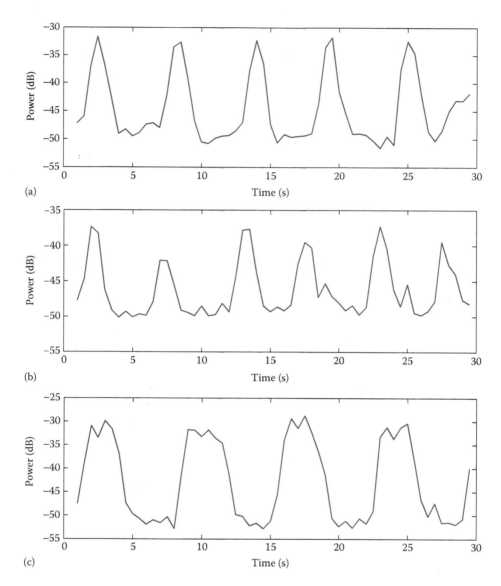

FIGURE 10.20
Motion powers over time for different human motions: (a) lifting an arm, (b) bending a leg, and (c) turning over.

and are used to indicate that more than one target can be monitored using the proposed method. First, the motion map of two moving targets is shown in Figure 10.22a, in which there are two human targets moving in beds 2 and 3. It can be seen that the strongest target can be detected/localized in the same way as it is discussed earlier, whereas the second target is harder to detect in the presence of the first target, due to ghost returns from the first target. Therefore, an iterative motion mapping method is proposed. Figure 10.22b and c shows the motion maps obtained by ignoring the first target and by removing returns from the first target (iterative motion mapping). It can be seen that the ghost return due to the first target is mitigated using the iterative method, which reduces the number of false alarms in detecting the second target.

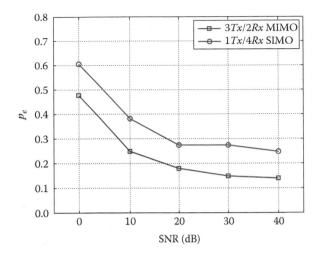

FIGURE 10.21
Probability of error versus SNR for motion localization of the first target using motion maps obtained from SIMO and MIMO.

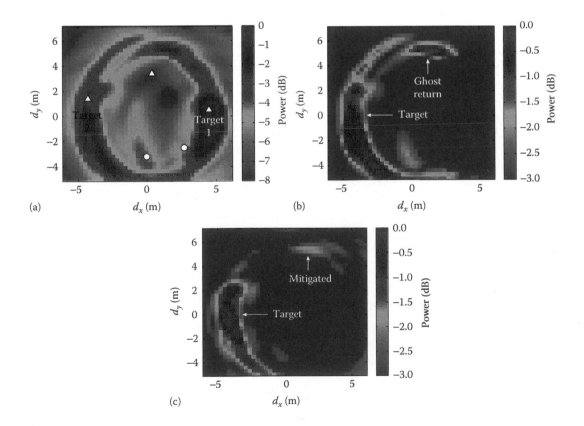

FIGURE 10.22
Iterative motion mapping is proposed for detecting the motions of multiple targets in a wide area. (a) Motion map of two moving targets, (b) motion map by ignoring the first target, and (c) motion map by removing the first target.

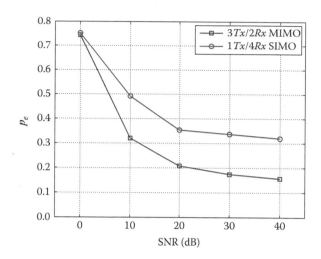

FIGURE 10.23

Probability of error versus SNR for motion localization of the second target using motion maps obtained from SIMO and MIMO.

Also, motion localization performances of SIMO and MIMO are compared for the second target, which is shown in Figure 10.23. It can be seen that MIMO again outperforms SIMO because of more sensor pairs for correctly localizing the second target. At 40 dB SNR, the probability of error using MIMO is about 16%.

Last, the ability of the noncoherent RF system to classify different types of human motion is evaluated, assuming that the motion of each target is detected and the location of each target is obtained correctly. It can be seen in Figure 10.20 that a bed turnover has a longer duration than those of arm and leg movements. This observation leads to the first feature we extract for motion classification, that is, the time duration. In addition to the time duration, more features are extracted from the velocity components for each motion. Figure 10.24 shows the velocity components of each type of motion at different beds. It is shown that the velocity distributions roughly form a tilted line in each plot, which is due to the velocity ambiguity shape and a limited number of sensor pairs. This validates that velocity ambiguities form a line. In order to classify human motions that move along different directions, are at different locations, and are monitored by different sensor pairs, rotational invariant features need to be extracted. It can be seen that the velocities of turning over extend to all directions from (0,0), whereas the velocities of arm/leg movements extend to only one side of all directions, which leads to the second feature we extract, which is the total area covered on the opposite to the strongest velocity. In addition, the total motion power (after normalizing the power of the strongest velocity to unit) on the opposite to the strongest velocity is extracted as the third feature. Also, we extract two more features to classify arm and leg movements, which are the total area that the velocity components cover and the maximum velocity of the motion. Therefore, a vector of five features is extracted for each motion.

We use the logistic discriminant classifier, naive Bayes classifier, k-nearest neighborhood classifier, and SVM classifier implemented in the PRT toolbox for classifying the motion. In total, there are 138 vectors of lifting an arm, 144 vectors of bending a leg, and 108 vectors of turning over. The 10-fold cross-validation is performed, and

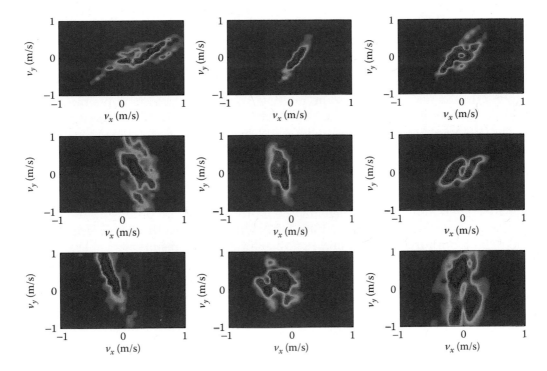

FIGURE 10.24
Velocity components of different human motions. The first row includes results from bed 1, the second row includes results from bed 2, and the third row includes results from bed 3. The first column includes the results from lifting an arm, the second column includes the results from bending a leg, and the third column includes the results from turning over. The dynamic range of all plots is 8 dB.

TABLE 10.3

Averaged 10-Fold Cross-Validation Rates of All Data Using SVM (the Noncoherent System)

Decision Truth	"Lift Arm" (%)	"Bend Leg" (%)	"Turnover" (%)
Lift arm	88.8	11.1	0.1
Bend leg	5.7	93.0	1.3
Turnover	0.0	5.6	94.4

the classification rates are averaged over 20 independently random data partitions. We notice again that SVM performs the best. Table 10.3 shows the confusion matrix of classifying the three motions.

10.5 Conclusion

This chapter addresses the problem of monitoring the motions of multiple human targets in wide-area indoor environments using WiFi-band signals. In particular, two RF systems have been proposed here for solving the problem: the coherent system using a monostatic MIMO array and the noncoherent system using multistatic transmitters and receivers. In the coherent system, a joint adaptive processing has been proposed to resolve the motion characteristics at each location in a wide area. The downrange resolution is limited by the signal bandwidth, whereas the crossrange resolution is limited by the array aperture. MIMO is used for mitigating ghost returns due to indoor multipath scattering. In the noncoherent system, multiple transmitters and multiple receivers are distributed in a wide area to estimate the locations and velocities of motions from multiple human targets. Ghost returns in the elliptical localization and due to the multipath propagation are suppressed based on the fact that these are less likely to be consistent with multiple $Tx–Rx$ pairs. Akin to the coherent system, we assume that the received signals from different transmitters are separable at the receiver. Different from the coherent system, the sensors used in the noncoherent system have independent clocks and different phases. Therefore, the noncoherent system provides a more promising solution due to its low complexity and high flexibility. To demonstrate the ability to recognize human motion using both the coherent and noncoherent systems, results with real data have been presented with measurements of different types of human motion at several locations in realistic indoor environments. Given resolved motion characteristics from each target, well-known motion classification methods have been used to evaluate the ability to classify bed turnovers from other movements such as lifting an arm and bending a leg, which is useful in long-term care facilities to prevent bed-ridden patients from developing pressure ulcers.

References

1. T. B. Moeslund, A. Hilton, and V. Krüger, A survey of advances in vision-based human motion capture and analysis, *Computer Vision and Image Understanding*, 104 (2), 90–126, 2006.
2. B. Najafi, K. Aminian, A. Paraschiv-Ionescu, F. Loew, C. J. Büla, and P. Robert, Ambulatory system for human motion analysis using a kinematic sensor: Monitoring of daily physical activity in the elderly, *IEEE Transactions on Biomedical Engineering*, 50 (6), 711–723, 2003.
3. K. Aminian and B. Najafi, Capturing human motion using body-fixed sensors: Outdoor measurement and clinical applications, *Computer Animation and Virtual Worlds*, 15 (2), 79–94, 2004.
4. W. Spillman Jr, M. Mayer, J. Bennett, J. Gong, K. Meissner, B. Davis, R. Claus, A. Muelenaer Jr, and X. Xu, A smartbed for non-intrusive monitoring of patient physiological factors, *Measurement Science and Technology*, 15 (8), 1614, 2004.
5. A. Bourke, J. Obrien, and G. Lyons, Evaluation of a threshold-based tri-axial accelerometer fall detection algorithm, *Gait & Posture*, 26 (2), 194–199, 2007.
6. C. Rougier, J. Meunier, A. St-Arnaud, and J. Rousseau, Fall detection from human shape and motion history using video surveillance, in *21st International Conference on Advanced Information Networking and Applications Workshops, 2007, AINAW'07*, vol. 2, IEEE, 2007, pp. 875–880.
7. F. Adib, Z. Kabelac, D. Katabi, and R. C. Miller, 3D tracking via body radio reflections, in *11th USENIX Symposium on Networked Systems Design and Implementation (NSDI 14)*, USENIX Association, Seattle, WA, 2014, pp. 317–329.
8. H. Fujiyoshi, A. J. Lipton, and T. Kanade, Real-time human motion analysis by image skeletonization, *IEICE Transactions on Information and Systems*, 87 (1), 113–120, 2004.
9. M. S. Ryoo and J. K. Aggarwal, Spatio-temporal relationship match: Video structure comparison for recognition of complex human activities, in *2009 IEEE 12th International Conference on Computer Vision*, IEEE, 2009, pp. 1593–1600.

10. M. Otero, Application of a continuous wave radar for human gait recognition Ed. Ivan Kadar, in *Defense and Security*. International Society for Optics and Photonics, Bellingham, WA, 2005, pp. 538–548.

11. L. Liu, M. Popescu, M. Skubic, M. Rantz, T. Yardibi, and P. Cuddihy, Automatic fall detection based on Doppler radar motion signature, in *2011 5th International Conference on Pervasive Computing Technologies for Healthcare (PervasiveHealth)*, IEEE, 2011, pp. 222–225.

12. T. L. Yap, S. M. Kennerly, M. R. Simmons, C. R. Buncher, E. Miller, J. Kim, and W. Y. Yap, Multidimensional team-based intervention using musical cues to reduce odds of facility-acquired pressure ulcers in long-term care: A paired randomized intervention study, *Journal of the American Geriatrics Society*, 61 (9), 1552–1559, 2013.

13. E. A. Ayello, K. L. Capitulo, C. E. Fife, E. Fowler, D. L. Krasner, G. Mulder, R. G. Sibbald, and K. W. Yankowsky, Legal issues in the care of pressure ulcer patients: key concepts for health care providers: A consensus paper from the international expert wound care advisory panel, *Journal of Palliative Medicine*, 12 (11), 995–1008, 2009.

14. E. Keelaghan, D. Margolis, M. Zhan, and M. Baumgarten, Prevalence of pressure ulcers on hospital admission among nursing home residents transferred to the hospital, *Wound Repair and Regeneration*, 16 (3), 331–336, 2008.

15. E. Park-Lee and C. Caffrey, Pressure ulcers among nursing home residents: United States, 2004. NCHS data brief, National Center for Health Statistics, Hyattsville, MD, no. 14, pp. 1–8, 2009.

16. K. Zulkowski, D. Langemo, M. E. Posthauer, and the National Pressure Ulcer Advisory Panel, Coming to consensus on deep tissue injury, *Advances in Skin & Wound Care*, 18 (1), 28–29, 2005.

17. D. Norton, R. McLaren, and A. N. Exton-Smith, *An Investigation of Geriatric Nursing Problems in Hospital*. Edinburgh, Scotland: Churchill Livingstone, 1962.

18. T. L. Yap, M. P. Rapp, S. Kennerly, S. G. Cron, and N. Bergstrom, Comparison study of Braden scale and time-to-erythema measures in long-term care, *Journal of Wound Ostomy & Continence Nursing*, 42 (5), 461–467, 2015.

19. N. Patwari and J. Wilson, RF sensor networks for device-free localization: Measurements, models, and algorithms *Proceedings of the IEEE*, 98 (11), 1961–1973, 2010.

20. R. Poppe, A survey on vision-based human action recognition, *Image and Vision Computing*, 28 (6), 976–990, 2010.

21. V. C. Chen, F. Li, S.-S. Ho, and H. Wechsler, Micro-Doppler effect in radar: Phenomenon, model, and simulation study, *IEEE Transactions on Aerospace and Electronic Systems*, 42 (1), 2–21, 2006.

22 V. C. Chen, Analysis of radar micro-Doppler with time-frequency transform, in *Proceedings of the Tenth IEEE Workshop on Statistical Signal and Array Processing*, IEEE, 2000, pp. 463–466.

23. I. Bilik and J. Tabrikian, Radar target classification using Doppler signatures of human locomotion models, *IEEE Transactions on Aerospace and Electronic Systems*, 43 (4), 1510–1522, 2007.

24. Y. Kim and H. Ling, Human activity classification based on micro-Doppler signatures using a support vector machine, *IEEE Transactions on Geoscience and Remote Sensing*, 47 (5), 1328–1337, 2009.

25. F. H. C. Tivive, A. Bouzerdoum, and M. G. Amin, A human gait classification method based on radar Doppler spectrograms, *EURASIP Journal on Advances in Signal Processing*, 2010 (1), 1–12, 2010.

26. S. S. Ram and H. Ling, Through-wall tracking of human movers using joint Doppler and array processing, *Geoscience and Remote Sensing Letters, IEEE*, 5 (3), 537–541, 2008.

27. F. Ahmad and M. G. Amin, Through-the-wall human motion indication using sparsity-driven change detection, *IEEE Transactions on Geoscience and Remote Sensing*, 51 (2), 881–890, 2013.

28. M. Malanowski and K. Kulpa, Two methods for target localization in multistatic passive radar, *IEEE Transactions on Aerospace and Electronic Systems*, 48 (1), 572–580, 2012.

29. S. Subedi, Y. D. Zhang, M. G. Amin, and B. Himed, Motion parameter estimation of multiple targets in multistatic passive radar through sparse signal recovery, in *2014 IEEE International Conference on Acoustics, Speech and Signal Processing (ICASSP)*, IEEE, 2014, pp. 1454–1457.

30. G. Frazer, Y. I. Abramovich, and B. A. Johnson, Multiple-input multiple-output over-the-horizon radar: Experimental results, *Radar, Sonar & Navigation, IET*, 3 (4), 290–303, 2009.

31. J. Li and P. Stoica, MIMO radar with colocated antennas, *Signal Processing Magazine, IEEE*, 24 (5), 106–114, 2007.

32. D. Bliss, K. Forsythe, S. Davis, G. Fawcett, D. Rabideau, L. Horowitz, and S. Kraut, GMTI MIMO radar, in *2009 International Waveform Diversity and Design Conference*, IEEE, 2009, pp. 118–122.

33. D. Bliss and K. Forsythe, Multiple-input multiple-output (MIMO) radar and imaging: Degrees of freedom and resolution, in *Conference Record of the Thirty-Seventh Asilomar Conference on Signals, Systems and Computers, 2004*, vol. 1, IEEE, 2003, pp. 54–59.

34. C. Xu and J. Krolik, MIMO RF probe for wide-area indoor human motion monitoring, in *2014 IEEE International Conference on Acoustics, Speech and Signal Processing (ICASSP)*, IEEE, 2014, pp. 6047–6051.

35. Y.-S. Yoon, M. G. Amin, and F. Ahmad, MVDR beamforming for through-the-wall radar imaging, *IEEE Transactions on Aerospace and Electronic Systems*, 47 (1), 347–366, 2011.

36. L. Li and J. L. Krolik, Vehicular MIMO SAR imaging in multipath environments, in *2011 IEEE Radar Conference (RADAR)*, IEEE, 2011, pp. 989–994.

37. L. Li and J. L. Krolik, Simultaneous target and multipath positioning with MIMO radar, in *IET International Conference on Radar Systems (Radar 2012)*, IET, 2012, pp. 1–6.

38. L. Li and J. L. Krolik, Simultaneous target and multipath positioning, *IEEE Journal of Selected Topics in Signal Processing*, 8 (1), 153–165, 2014.

39. C. Xu and J. Krolik, Space-delay adaptive processing for MIMO RF indoor motion mapping, in *2015 IEEE International Conference on Acoustics, Speech and Signal Processing (ICASSP)*, IEEE, 2015, pp. 2349–2353.

40. J. Ward, Space-time adaptive processing for airborne radar, in *1995 IEEE International Conference on Acoustics, Speech, and Signal Processing (ICASSP)*, IEEE, 1995, vol. 5, pp. 2809–2812.

41. F. Ahmad and M. G. Amin, Noncoherent approach to through-the-wall radar localization, *IEEE Transactions on Aerospace and Electronic Systems*, 42 (4), 1405–1419, 2006.

42. Q. He, R. S. Blum, and A. M. Haimovich, Noncoherent MIMO radar for location and velocity estimation: More antennas means better performance, *IEEE Transactions on Signal Processing*, 58 (7), 3661–3680, 2010.

43. Y. Zhou, C. L. Law, Y. L. Guan, and F. Chin, Indoor elliptical localization based on asynchronous UWB range measurement, *IEEE Transactions on Instrumentation and Measurement*, 60 (1), 248–257, 2011.

44. C. Chang and A. Sahai, Object tracking in a 2D UWB sensor network, in *Conference Record of the Thirty-Eighth Asilomar Conference on Signals, Systems and Computers*, vol. 1, IEEE, 2004, pp. 1252–1256.

45. F. Adib, Z. Kabelac, and D. Katabi, Multi-person localization via RF body reflections, in *12th USENIX Symposium on Networked Systems Design and Implementation (NSDI 15)*, USENIX Association, Oakland, CA, 2015, pp. 279–292.

11

Activity Recognition and Localization Using Array Antenna: Array Sensor

Tomoaki Ohtsuki

CONTENTS

11.1 Introduction

Japan is known as a rapid aging country. There are many elderly people living at home alone in this country [1]. They need to be monitored for their safety. Of course, people do not desire to be monitored by cameras at home. Therefore, we need a system that can monitor people without invasion of privacy.

Several electrical wave-based security systems are reported such as in Reference 2 where an event such as intrusion is detected based on the change of received signal strength (RSS). Electrical waves arrive in every corner, and thus wide sensing range is realized. In addition, there is no need to be concerned about privacy invasion in these systems. However, RSS suffers from the effects of noise and signal level fluctuations, even in static environments. Thus, detection errors occur.

An approach for human activity classification technology based on micro-Doppler features of radar using support vector machine (SVM) was evaluated in Reference 3. Measured data of different activities were collected using the Doppler radar, and their features used for SVM to classify the activities were extracted from the Doppler spectrogram. The micro-Doppler radar achieves the classification accuracy above 90%. However, the detection range is small and not sufficient to monitor an entire room. In addition, general micro-Doppler radar is not applicable to non-line-of-sight (NLOS) environments.

In addition to human motion activity classifications, it is important to know not only whether a particular event happens but also where it happens, that is, its location. Thus, localization is particularly helpful for monitoring applications. Localization techniques are roughly classified into two classes: active localization and passive localization.

The majority of localization techniques fall in the class of active localization where a person being localized and/or tracked needs to carry tags/electric devices. In passive localization, a person is localized and/or tracked without the need of tags/electric devices being carried by him or her. In general, passive localization is preferred due to relief of stress brought in by carrying tags/devices; this makes this kind of localization feasible even in a bathroom environment. However, in general, localization accuracy of passive localization techniques is relatively low compared with that offered by their active counterparts.

This chapter introduces the activity recognition and localization using array antenna, referred to as an array sensor [4–12]. The array sensor exploits an antenna array on the receiver side and decomposes the received signals into eigenvectors and eigenvalues. It uses these components and possibly RSS, depending on the underlying applications, which include intrusion detection, monitoring, and passive localization. When an event (e.g., falling) occurs, the propagation environment changes, and thus, the eigenvector changes. Based on this change, we can detect an event more accurately than the system based on RSS. Using machine learning technique such as SVM based on those features obtained by array sensor, several more complex states and activities can be classified, such as sitting in a bathtub and falling in a bathroom. Because the array sensor does not rely on the exact direction of arrival (DOA) information but the changes in radio propagation, it does not need a precisely designed array antenna nor array calibration; just multiple antennas are needed so that the sensor becomes low cost and is easy to install. Of course, general array antennas can be used as array sensors, and sometimes, we can expect better performance at the expense of cost. The array sensor can also realize passive localization using SVM based on those features in a fingerprinting manner.

11.2 Array Data Model

Consider the L-element linear array shown in Figure 11.1 as an example. We assume one source signal $s(t)$. DOA θ is defined clockwise relative to the broadside. The source signal $s(t)$ is a plane wave owing to the far field assumption. The noise $\mathbf{n}(t)$ is an additive white

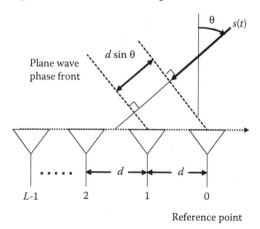

FIGURE 11.1
L-element uniform linear array with a source signal $s(t)$ from DOA θ.

Gaussian noise of zero mean and variance σ^2. The received signal vector $x(t)$ from DOA θ is represented as

$$x(t) = \mathbf{a}(\theta)s(t) + \mathbf{n}(t) \tag{11.1}$$

where $\mathbf{a}(\theta)$ is referred to as a steering vector, which is a complex vector defined by a phase shift of a source signal at each antenna relative to the first antenna (reference point). For a uniform linear array, the steering vector is represented by

$$\mathbf{a}(\theta) = \left[1, e^{-j\frac{2\pi}{v}d\sin\theta}, \ldots, e^{-j\frac{2\pi}{v}(L-1)d\sin\theta} \right]^T \tag{11.2}$$

where:
  v is the wavelength
  $[\cdot]^T$ is the transposition

In the indoor wireless scenario, the antenna array receives not only signals from the direct path but also many reflected multipath components with different DOAs. Thus, the signal vector that includes the total signal received by the antenna array is expressed as follows:

$$x(t) = \sum_{i=1}^{M} \alpha_i \mathbf{a}(\theta_i)s(t) + \mathbf{n}(t) = \mathbf{a}'s(t) + \mathbf{n}(t) \tag{11.3}$$

where:
  M is the total number of direct path and multipath components
  α_i is the phase difference and amplitude decay between the direct path and the ith multipath

To analyze wave propagation, we use the data correlation matrix estimated from the received signal vector. The data correlation matrix $\mathbf{R}_{xx}$ is defined as follows:

$$\mathbf{R}_{xx} = E[x(t)x(t)^H] \tag{11.4}$$

where:
  $E[\cdot]$ is the ensemble average
  $[\cdot]^H$ is the conjugate transpose

In general, additive noise is uncorrelated with the source signal. The noise is independent in each array element. Therefore, the data correlation matrix can be simplified as follows:

$$\mathbf{R}_{xx} = E[\mathbf{a}'s(t)s(t)^H\mathbf{a}'^H] +$$

$$\underbrace{E[\mathbf{a}'s(t)\mathbf{n}(t)^H] + E[\mathbf{n}(t)\mathbf{a}'^H s(t)^H]}_{\to 0} + E[\mathbf{n}(t)\mathbf{n}(t)^H] \tag{11.5}$$

$$= \mathbf{a}'S\mathbf{a}'^H + \sigma^2 I$$

where $S = E[s(t)s(t)^H]$ and $\mathbf{I}$ is the identity matrix. However, the data correlation matrix $\mathbf{R}_{xx}$ cannot be strictly obtained. In effect, based on ergodic hypothesis, the ensemble average of Equation 12.4 is replaced with the time average:

$$\hat{\mathbf{R}}_{xx} = \frac{1}{N_s} \sum_{k=1}^{N_s} x(t_k)x(t_k)^H \tag{11.6}$$

where N_s is the number of snapshots. The more the number of snapshots, the better the estimation accuracy of $\hat{\mathbf{R}}_{xx}$ becomes.

11.3 Subspace-Based Method

The subspace-based method [13] decomposes the data correlation matrix into orthogonal signal and noise subspaces via the eigenvalue decomposition (EVD). By the EVD of $\mathbf{R}_{xx}$, we obtain the eigenvalue λ_i and the eigenvector $\mathbf{v}_i$, which satisfies the following equation:

$$\mathbf{R}_{xx}\mathbf{v}_i = (\mathbf{a}'S\mathbf{a}'^H + \sigma^2\mathbf{I})\mathbf{v}_i = \lambda_i\mathbf{v}_i, i = 1,2,\cdots,L \tag{11.7}$$

The EVD of the L-element data correlation matrix $\mathbf{R}_{xx}$ is as follows:

$$\mathbf{R}_{xx} = \mathbf{a}'S\mathbf{a}'^H + \sigma^2\mathbf{I}$$

$$= \sum_{i=1}^{L} \lambda_i\mathbf{v}_i\mathbf{v}_i^H \tag{11.8}$$

$$= \mathbf{V}\mathbf{\Lambda}\mathbf{V}^H \tag{11.9}$$

$$\mathbf{V} = [\mathbf{v}_1, \mathbf{v}_2, \cdots, \mathbf{v}_L] \tag{11.10}$$

$$\mathbf{\Lambda} = \text{diag}\{\lambda_1, \lambda_2, \cdots, \lambda_L\} \tag{11.11}$$

Because $\mathbf{R}_{xx}$ is a positive definite Hermitian matrix, the eigenvalue λ is a nonnegative real number and is sorted in the descending order: $\lambda_1 \geq \lambda_2 \geq \cdots \geq \lambda_L(> 0)$. Then, we can write

$$\mathbf{a}'S\mathbf{a}'^H\mathbf{v}_i = (\lambda_i - \sigma^2)\mathbf{v}_i = \lambda_i' \mathbf{v}_i, i = 1,2,\cdots,L \tag{11.12}$$

$$\lambda_i' = \lambda_i - \sigma^2 \tag{11.13}$$

because rank $[\mathbf{a}'S\mathbf{a}'^H] = 1$

$$\lambda_1' > \lambda_2' = \cdots = \lambda_L' = 0 \tag{11.14}$$

Thus, the eigenvalue distribution of the data correlation matrix is

$$\lambda_1 > \lambda_2 = \cdots = \lambda_L = \sigma^2 \tag{11.15}$$

Therefore, the eigenvalue matrix Λ is decomposed into signal and noise eigenvalues. The space spanned by the eigenvector matrix $\mathbf{V}$ is decomposed into the orthogonal signal and noise subspaces via the EVD. The first eigenvector $\mathbf{v}_1$ spans the signal subspace, and all the eigenvectors are mutually orthogonal.

11.4 Array Sensor

Figure 11.2 shows an example of an array sensor. When a pair of transmitter and receiver is fixed, the signal subspace spanned by eigenvector changes when the indoor environment of interest changes. For detecting simple events, such as intrusion, we can use a simple threshold detection based on the change of the first eigenvector. For detecting and classifying more complex states and activities, such as sitting in a bathtub and falling in a bathroom, we use machine learning, such as SVM. We explain how these methods are implemented in the array sensor.

A signal subspace spanned by an eigenvector is obtained as the first eigenvector by the EVD of the data correlation matrix. Thus, the signal subspace spanned by the eigenvector consists of a linear coupling of the steering vectors from incident multipath signals. The incident multipath signals go through every indoor point of interest. Therefore, the signal subspace spanned by the eigenvector represents a wave propagation. When the environment of interest changes, the wave propagation changes, and thus, the eigenvector spanning signal subspace changes. Consequently, the signal subspace spanned by the eigenvector is inherent in each environment of interest.

In the array sensor, we use cost functions based on the eigenvector and eigenvalue to detect events. Each cost function is obtained from N_t received signal vectors. The cost function $P(t)$ based on the eigenvector is defined as

$$P(t) = | \mathbf{v}_1 (u_{\mathrm{no}})^H \mathbf{v}_1 (t) |, \qquad (0 \le P(t) \le 1) \tag{11.16}$$

FIGURE 11.2
An example of array sensor.

where:

$\mathbf{v}_1(u_{no})$ is the first eigenvector obtained in advance, the reference vector
$\mathbf{v}_1(t)$ is the first eigenvector obtained at the observation time u

Both eigenvectors are normalized to unity. $P(t)$ means the correlation between the indoor environment at the reference time and the observation time u. Therefore, the closer the value of $P(t)$ to 1, the smaller the change of environment, and the smaller the value of $P(t)$, the larger the change of environment. The eigenvector is stationary even in the noise and fading environment, because it does not include RSS information.

The cost function $Q(t)$ based on the eigenvalue is defined as

$$Q(t) = 1 - \frac{|\lambda_1(t) - \lambda_1(t_{red})|}{\lambda_1(r_{ref})}, \qquad (Q(t) \le 1) \tag{11.17}$$

where:

$\lambda_1(t_{ref})$ is the reference value, the first eigenvalue obtained in advance
$\lambda_1(t)$ is the first eigenvalue obtained at the observation time t

Like $P(t)$, the closer $Q(t)$ is to 1, the smaller the change of environment, and the smaller the value of $Q(t)$, the larger the change of environment. The eigenvalue is less stationary than the eigenvector, but $Q(t)$ can detect even the small events. Then, we use both $P(t)$ and $Q(t)$ as the situation demands [7].

To detect simple events, we just set the threshold P_{th} to the cost function to detect an event. For detecting and classifying more complex states and activities, such as sitting in a bathtub and falling in a bathroom, we use machine learning, such as SVM [14].

11.5 Activity Recognition Using Array Sensor

First, we show the change of the values of the cost functions for some simple state and activities: no one in the room, walking and stopping in a room. Figure 11.3 shows the cost functions for simple activities in a typical office room. When no one is in the room, both the cost functions take value of about one. The cost function of $P(t)$ is more stable than that of $Q(t)$, as we can expect. When the person is walking in the room, fluctuations of the cost functions become large, owing to the changes in the propagation environment that reflect on the first eigenvector and its corresponding eigenvalue. When the person is standing still, the fluctuations of the cost functions become small, compared with that when the person is walking; in particular, the cost function $P(t)$ is more stable. Note that $P(t)$ assumes smaller values than one, because the propagation environment is different from that when no one is in the room so that the correlation becomes smaller. Therefore, we can easily classify some simple activities such as those in the above example. When classification of more complex states is in order, we need to use machine learning algorithms, which is shown in Figure 11.4.

Figure 11.4 also shows the change of values of the cost function $P(t)$ for falling and lying, and walking both after standing still. Around the changes of the movements, from standing still to falling and lying, and walking, we can observe the rapid changes of the cost function $P(t)$. Based on the mere changes of the cost function $P(t)$, it is difficult to classify

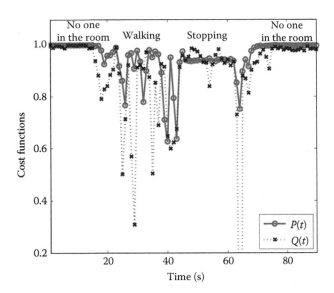

FIGURE 11.3
Cost functions for some simple activities in a typical office room.

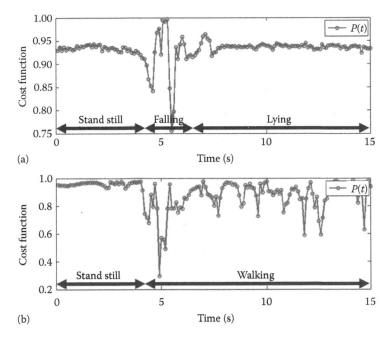

FIGURE 11.4
Change of values of the cost function $P(t)$ for (a) falling and lying and (b) walking both after standing still. (From Hino, Y. et al., *IEEE International Conference on Communications (ICC)*, 507–511, June 2015.)

the activities. Thus, we resort to machine learning techniques based on the features obtained by the array sensor.

The classification performance of array sensor under real-life scenarios in a bathroom is reported in Reference 9, where seven states are considered: (1) *No event*, (2) *Walking*, (3) *Entering into a bathtub*, (4) *Standing while showering*, (5) *Sitting while showering*, (6) *Falling down*,

TABLE 11.1

Six Scenarios of the Bathroom Experiments

S1	State (1) → (2) → (3)
S2	State (1) → (2) → (7)
S3	State (1) → (2) → (6)
S4	State (1) → (2) → (3) → (6)
S5	State (1) → (2) → (4) → (6)
S6	State (1) → (2) → (5) → (6)

Source: Hong, J. and Ohtsuki, T., *IEICE Trans. Commun.*, E95-B, 3088–3095, 2012.

TABLE 11.2

Accuracy Comparison between RSS-Based and Signal Subspace Feature-Based Using Array Sensor

	Average Classification Accuracy (%)	
	RSS Based	Signal Subspace Feature Based
S1	80.0	100
S2	93.3	96.7
S3	80.0	100
S4	57.5	95.0
S5	52.5	92.5
S6	50.0	95.0
Average	68.9	96.5

Source: Hong, J. and Ohtsuki, T., *IEICE Trans. Commun.*, E95-B, 3088–3095, 2012.

and (7) *Passing out.* In the experiments, six scenarios combining the above seven states are defined in Table 11.1. For the state classification, a feature vector whose elements are the cost functions based on the eigenvector and eigenvalue over a specific length are used. SVM was used as a classifier.

By comparison, we show the performance of the activity recognition based on RSS. Table 11.2 shows the accuracy comparison results between the method based on RSS and the array sensor. From this table, it is evident that the array sensor achieves much better classification accuracy than the method based on only RSS. This is because RSS variation over time is large due to fading and noise, which results in worse classification accuracy of the method based on only signal strength.

11.6 Localization Using Array Sensor

The array sensor can localize a person based on a fingerprinting technique. In the fingerprinting, first we create a map (database) of a given area based on the feature values for a given location. The observed feature values are then compared to the fingerprint

TABLE 11.3

Experimental Parameters

Transmitter	Dipole antenna
Modulation method	No modulation
Transmission power	0.5 dBm
Frequency $(T \times 1)$	2.412 GHz
Frequency $(T \times 2)$	2.417 GHz
Frequency $(T \times 3)$	2.422 GHz
Array geometry	Linear
Number of array elements	8
Array aperture	0.8 m
Sampling rate	60 MHz
Number of snapshots	1024
Room size	7 m × 7 m

(database) to find the closest match and generate a predicted location. As a localization using fingerprinting, WiFi localization using fingerprinting is well known [15]. Here, we explain the localization using array sensor [10–12].

Assume that we classify N_p positions. In the training phase, we obtain the received signals $\mathbf{x}_p(t) (p = 1, \ldots, N_P)$ when a person stands at position p for T_N observation times. From the signals, we compute the cost functions $P_i(t), Q_i(t)$, where $u = 1, \ldots, T_N$. That is, we have $N_P T_N$ training samples. Next, all the cost functions are combined into one feature vector. Then, the feature vector is mapped into high-dimensional space by a kernel such as the radial basis function kernel and the training model is obtained.

In the localization algorithm using array sensor, in the testing phase, although we get the cost functions and the feature vector in the same way as in the training phase, we do not know what position this feature vector is classified to. However, once the SVM has been trained, all future unknown samples can be classified in real time.

We show one of the experimental results of localization using array sensor. The experimental parameters are summarized in Table 11.3 [8]. The experimental room is a usual class room constructed of ferro-concrete walls and glass windows on our campus. There are obstacles in front of each transmitter, and then the transmitters and receiver are set based on the NLOS conditions. In this experiment, the localization using array sensor achieves a high accuracy of 76.47% and a root mean squared error of 1.61 m, even though it is a passive localization technique.

11.7 Conclusion

This chapter introduces the activity recognition and localization method using array antenna, referred to as an array sensor. The array sensor exploits an antenna array on the receiver side and decomposes the received signals into eigenvectors and eigenvalues. The array sensor exploits these components depending on its applications, such as monitoring and passive localization. Using machine learning based on these components, the array sensor can classify several more complex states and activities without invasion of privacy. Although the name is array sensor, it does not necessarily need a precisely designed array antenna, nor does it require array calibration. We presented some experimental results, such as classifying a person's activities in the bathroom, and localization performance of a person's position in the office room. The proposed array sensor can be useful for monitoring an elderly person living alone and monitoring a person in a bathroom and a restroom where the camera cannot be installed. The ability of monitoring a person without invasion of privacy is a key advantage of array sensor. The array sensor would be particularly helpful to an aging society like Japan.

References

1. Cabinet Office, Government of Japan, Annual Report on the Aging Society: 2015.
2. M. Nishi, S. Takahashi, and T. Yoshida, Indoor human detection systems using VHF-FM and UHF-TV broadcasting waves, *IEEE International Symposium on Personal Indoor and Mobile Radio Communications*, pp. 1–5, September 2006.
3. Y. Kim and H. Ling, Human activity classification based on micro-Doppler signatures using a support vector machine, *IEEE Trans. Geosci. Remote Sens.*, 47 (5), 1328–1337, 2009.
4. S. Ikeda, H. Tsuji, and T. Ohtsuki, Indoor event detection with eigenvector spanning signal subspace for home or office security, *IEEE 68th Vehicular Technology Conference*, Calgary, Canada, pp. 1–5, September 2008.
5. S. Ikeda, H. Tsuji, and T. Ohtsuki, Indoor event detection with signal subspace spanned by eigenvector for home or office security, *IEICE Trans. Commun.*, E92-B (7), 2406–2412, 2009.
6. S. Ikeda, T. Ohtsuki, and H. Tsuji, Signal-subspace-partition event filtering for eigenvector-based security system using radio waves, *IEEE International Symposium on Personal Indoor and Mobile Radio Communications (PIMRC2009)*, Tokyo, Japan, pp. 2792–2796, September 2009.
7. S. Kawakami, K. Okane, and T. Ohtsuki, Detection performance of security system using radio waves based on space-time signal processing, *Procedia–Social and Behavioral Sciences*, 2 (1), 171–178, 2010.
8. T. Ohtsuki, (Invited paper) Wireless security and monitoring system using array antenna: Array sensor, *IEEE International Conference on Computing, Networking and Communications (ICNC2012)*, pp. 551–555, Maui, HI, January/February 2012.
9. J. Hong and T. Ohtsuki, State classification with array sensor using support vector machine for wireless monitoring systems, *IEICE Trans. Commun.*, E95-B (10), 3088–3095, 2012.
10. J. Hong and T. Ohtsuki, Signal eigenvector-based device-free passive localization using array sensor, *IEEE Trans. Vehicular Tech.*, 64 (4), 1354–1363, 2015.
11. J. Hong and T. Ohtsuki, Ambient intelligence sensing using array sensor: Device-free radio based approach, *The 4th Workshop on Context-Systems Design, Evaluation and Optimisation: Device-Free Radio-Based recognition*, Zurich, Switzerland, September 2013.
12. J. Hong, S. Kawakami, and T. Ohtsuki, Passive localization using array sensor with support vector machine, *9th Workshop on Positioning, Navigation and Communication 2012*, Dresden, Germany, March 2012.
13. H. Krim and M. Viberg, Two decades of array signal processing research, *IEEE Sig. Proc. Mag.*, 13 (4), 67–94, 1996.
14. C. J. C. Burges, A tutorial on support vector machines for pattern recognition, *Data Min. Knowl. Disc.*, 2 (2), 1–47, 1998.
15. K. Kaemarungsiand and P. Krishnamurthy, Modeling of indoor positioning systems based on location fingerprinting, *Proceedings of IEEE INFOCOM2004*, pp. 1012–1022, March 2004.
16. Y. Hino, J. Hong, and T. Ohtsuki, *IEEE International Conference on Communications (ICC)*, pp. 507–511, June 2015.

12

Radar Monitoring of Humans with Assistive Walking Devices

Ann-Kathrin Seifert, Moeness G. Amin, and Abdelhak M. Zoubir

CONTENTS

12.1 Introduction

Worldwide, the elderly population aged more than 65 years is growing (World Health Organization 2007). In particular, its ratio to the population aged more than 20 years is predicted to be 50% by 2050 (United Nations 2015). Older adults desire to stay at their own homes as long as possible. Also from a clinical point of view, aging in place has been shown to be advantageous over moving to assisted living facilities or nursing homes

(Marek et al. 2005). However, a major challenge for seniors, when living independently, is the risk of falling. Falls can lead to injuries and long-term health complications and are the second leading cause of death due to accidental or unintentional injuries (World Health Organization 2007). According to the WHO, the greatest number of fatal falls occurs among seniors aged 65 years and older (World Health Organization 2007).

Age is reported to be one of the key risk factors for falls (World Health Organization 2007). With aging, the human body's physical (muscles), sensory (seeing, hearing), and cognitive (mental) ability and strength naturally decrease, which in turn increase the risk of falling. Moreover, due to the higher susceptibility of the elderly to injury, falls are more consequential at an old age—not least because the ability to recover from a fall deteriorates. The fear of falling also increases with age, which constitutes another key risk factor for falls (Gell et al. 2015).

In order to compensate for decrements in balance, gain mobility, and overcome the fear of falling, a great number of seniors resort to assistive walking devices, such as a cane or a walker (Gell et al. 2015). In 2011, 8.5 million U.S. seniors aged 65 years and older reported having assistive walking devices, with a cane being the most commonly used by two-thirds of the elderly (Gell et al. 2015). The correct use of mobility devices is essential to guarantee optimal support and avoid postural deformities, in the case of both the elderly gaining mobility and patients recovering from injuries or physical impairment with the purpose of reestablishing a normal gait. However, assessing proper handling of walking aids is often difficult for health-care providers and nursing staff. The information on the elderly resorting to frequent or continuous use of a cane inside his/her home can be valuable in designing proper treatment and a recovery course.

Further, constant monitoring of changes in gait enables early diagnosis of different diseases, including Parkinson's, cardiopathies, and strokes, and facilitates studying disease progressiveness for designing an adequate course of treatment (Muro-de-la-Herran et al. 2014). Changes in gait characteristics have also been shown to associate with the risk of falling (Barak et al. 2006).

For these reasons, it is essential to detect gait abnormalities and monitor alterations in walking patterns over time. However, it is important that fall detection and gait remote monitoring systems for assisted living recognize the use of walking aids and account for their impact on the respective motion articulations.

As personal emergency response systems become more and more popular, fall detection technologies have been widely investigated. Besides classical indoor monitoring modalities, such as cameras, audio, or wearable devices, radar has drawn much attention, specifically over the last decade, due to its attractive attributes, and it is promising to become a leading technology in assisted living in the near future (Chen et al. 2014, Amin et al. 2016). Radar is a reliable, nonintrusive, and safe sensing modality. Radar sensing of human motions is insensitive to lighting conditions, and its signals are invisible and transmitted at low electromagnetic energy (3 dBm compared to a phone 27 dBm). Radar human monitoring preserves privacy, which establishes one of its major attractions when continuously used in homes and retirement facilities. Operating in the L- and S-band electromagnetic waves, unlike optics, can penetrate interior walls and are not occluded by many furniture items, giving radar the ability to operate in wide areas. Besides these favorable characteristics, radar systems are able to sense nuances in target motions and are thus well suited to detect falls and screen changes in gait characteristics.

This chapter examines the human gait with and without the use of a cane as an assistive walking device. It analyzes various attributes of the motions of the feet, and the upper and lower parts of the leg, and shows how to separate these motions from that of the cane.

In particular, we point out the saddle differences in the gait time–frequency signal representations when the person is walking toward and away from the radar. We use both the spectrograms and the cadence velocity diagram (CVD) for extracting the features that reveal these differences. Although still preliminary, the study offered by this chapter is poised to improve knowledge and practice of geriatricians and rehabilitation physicians, and aid them in their research and pedagogy, thus benefiting a large segment of the population. The outline of the chapter is as follows: Section 12.2 introduces the fundamentals of human gait analysis. First, the radar signal model for human monitoring is presented. Then, Doppler and micro-Doppler frequencies associated with human gait are delineated and shown to correspond to a clear time–frequency signature when applying time–frequency signal representations to the backscattered radar data. Further, the analysis of human gait motion signature in the cadence-velocity domain is provided, along with the general framework for micro-Doppler gait signature classification. Using experimental data, Section 12.3 reveals the distinct human gait patterns for different walking styles in both time–frequency and cadence-velocity signal representation. These distinctions permit extractions of relevant features for identifying gait abnormalities and perform gait classification in cane-assisted walks. Finally, Section 12.4 highlights the main chapter contributions.

12.2 Fundamentals of Human Gait Analysis

12.2.1 Radar Signal Model

Suppose a monostatic continuous-wave (CW) radar transmits an electromagnetic wave at radiofrequency f_c, the transmitted signal can be expressed as

$$s(t) = e^{j2\pi f_c t} \tag{12.1}$$

where $0 \le t \le T$ and T is the time duration of the signal. Consider a point target whose initial position at time instant $t = 0$ is defined by the distance R_0 from the radar and moves with a velocity $v(t)$ in a direction forming an angle θ with the radar line-of-sight (LOS) (Chen and Ling 2001, Amin et al. 2015b). Then, the distance between the radar and the target at time instant t is given by

$$R(t) = R_0 + \int_0^t v(u) \cos \theta \, du \tag{12.2}$$

The received radar signal reflected from the point target is a delayed and attenuated version of the transmitted signal. The round-trip delay, which refers to the time that the wave needs to travel to the target and back to the receiver, is given by

$$\tau(t) = \frac{2R(t)}{c} \tag{12.3}$$

where c denotes the propagation speed of electromagnetic waves in free space. Hence, for a moving target, the round-trip delay is a time-varying delay $\tau(t)$. Given the target scatters the radar signal with a reflectivity coefficient ρ, the radar return signal can be expressed by

$$x_a(t) = \rho \, \exp\left(j2\pi f_c(t - \tau(t))\right) = \rho \exp\left(j2\pi f_c\left(t - \frac{2R(t)}{c} \right) \right) \tag{12.4}$$

12.2.2 Micro-Doppler Signatures of Human Gait

The Doppler effect identifies the moving targets by observing a frequency shift in the radar signal reflected from the target. The frequency shift f_D, also referred to as Doppler frequency or Doppler shift, is related with the targets radial velocity v as

$$f_D = \frac{2v}{c - v} f_c \cos\theta \tag{12.5}$$

where:
  f_c is the radar carrier frequency
  c is the speed of light
  θ is the angle between the motion direction and the radar LOS

Assuming that c is much larger than the target velocity v of a moving target, the Doppler frequency that corresponds to $x_a(t)$ in Equation 12.4 is well approximated by

$$f_D \approx \frac{2v}{c} f_c \cos\theta = \frac{2v}{\lambda} \cos\theta \tag{12.6}$$

Note that as the velocity of the target changes with time, the Doppler frequency shift becomes time dependent. That is,

$$f_D(t) \approx \frac{2v(t)}{\lambda} \cos\theta \tag{12.7}$$

This is generally the case for human movements as, for example, the body speeds up and slows down alternately throughout a gait cycle. Clearly, the observed Doppler frequency shift depends on the look angle θ and decreases to zero when this angle becomes perpendicular to the radar LOS.

The micro-Doppler effect describes the observation of signal components in the Doppler domain, which do not stem from the primary motion of the target, but rather from its individually moving components (Chen et al. 2014). Typically, these micro-Doppler components occur as modulations, around the carrier frequency of the transmitted radar signal, and form patterns that are characteristic for an object or a movement. These patterns are referred to as micro-Doppler signatures. By analyzing a target's micro-Doppler signature, its motion can be inferred.

In the case of a walking person, the overriding Doppler signature is the Doppler shift caused by the torso's motion, whereas the micro-Doppler modulations around that frequency stem from the swinging of arms and legs (Chen et al. 2014). Highly articulated and flexible targets, such as the human body, can be considered as a collection of individual point scatterers. Hence, the received radar signal $x(t)$ can be described by the integration over the target region Ω such that

$$x(t) = \int_\Omega x_a(t)\mathrm{d}a \tag{12.8}$$

which comprises a superposition of delayed, attenuated, and Doppler-shifted versions of the transmitted radar signal. Here, we are particularly interested in the individual Doppler shifts induced by different moving body parts, all of which are contained in $x(t)$. An appropriate method to analyze these micro-Doppler components in the radar return signal is the time–frequency representation of $x(t)$, which reveals the time dependency of the Doppler shifts in the joint Doppler frequency and time domain.

12.2.3 Time–Frequency Representations of Human Gait Signatures

When observing a nonrigidly moving target, the radar signal return is composed of multiple time-varying Doppler and micro-Doppler components. In the joint time–frequency domain, these different components form an intricate signature that is characteristic of the observed motion. These captured time–frequency signatures describe the signal's local frequency behavior and are considered instrumental to signal analysis and feature extraction (Amin et al. 2015a).

12.2.3.1 Short-Time Fourier Transform and Spectrogram

The most common choice of the time–frequency signal representation of radar signals scattered by humans in motions is the spectrogram (Chen and Ling 2001, Amin et al. 2016). In particular, the spectrogram is well suited to extract the Doppler and micro-Doppler signatures of human gait. It belongs to the class of quadratic time–frequency signal representations and depicts how the signal power varies with time and frequency. The spectrogram is obtained by calculating the magnitude squared of the short-time Fourier transform (STFT). The STFT is defined for a discrete-time signal $x(n) = x(n\Delta t)\Delta t$, sampled at an interval of Δt, as (Stanković et al. 2013)

$$\text{STFT}(n,\omega) = \sum_{m=-\infty}^{+\infty} w(m)x(n-m)e^{-j\omega m} \qquad (12.9)$$

where:
  n is the discrete-time index
  ω is the normalized frequency
  $w(\cdot)$ is a discrete-time window function

From the STFT, the spectrogram is then obtained as

$$S(n,k) = \left| \text{STFT}(n,\omega) \right|_{\omega=\frac{2\pi}{M}k}^2 = \left| \sum_{m=0}^{M-1} w(m)x(n-m)e^{-j2\pi\frac{mk}{M}} \right|^2, n = 0,\dots,N-1 \qquad (12.10)$$

where:
  k is the discrete frequency index
  M is the length of the window function
  N is the number of samples in $x(n)$

Here, we assume that the number of discrete frequency points is equal to the window length.

When calculating the spectrogram, the choice of the window function is eminent as it trades off the time and frequency resolutions. A larger window length will typically degrade the time resolution as signal components with a shorter duration than the window length may get smeared out. However, a smaller window length amounts to convolution with a wideband signal limiting the frequency resolution. As the time window lengths in time and frequency are inversely proportional to each other by the uncertainty principle, using the STFT has the effect of trading off time resolution against frequency resolution (Chen and Ling 2001). The multiwindow spectrogram can be used to partially mitigate this effect and has been successfully used for fall detection (Jokanović et al. 2015).

A typical spectrogram of a human walking toward the radar system is shown in Figure 12.1a, in which the color indicates the received power in dB. The strongest Doppler component, which is identified by the highest energy in the spectrogram, is due to the torso as it has the largest radar cross section (RCS). The torso time–frequency signature reveals the acceleration and deceleration phases during the human walking cycle. Here, it periodically varies between approximately 50 and 150 Hz Doppler frequency. Additionally, when taking a step, the swinging of the foot leads to a sinusoidal-shaped micro-Doppler signature in the spectrogram, which is simply referred to as stride signature hereafter. The depicted spectrogram reveals five stride signatures during the 5 s measurement, indicating that the observed walk is relatively slow. Note that there is no arm swinging involved in this motion.

In contrast, the micro-Doppler signature of a person walking away from the radar system is shown in Figure 12.1b. Clearly, the micro-Doppler stride signatures are different compared to the toward radar measurement. Figure 12.2 shows an excerpt of the spectrogram in Figure 12.1b, which corresponds to the swing phase of one leg during the human walking cycle.

The swing phase consists of acceleration, mid-swing, and deceleration phases (Chen 2011). In the acceleration phase, marked with (I) in Figure 12.2, the heel comes off the ground and the thigh swings forward. The latter is considered a pendulum-like motion and, as such, it reveals a sinusoidal-shaped micro-Doppler component, which can be observed between 0.2 and 0.5 s with a maximum Doppler frequency of 200 Hz.

In the mid-swing phase, marked with (II) in Figure 12.2, the swinging of the foot causes the highest Doppler frequency, that is, it has the highest velocity, here up to 350 Hz between 0.5 and 0.7 s. Again, due to its pendulum-like motion, the micro-Doppler component has a sinusoidal shape in the spectrogram. Embraced in the foot signature, the lower leg signature becomes visible in the form of a spike, that is, an impulse-like behavior in the

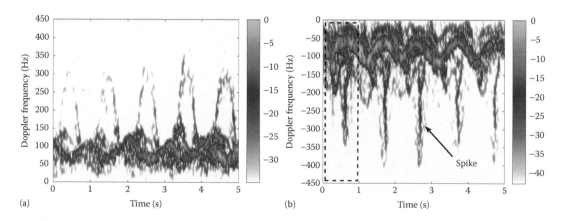

FIGURE 12.1
Typical spectrogram of a person walking slowly (a) toward and (b) away from the radar. (From Seifert, A.-K. et al., New analysis of radar micro-Doppler gait signatures for rehabilitation and assisted living, in *Proceedings of the 2017 IEEE International Conference on Acoustics, Speech and Signal Processing* © 2017 IEEE. With Permission.)

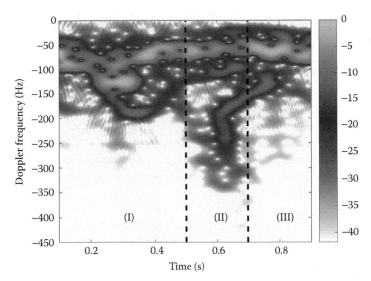

FIGURE 12.2
Excerpt of Figure 12.1b revealing the characteristics of a micro-Doppler stride signature in away from radar walks. (From Seifert, A.-K. et al., New analysis of radar micro-Doppler gait signatures for rehabilitation and assisted living, in *Proceedings of the 2017 IEEE International Conference on Acoustics, Speech and Signal Processing* © 2017 IEEE. With Permission.)

time–frequency domain. The spike appears during the mid-swing phase when the swinging foot causes the highest Doppler shift corresponding to its highest velocity. However, due to the larger cross-sectional area in comparison with the foot, the calf reveals a higher energy in the time–frequency representation and is eclipsing the foot signature. Hence, the foot signature may not be noticeable by standard feature extraction techniques.

Progressing in time, the spike passes into a half sinusoidal-shaped micro-Doppler signature. This phase represents the straight leg swinging to the front of the body during the deceleration phase and is marked with (III) in Figure 12.2. This part of the signature can be attributed to reflections from the upper calf. It experiences the same deceleration as the foot, for which reason the signature is in parallel with the foot's signature. However, the calf's motion leads to a smaller Doppler frequency as the swinging angle with respect to the knee joint is smaller compared to that of the foot.

It is fair to conclude that the micro-Doppler signatures of a person walking toward the radar are substantially different from those where the radar has a back view. This has to be taken into account when extracting the features from the spectrogram for classification of human gait.

12.2.3.2 Other High-Resolution Quadratic Time–Frequency Distributions

Besides the spectrogram, there are a large number of quadratic time–frequency representations defined within Cohen's class (Cohen 1989, 1995) and given by

$$D(t,f) = \sum_{v=-\infty}^{\infty} \sum_{\tau=-\infty}^{\infty} \Phi(v,\tau)A(v,\tau)e^{j2\pi t v - j2\pi f \tau} \tag{12.11}$$

where:
$\Phi(v,\tau)$ is the time–frequency kernel
$A(v,\tau)$ is the ambiguity function

The latter is defined in the Doppler–lag domain as

$$A(v,\tau) = \sum_{t=-\infty}^{\infty} x\left(t+\frac{\tau}{2}\right) x^*\left(t-\frac{\tau}{2}\right) e^{-j2\pi tv} \tag{12.12}$$

where:
  v is the Doppler frequency and τ is the time lag.

High-resolution quadratic time–frequency distributions (QTFDs) can be designed by carefully choosing time–frequency kernels that exhibit a two-dimensional low-pass filter characteristic in the Doppler–lag domain. An example of a high-resolution QTFD is the B-distribution, whose kernel is defined as

$$\Phi(v,\tau) = g(\tau)G(v) = |\tau|^{\beta} \frac{\left|\Gamma(\beta + j\pi v)\right|^2}{2^{1-2\beta}\Gamma(2\beta)}, \tag{12.13}$$

with $|v| \leq 0.5, |\tau| \leq 0.5$, and $0 \leq \beta \leq 1$.

  Improved time–frequency representation capabilities are provided by the extended modified B-distribution (EMBD), which is based on the B-distribution and defined as

$$\Phi(\theta,\tau) = \frac{\left|\Gamma(\beta + j\pi\theta)\right|}{\Gamma^2(\beta)} \frac{\left|\Gamma(\alpha + j\pi)\right|}{\Gamma^2(\alpha)} \tag{12.14}$$

where $|v| \leq 0.5, |\tau| \leq 0.5, 0 \leq \beta \leq 1, 0 \leq \alpha \leq 1$, and $\Gamma(\cdot)$ is the standard Gamma function (Boashash et al. 2015). Here, the parameters α and β control the length of the Doppler and lag window, respectively. This means that the EMBD allows for adjusting the length of the window in both Doppler and lag domains separately. Examples of TFRs of gait signatures utilizing EMBD are shown in Figure 12.3a and b, which correspond to the same measurements as depicted in Figure 12.1 using the spectrogram.

12.2.3.3 Wavelet Transform

The legs' movement reveals high-frequency components of short bursts. The Fourier transform is unsuitable for analyzing signals of this kind owing to its limitations of offering constant time and frequency resolutions for both low and high frequencies. Using the wavelet transform (WT), these resolutions vary with scale, which is proportional to frequency. Accordingly, higher time resolution can be achieved for high-frequency components. This is particularly relevant for analyzing the salient impulsive-like micro-Doppler component in the stride motion underlying the micro-Doppler signatures of humans walking away from the radar system. The continuous WT is defined as (Stanković et al. 2013)

$$\mathrm{WT}(t,a) = \frac{1}{\sqrt{|a|}} \int_{-\infty}^{\infty} x(\tau) h^*\left(\frac{\tau-t}{a}\right) d\tau \tag{12.15}$$

where:
  $x(\cdot)$ is the time-domain radar signal
  $h(\cdot)$ is the wavelet function
  $*$ denotes the complex conjugate
  a is the scale

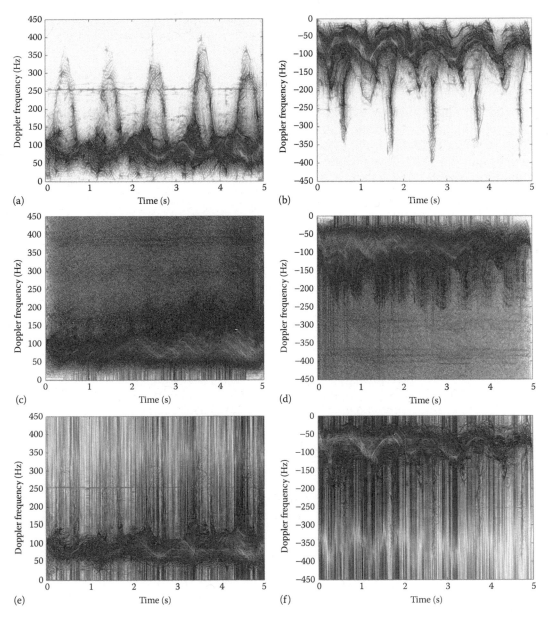

FIGURE 12.3
Comparison of different time–frequency distributions for a radar return of a person walking toward (left) and away from (right) the radar. (a–b) EMBD ($\alpha = 0.04$, $\beta = 0.2$). (c–d) Wigner–Ville distribution. (e–f) Choi–Williams distribution.

That is, the signal is projected onto scaled and translated versions of the wavelet function, where the choice of the wavelet function depends on the application.

Similar to the definition of the spectrogram, we find the scalogram as the squared magnitude of the WT as

$$W(t,a) = \left| WT(t,a) \right|^2 \tag{12.16}$$

The scalogram represents the energy for each wavelet coefficient at each scale. It is noted that discussion of QTFDs and linear multiresolution transforms and their use in fall detection are given in Chapter 4.

12.2.4 CVD for Human Gait Analysis

The human walk is periodic with each stride or half of the gait cycle. In order to analyze these periodicities in the spectrogram, we generate the CVD (Otero 2005, Björklund 2012, Björklund et al. 2015, Clemente et al. 2015). The CVD is obtained by taking the Fourier transform of the spectrogram along frequency bins. Using the definition of the spectrogram in Equation 12.10, the CVD is thus calculated as

$$C(\mathcal{E},k) = \mathcal{F}_n\{S(n,k)\} = \left| \sum_{n=0}^{N-1} S(n,k)e^{-j2\pi\frac{\mathcal{E}n}{N}} \right| \tag{12.17}$$

where:
  $\mathcal{E}$ is the cadence frequency
  N is the number of samples in a frequency slice of the spectrogram

For a walking person, the CVD reveals the average walking speed as well as the step rate in a joint Doppler-frequency and cadence-frequency representation. Figure 12.4 shows the CVD of a person walking slowly toward the radar, with Figure 12.1a showing the corresponding spectrogram.

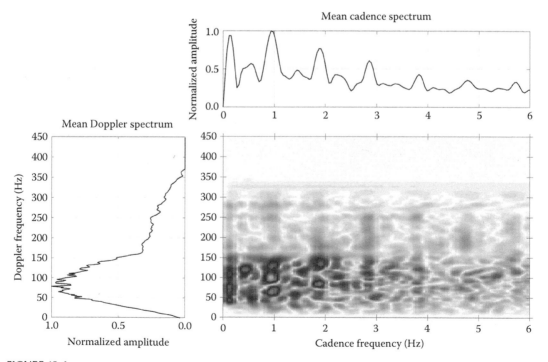

FIGURE 12.4
CVD for a normal walk toward the radar (see Figure 12.1a for the corresponding spectrogram), along with its mean Doppler spectrum and its mean cadence spectrum.

Taking the mean over all cadence frequencies for each Doppler frequency results in the mean Doppler spectrum. The Doppler frequency corresponding to the maximum in the mean Doppler spectrum $f_{D,\max}$ is proportional to the average walking speed v_0 of a walking person, that is,

$$v_0 = \frac{\lambda}{2} \cdot f_{D,\max} \tag{12.18}$$

where λ is the wavelength of the transmitted radar signal. Here, the maximum of the mean Doppler spectrum is approximately 80 Hz, which corresponds to an average walking speed of 0.5 m/s given a transmitting frequency of 24 GHz.

From the projection of the CVD on the cadence-frequency axis, that is, taking the mean over all Doppler frequencies for each cadence frequency, one can extract the step rate by finding the cadence frequency that reveals the highest amplitude in the mean cadence spectrum. The latter is herein referred to as fundamental cadence frequency. From Figure 12.4, we observe a fundamental cadence frequency of 1 Hz, which is consistent with five strides in a 5 s data measurement in Figure 12.1a.

From the stride rate estimate f_0 and the mean velocity estimate v_0, we can calculate the average length of a stride l_s as

$$l_s = \frac{v_0}{f_0} \tag{12.19}$$

The average walking speed and stride rate are commonly referred to as physical features, as they are easily interpretable. Other physical features include, for example, the torso Doppler frequency, the total bandwidth of the Doppler signal, and the offset of total Doppler (Kim and Ling 2009). A detailed overview of feature extraction techniques for human gait classification is given in Section 12.2.5.

12.2.5 Human Micro-Doppler Signature Classification

Classifications of different human activities require feature extractions from the observed radar signal return. The general processing steps for performing radar-based micro-Doppler classification are depicted in Figure 12.5. First, an appropriate domain must be chosen which best reveals the characteristics of the different motions. The most common signal domains for analyzing the radar return signal of human motions are TFRs and CVDs. Once meaningful features are extracted from one or multiple domains, feature selection can proceed to retain the minimum number of significant features. The extracted features are then used to train a classifier and perform the classification of test data.

Table 12.1 gives an overview of a variety of micro-Doppler features that have been used to perform human motion classification, including gait recognition, which identifies different walking styles such as normal walking, running, jogging, assisted, and unassisted walks.

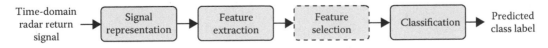

FIGURE 12.5
Typical processing steps for radar-based micro-Doppler signature classification.

TABLE 12.1

Overview of Micro-Doppler Features Used for Human Motion Classification and Gait Recognition

Features	Method	Description
Physical	Spectrogram	Doppler frequency profile (Clemente et al. 2013); torso Doppler frequency, total bandwidth of Doppler signal, offset of total Doppler, bandwidth without micro-Doppler, normalized standard deviation of Doppler signal strength, period of the limb motion (Kim and Ling 2009); base velocity, step time (Wang et al. 2014)
	Hermite S-method	Envelope function of micro-Doppler and time offset between the corresponding minima and maxima (Orović et al. 2011)
	EMBD	Width ratio, mean magnitude of intensity differences of stride signatures (Amin et al. 2015b)
	CVD	Base velocity, stride rate (Yardibi et al. 2011); base velocity, stride length, appendage/torso ratio (Otero 2005); CVD frequency profile (Clemente et al. 2013); cadence frequencies, velocity profiles, base velocity (Björklund et al. 2015); step repetition frequency, mean Doppler spectrum (features) (Ricci and Balleri 2015)
	Empirical mode decomposition (EMD)	Energy of intrinsic mode functions (Fairchild and Narayanan 2014)
Transform-based	Discrete cosine transform (DCT)	DCT coefficients (Molchanov et al. 2011)
	Fourier transform (FT)	Highest peak of FT, FT's power, highest peak of FT's derivative (Philips et al. 2012)
	Image moments	Pseudo-Zernike moments of CVD (Clemente et al. 2015)
Speech inspired	Cepstrum/linear prediction coding (LPC)	Cepstrum (Bilik and Khomchuk 2012); LPC coefficients (Yessad et al. 2011, Javier and Kim 2014); linear predictive cepstral coefficients (LPCCs) (Yessad et al. 2011, Bilik and Khomchuk 2012); Mel frequency cepstral coefficients (MFCCs) (Yessad et al. 2011, Bilik and Khomchuk 2012)
Nonparametric	Subspace representations of TFRs	PCA of spectrogram (Mobasseri et al. 2009); PCA of spectrogram frequency profile, CVD frequency profile (Clemente et al. 2013); two-directional 2D PCA and two-directional 2D LDA on spectrogram (Li et al. 2012); two-directional 2D PCA and two-directional 2D LDA of log-Gabor-filtered TFRs utilizing the STFT and the S-method (Tivive et al. 2015)
	TFRs	Distance between training and testing spectrogram (Lyonnet et al. 2010); hierarchical image classification architecture with nonlinear directional and adaptive 2D filters utilizing the spectrogram (Tivive et al. 2010)

Physical features are best interpretable as they describe classical gait parameters, such as the average gait velocity, the average stride rate, and the stride variability. They are typically extracted from the TFRs or CVDs of the data. Other approaches utilize well-known feature extraction methods from other disciplines, for example, speech or image processing. The importance of feature selection for human micro-Doppler classification and the impact of the radar's parameters and test scenario on the classification performance are discussed in the works of Gürbüz et al. (2013), Tekeli et al. (2016), and Gürbüz et al. (2015).

12.3 Human Gait Recognition

Considering human gait with and without assistive walking devices, we aim at discerning different variations of the same motion category, which is referred to as intraclass motion discrimination. For example, within the class of human walk, it is important to distinguish between normal and abnormal walks. Abnormal gait patterns arise, for example, from limping, which is typically characterized by the inability to normally bend one of the knees. Further, the use of assistive walking devices, such as a cane or a walker, leads to distinct gait characteristics.

The challenge in intraclass motion discrimination is selecting features that capture the underlying gait characteristics of different walking styles. After all, depending on the specific motion, the intraclass variability due to different human test subjects may be significant. This implies that features should to be chosen such that they represent a generic person and are invariant to the different population.

Focusing on human gait recognition, Wang et al. (2014) utilize the stride rate information and walking velocity information to assess human walking characteristics (see also Otero 2005, Yardibi et al. 2011). Orović et al. (2011) propose a human gait classification method that relies on the motion signature from arm and leg movements.

Analysis of human gait characteristics using assistive walking devices can be found in the works of Amin et al. (2015a, 2015b). It is shown that walking aids, such as a cane or a walker, can be detected by carefully analyzing the backscattered radar signal in the time–frequency domain. Moreover, they observe changes in gait patterns when using assistive walking devices compared to an unassisted gait. These changes were analyzed and revealed the following fundamental TFR characteristics. First, the stride that is aligned with the cane movement results in a strengthened radar return due to the combined scattering of the leg and the cane. In essence, the cane increases the RCS of the leg when moved in concert. Second, the stride signature, which encompasses the signature of the cane, is dispersed in time. This is attributed to the incremental time difference between the leg motion and the swinging of the cane, which are typically not perfectly aligned. The EMBD is used to analyze the radar signal return in the time–frequency domain, where the parameters are chosen as $\alpha = 0.04$ and $\beta = 0.2$. Elderly gaits with and without the use of a cane are analyzed.

To discriminate between these two gaits, two distinguishing features are extracted from the TFR. The mean magnitude of intensity difference reveals the use of the cane by returning high values for an assisted walk compared to a normal gait, which is motivated by the stronger scattering during the stride with a cane. From a time slice obtained from the time–frequency domain at 60% (found empirically) of the maximum Doppler frequency, the mean magnitude of intensity difference is calculated as

$$\overline{\Delta} = \frac{|\delta_1| + |\delta_2|}{2} \tag{12.20}$$

where δ_1 and δ_2 are the intensity differences between three consecutive peaks in the time slice. Using real data, values range from 0.58 to 0.76 and from 0.21 to 0.28 for a walk with and without a cane, respectively. That is, the mean magnitude of intensity difference reveals a higher value for assisted walks compared to normal walks.

To exploit the fact that the cane and the leg are typically not perfectly aligned, the width ratio measures the time dispersion of the stride signature with a cane. Using the time slice as above, the width ratio is defined as

$$R = \frac{\overline{T_{odd}}}{\overline{T_{even}}}. \tag{12.21}$$

Here, $\overline{T_{odd}}$ and $\overline{T_{even}}$ denote the mean temporal widths of the odd and even leg cycles of the time–frequency signatures, respectively. When the odd leg cycles are those associated with a cane, the width ratio is expected to reveal a much higher value for assisted walks compared to normal, in which case the width ratio would exhibit values close to 1. Using real data based on laboratory experiments, the width ratio varies from 2.61 to 7.89 for walks with a cane, and from 1.24 to 1.67 for walks without a cane.

The work by Gürbüz et al. (2016) investigates the effect of different walking aids on the characteristics of radar return time–frequency signatures of human walk. Different degrees of mobility are compared: a normal, that is, unaided, walk, walking with a limp, walking with a cane or tripod, walking with a walker, and using a wheelchair. Besides using different radar systems (ultra high frequency (UHF), L-band, K-band, 24/77 GHz), ultrasound systems (20–80 kHz) are examined. Micro-Doppler features are obtained from the most commonly used TFR, the spectrogram. Machine learning and pattern recognition techniques are applied to classify the different walking styles. Different types of micro-Doppler features are investigated. Traditionally, physical features are extracted from the spectrogram and relate to, for example, the mean gait velocity, mean stride frequency, stride variability, and bandwidth of Doppler signal. Moreover, features are extracted by making use of speech processing algorithms such as linear prediction coding (LPC) coefficients, cepstrum coefficients, and mel-frequency cepstrum coefficients (MFCCs). Feature selection is performed utilizing the mutual information feature selector under uniform information distribution algorithm.

Using the 24 GHz, pulse-Doppler radar system, features were extracted utilizing the pseudo-Zernike moments, as suggested by Clemente et al. (2015), and a support vector machine (SVM) for classification. Using measurements of 5 and 10 s duration, a classification accuracy of 72.6% and 79.7% was achieved, respectively. Unaided walks were correctly identified as such in 73.3% (81.3%) of the cases, whereas the cane-assisted walk was correctly classified in 55.4% (61.9%) of the cases.

12.3.1 Experimental Setup

The measurements presented in this work were obtained using an ultra-wideband radar system (Ancortek Inc. 2016). The radar was set to frequency-modulated CW (FMCW) mode with linear frequency modulation sweeps and a carrier frequency of 24 GHz. Doppler filtering was applied to obtain the velocity information of the target by utilizing the phase shift between different sweeps. All measurements were conducted in a semicontrolled environment at the Radar Imaging Lab at Villanova University, Villanova, Pennsylvania. In total, seven subjects, both males and females, were asked to walk slowly back and forth between two points in front of the radar, approximately 4.5 and 1 m from the antenna feed point. None of the subjects were disabled or elderly, but were instructed to walk like an elderly, that is, very slowly and with small lifting of the feet. The feed point of the antenna was positioned 1.15 m above the floor. Data were collected with a non-oblique view to the targets and at a 0° angle relative to the radar LOS.

12.3.2 Human Gait Patterns in the Time–Frequency and Cadence-Velocity Domains

In order to understand the composition of micro-Doppler signatures of different gaits, we carefully analyze the corresponding characteristics in the time–frequency and cadence-velocity domains. From the spectrogram, we can deduce the course of motion and provide a biomechanical interpretation of the observed data. The CVDs reveal the periodicities in the spectrogram underlying specific walking styles, for example, whether the gait is considered normal or whether a walking aid is used.

Figure 12.6 shows the spectrograms of different walking styles when moving toward the radar system. A person walking slowly toward the radar is depicted in Figure 12.6a. Here, the swinging of the feet leads to clear sinusoidal-shaped micro-Doppler signatures around the torso's radar return, which can be identified by the highest energy in the spectrogram. Note that there is no arm swinging involved in all presented measurements. The use of assistive walking devices affects the walking style, that is, the gait pattern, and thus leads to distinct micro-Doppler frequencies. Figure 12.6c–f shows the spectrograms of a person walking with one or two canes. In Figure 12.6c and e, the cane is moved in a complete sync with the opposite-side leg. Being aligned with one leg, the cane's micro-Doppler signature combines with the foot signature and leads to higher energy values in the spectrogram. This can, for example, be seen in the first and third micro-Doppler cycles presented in Figure 12.6c. Also, we observe that the resulting micro-Doppler signature is more filled compared to a normal stride signature because the movement of the (upper) arm contributes to the signature. In Figure 12.6e, each cycle is composed of the overlaying foot and cane signature. However, Figure 12.6d and f depicts the cases where the cane is moved alone, that is, not aligned with either leg. In these cases, the leg and cane micro-Doppler signatures are nonoverlapping such that distinct leg and cane micro-Doppler signatures are revealed. In Figure 12.6d, the first, fourth, and seventh cycles stem from the cane movement; the remainders are normal stride signatures as in Figure 12.6a. In Figure 12.6f, alternating leg and cane signatures can be observed, that is, the second, fourth, and sixth sinusoidal signatures are attributed to the cane movement. In this example, the cane signatures exhibit a higher maximum Doppler frequency compared to the stride signatures. In general, we expect the latter whenever the cane is actually used to relieve a foot or leg, which typically results in a fast movement of the cane.

Figure 12.7 depicts the corresponding CVDs for the measurements in Figure 12.6, along with the mean cadence spectra. In the latter, the first peak in the lower cadence frequencies is due to the torso's motion. When walking normally toward the radar system, as depicted in Figure 12.7a, we additionally observe a strong peak at 0.98 Hz cadence frequency, which indicates the stride rate. Here, the person takes about one step per second, which is equivalent to a walking speed of about 3.6 km/h and which is recognized as a slow walk. Further, we observe strong harmonics of the fundamental cadence frequency at integer multiples of 0.98 Hz. As outlined previously, the use of assistive walking devices changes the periodicities in the TFR. Thus, the pattern in the corresponding CV domain is different compared to that of an unassisted walk. When the cane is moved in alignment with one leg, the cane alters every second stride signature, which results in a peak in the mean cadence spectrum at half the stride rate. This can be seen in Figure 12.7c, where the stride rate is 0.78 Hz, but the highest peak in the mean cadence spectrum appears at approximately 0.4 Hz. Similarly, walking with a cane, where the cane is moved while both legs are stationary, leads to specific sequences of leg and cane micro-Doppler signatures and thus distinct patterns in the CVD. This case is depicted in Figure 12.7d, where the mean cadence spectrum peaks at 1.48 Hz, which describes

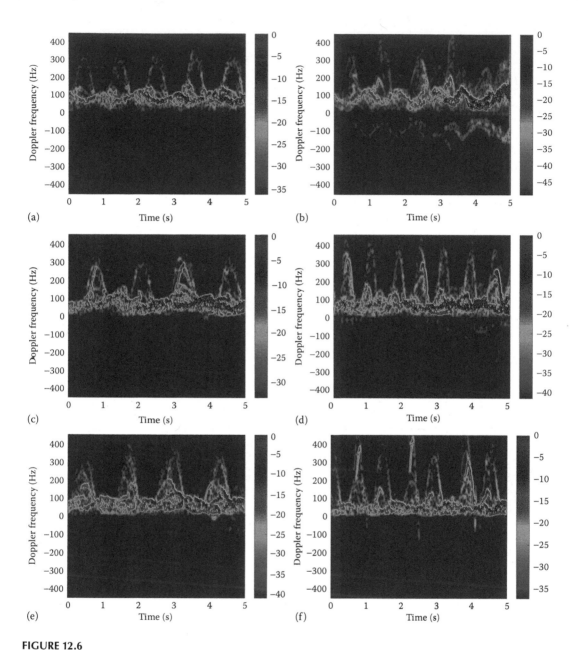

FIGURE 12.6

(a–f) Different walking styles moving toward the radar system. ([a, c, and d] From Seifert, A.-K. et al., Radar-based human gait recognition in cane-assisted walks, in *Proceedings of the 2017 IEEE Radar Conference* © 2017 IEEE. With Permission.)

the rate of adjacent micro-Doppler signatures. Additionally, the mean cadence spectrum shows a peak at one-third of the fundamental cadence frequency, that is, approximately 0.5 Hz. The latter relates to the similarity of every third micro-Doppler signature, which is a pair of cane or leg signatures. Moreover, a strong harmonic is present at two-thirds of the fundamental cadence frequency. Note that we cannot define a stride rate in this case because the strides are not periodic. Figures 12.7e and f shows the CVDs and mean

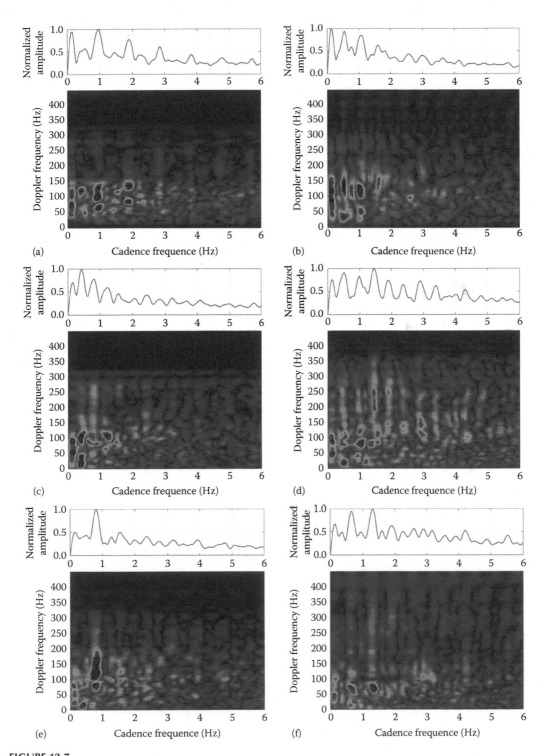

FIGURE 12.7
(a–f) CVD analysis of walking with assistive walking devices toward the radar. ([a, c, and d] From Seifert, A.-K. et al., Radar-based human gait recognition in cane-assisted walks, in *Proceedings of the 2017 IEEE Radar Conference* © 2017 IEEE. With Permission.)

cadence spectra for a person walking with two canes. Moving both canes in alignment with the legs, as shown in Figure 12.7e, the mean cadence spectrum reveals a dominant peak at approximately 0.8 Hz, which is the fundamental cadence frequency and corresponds to the stride rate. When both canes are moved independently of the legs, the mean cadence spectrum shows an additional peak at half the fundamental cadence frequency, as depicted in Figure 12.7f. Here, the fundamental cadence frequency is 1.2 Hz, which corresponds to the stride rate and indicates a rather slow walk. An additional peak in the mean cadence spectrum can be found at approximately 0.6 Hz, which relates to the period of every other stride signature, that is, the periodicity of leg- or cane-only micro-Doppler components.

Figure 12.8 shows the spectrograms of the same walking styles, but with the radar having a back view of the person. The micro-Doppler signature of a person walking slowly away from the radar is shown in Figure 12.8a. Here, the salient characteristic of a micro-Doppler stride signature is the spike, as outlined in Section 12.2.3. This feature is also visible in every other stride signature in Figure 12.8b, which shows a person walking with a limp, that is, only one knee is bent, and the other is not. Here, the second, fourth, and sixth micro-Doppler stride signatures are abnormal. Clearly, the salient spike does not show in abnormal stride signatures. Instead, we observe two overlaying sinusoidal-shaped components, which arise from the upper calf and the foot swinging jointly forward in a pendulum-like manner. Figure 12.8c shows the spectrogram of a person walking with one cane, where the cane is moved in alignment with the opposite side leg. The movement of the cane alters every other micro-Doppler signature, and its signature superimposes on the stride's signature. The first, third, and fifth stride signatures in Figure 12.8c are due to taking a step and moving the cane forward synchronously. They are visually distinguishable from normal stride signatures that appear in the second and fourth micro-Doppler cycles. Combined stride and cane micro-Doppler signatures can also be observed in Figure 12.8e, where a person is walking with two canes using both in a synchronized manner. Again, the micro-Doppler signatures are visually distinguishable from those in the spectrogram in Figure 12.8a. Cases in which the cane is moved alone are shown in Figure 12.8d and f. Here, distinct cane and leg signatures become visible. In Figure 12.8d, the second and fifth micro-Doppler signatures correspond to the cane being moved forward. The remainders are normal stride signatures as in Figure 12.8a. In Figure 12.8f, the second, fourth, and sixth micro-Doppler signatures correspond to the cane's movement. Note again that the salient spike is present in the remaining stride signatures, but not in the cane's signatures.

The corresponding CVDs are presented in Figure 12.9. In general, the mean cadence spectra reveal stronger peaks at the fundamental cadence frequency compared to the toward radar measurements. However, the effects of the cane on the CVD structure are less prominent.

12.3.3 Detection of Gait Abnormalities

First, we aim to detect gait abnormalities in order to distinguish between a normal and an abnormal walk. Here, a gait is considered abnormal if the person is, for example, limping, which is typically characterized by unbent or not fully bent knees compared to a normal walk. Further, the use of assistive walking devices is considered an abnormal walk, indicating a level of physical or cognitive impairment. Thus, we want to access from the radar return signal whether the person has a healthy gait or whether the gait shows abnormalities.

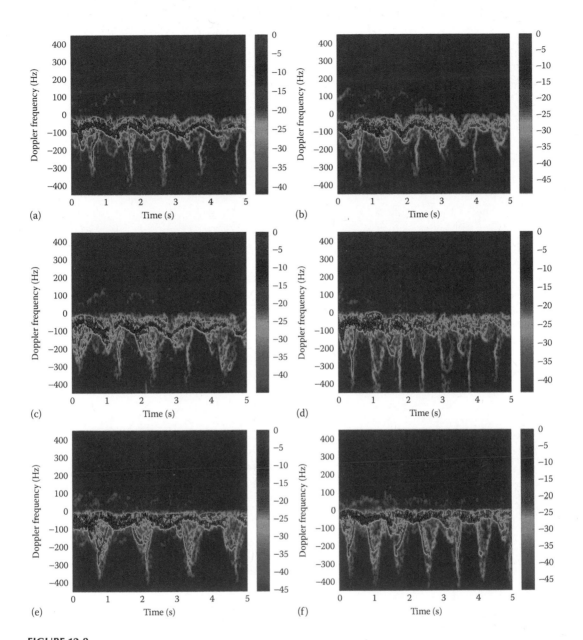

FIGURE 12.8
(a–f) Different walking styles receding from the radar system. ([a, c, and d] From Seifert, A.-K. et al., Radar-based human gait recognition in cane-assisted walks, in *Proceedings of the 2017 IEEE Radar Conference* © 2017 IEEE. With Permission.)

12.3.3.1 Feature Extraction

A major challenge in human gait classification is the choice of features to discriminate different walking styles. Many works have presented extraction methods for human motion classification and recognition (Kim and Ling 2009, Björklund et al. 2015, Gürbüz et al. 2016). In Sections 12.3.3.1.1 and 12.3.3.1.2, we use timescale features from the WT of the backscattered radar signal to detect abnormalities in human walks.

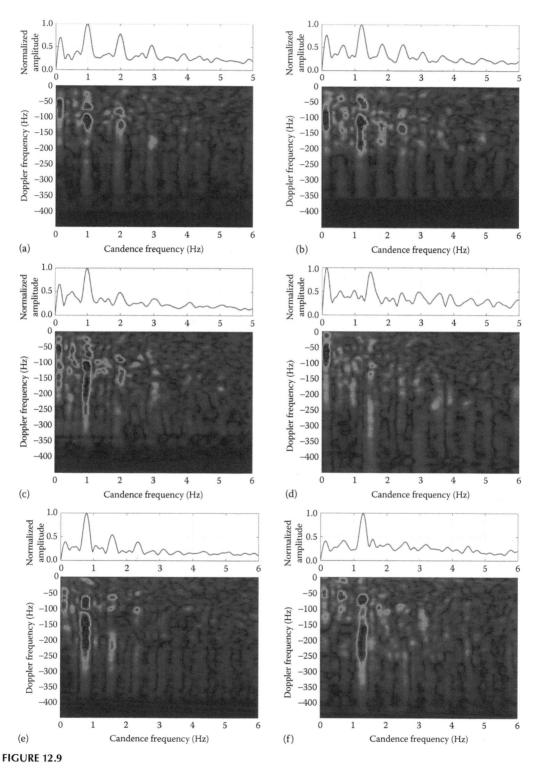

FIGURE 12.9
(a–f) CVD analysis of walking with assistive walking devices away from the radar. ([a, c, and d] From Seifert, A.-K. et al., Radar-based human gait recognition in cane-assisted walks, in *Proceedings of the 2017 IEEE Radar Conference* © 2017 IEEE. With Permission.)

12.3.3.1.1 Prescreening

In order to find the time location of a micro-Doppler stride signature, the WT coefficients of the radar return signal are utilized, where the wavelet function is the reverse biorthogonal 3.3 (Su et al. 2015). Here, we use the WT coefficients at scale four, which corresponds to a Doppler frequency range of 180–220 Hz and was found to best reveal the impulsive-like behavior of the spike in the frequency domain.

The stride signature location in time is determined by detecting maxima in the short-time energy of the corresponding wavelet coefficients WT$(t,4)$ obtained using Equation 12.15 as

$$D(k) = \sum_{t=1}^{N} \left\{ w(t) \mathrm{WT}\left(t + k \cdot N/5, 4\right) \right\}^2 \qquad (12.22)$$

where:
  $w(\cdot)$ is a Hamming window of length 0.5 s
  N is the window length in samples
  k denotes the frame index

The windows overlap by 80%, that is, the resolution in time for detecting a stride signature is 0.1 s.

Figure 12.10 shows examples of micro-Doppler stride signatures that are detected by the prescreener. Note that the depicted spectrograms are just for illustration purposes, as the prescreener actually utilizes the energy of the wavelet coefficients. In Figure 12.10a, each stride signature of a normal walk is detected, which is indicated by the black boxes. Clearly, the first detection is a false alarm (FA) in this case. In Figure 12.10b, the performance of the prescreener in case of an abnormal walk is shown. Again, all stride signatures are correctly detected, where the last stride signature is missed, as it is only partly present.

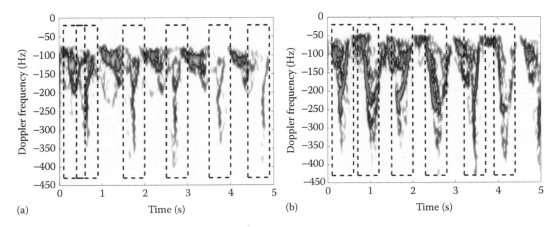

FIGURE 12.10
(a,b) Examples of micro-Doppler stride signatures detected by the prescreener.

12.3.3.1.2 *Timescale Features*

Next, we extract the features from the timescale domain by forming a feature vector for each detected frame during prescreening as follows: Let $E(t,a)$ be the relative energy of a wavelet coefficient at scale a and time lag t, we calculate an energy profile as

$$F(t) = \sum_{a=1}^{M} E(t,a) \tag{12.23}$$

where $M = 8$ is the number of scales used.

Next, we determined the time span occupied by the signature by thresholding the energy profile $F(t)$ at 20% of its maximum. The rationale behind it is that due to its impulsive-like behavior, the normal stride signature spans a much shorter interval than abnormal ones. Thus, the first feature is defined as the time duration t_{span} that contains the wavelet coefficients with the highest energy over all scales (Figure 12.11).

Further, we observe from the TFRs that the spike manifests itself in higher frequency bands or scales. Hence, we define the sum of the relative energy of wavelet coefficients at scale a as

$$G(a) = \sum_{t=1}^{N} E(t,a) \tag{12.24}$$

where N is the number of times samples in the detected frame. The final feature vector for a detected micro-Doppler stride signature is then given by

$$\vec{z}_1 = \left[t_{\text{span}} \, G(1)G(2)\ldots G(M) \right]^T \tag{12.25}$$

where $M = 8$ is again the number of scales used such that $\vec{z}_1 \in \mathbb{R}^{9\times 1}$.

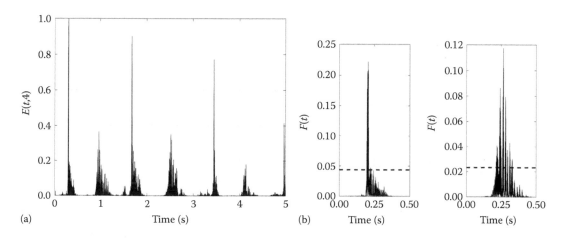

FIGURE 12.11
(a) Relative energy of WT coefficients at scale four (normalized) for the measurement depicted in Figure 12.10b. (b) The first and second stride signatures of (a) thresholded at 20% of their respective maximum.

12.3.3.2 Classification

From the measurements listed in Table 12.2, a total number of 216 micro-Doppler stride signatures were detected by the prescreener with a FA rate of 6% and a missed detection (MD) rate of 5%. After prescreening, there are 104 normal gait signatures and 89 abnormal signatures available for classification. Using the feature vector defined in Equation 12.25, we train a linear SVM (Boser et al. 1992, Cortes and Vapnik 1995) using 70% of the detected frames, whereas the remainder is kept for testing. Classification results are presented in Table 12.3. The numbers give the correct classification rate for each true class in percent. All rates are obtained by averaging 100 classification results, where training and test samples were randomly chosen. The overall detection rate is 76%.

Here, we only evaluated the measurements of 5 s duration with only three to seven stride signatures present. In practice, the observation time could be much longer, and, additionally, the classifier could be trained to be person specific, which would likely improve the classification performance.

12.3.4 Gait Classification in Cane-Assisted Walks

Next, we aim at distinguishing different walking styles with and without assistive walking devices, that is, normal walk, walking with a cane and walking with a cane out of sync. Here, it is important that the use of the cane does not affect the estimation of commonly used gait features such as the walking velocity and the stride rate.

TABLE 12.2

Data Set 1

Walking Style	Away
Normal walk	7
Walking with one cane	18
Walking with two canes	9
Walking without bending knees	2
Walking with bending only one knee	6
Total	42

Source: Seifert, A.-K. et al., New analysis of radar micro-Doppler gait signatures for rehabilitation and assisted living, in *Proceedings of the 2017 IEEE International Conference on Acoustics, Speech and Signal Processing* © 2017 IEEE.

TABLE 12.3

Classification Results for Micro-Doppler Stride Signatures

		Predicted Class	
		NW	ANW
True Class	Normal walk (NW)	80	20
	Abnormal walk (ANW)	29	71

Source: Seifert, A.-K. et al., New analysis of radar micro-Doppler gait signatures for rehabilitation and assisted living, in *Proceedings of the 2017 IEEE International Conference on Acoustics, Speech and Signal Processing* © 2017 IEEE.

As outlined in Section 12.3.2, CVDs are well suited to reveal the periodicities in human gait. In order to perform the classification of different human walking styles, both assisted and unassisted, we first extract classical gait features from the cadence velocity domain, that is, the average walking speed and the step rate (Otero 2005, Björklund et al. 2015, Clemente et al. 2015). Additionally, we perform a principal component analysis (PCA) of the CVDs to recognize the patterns in a subspace representation of the CVDs. A similar approach, that is, PCA of spectrograms, has been successfully applied to radar-based human gait recognition (Mobasseri et al. 2009) as well as fall motion detection (Jokanović et al. 2016).

12.3.4.1 Feature Extraction

The CVDs are obtained from the spectrograms of 6 s measurements sampled at 2.56 kHz. The STFT is calculated using a Hamming window of length $M = 255$, which corresponds to 0.1 s, and 2048 Doppler frequency bins. Thus, the spectrogram is given on a grid of 2048 Doppler frequency bins by 15,360 time samples. Then, frequency slices of the spectrogram are transformed to the cadence-frequency domain. Calculating the FT of the spectrogram along the time axis with $M = 2^{16}$ points leads to a cadence-frequency resolution of approximately 0.04 Hz. As we are expecting the stride rates in the range of 0.5–1.5 Hz for a normal human walk, the cadence-frequency resolution is sufficiently high for our purpose. Moreover, we find the maximum cadence frequency that can be detected is 1.28 kHz. Next, we extract the relevant part of the CVD, that is, Doppler frequencies between 0–450 and −450–0 Hz for toward and away from radar measurements, respectively, and cadence frequencies up to 3 Hz. The latter range is sufficiently large to capture the fundamental cadence frequency, which is typically the stride rate of the walk. Considering the CVD as an image, this excerpt of the CVD results in an image of dimension 361 × 78 pixels.

The aforementioned CVDs are used for extracting features as follows: First, the maximum of the mean Doppler spectrum is determined, which serves as an estimate of the average walking speed, also referred to as base velocity. Then, the maximum in the mean cadence spectrum is used as an estimate of the fundamental cadence frequency. Here, we exclude lower Doppler frequency components from the mean cadence spectrum calculation. This confines the mean cadence spectrum to showing the periodicities of leg and cane micro-Doppler signatures, which typically have absolute Doppler components larger than 150 Hz. In the case of unassisted walks, the fundamental cadence frequency corresponds to the stride rate. However, for assisted walks, this relation does not always hold. Particularly, if the cane is moved alone, the strides are usually nonperiodic, and thus, a stride rate cannot be defined. Further, to capture the profile of the mean cadence spectrum, we extract its dominant peaks and calculate the average peak distance in cadence frequency. The intuition is that the smaller the average peak distance, the more likely an assistive walking device is present.

Thus, the CV feature vector is defined as

$$\vec{f} = \left[v_0,\ f_0,\ \bar{d}_{\mathrm{pks}} \right]^T \tag{12.26}$$

where:
  v_0 is the estimated base velocity of the walk
  f_0 is the fundamental cadence frequency
  $\bar{d}_{\mathrm{pks}}$ is the average distance of detected peaks in the mean cadence spectrum

Next, we perform a PCA of the CVD images in order to obtain the features that describe the patterns inherent in the CVDs. For that, we form a $i \times j$ data matrix whose j columns contain vectorized CVD images with i pixels. Note that we arrange the CVD images row by row to a one-dimensional vector to take into account the relations of a given pixel to pixels in neighboring columns, which is more appropriate in this case. A vectorized CVD image is of size 28158×1, where the length of the vector describes the dimensionality of the vector space. Using the data matrix, we perform PCA to find the j eigenvectors of the vector space and the corresponding j eigenvalues. Principal components are those eigenvalues that explain most of the variance in the data. The eigenvectors of length i that correspond to the principal components span a subspace of the original vector space. Projections of data onto this subspace result in a data representation of much lower dimensionality and are used as features during classification. Choosing l principal components leads to a feature vector with l entries, which is denoted as

$$\vec{p} = \left[p_1, p_2, \ldots, p_l \right]^T \tag{12.27}$$

where $l < j \in \mathbb{N}$.

We obtain the final feature vector by concatenation of the CV feature vector and the PCA-based features as

$$\vec{z}_2 = \left[\vec{f}, \vec{p} \right]^T \tag{12.28}$$

such that $\vec{z}_2 \in \mathbb{R}^{(l+3) \times 1}$.

12.3.4.2 Feature Selection

Because of the inherent differences in micro-Doppler characteristics in toward and away from the radar walks, the performance of the classifier is sensitive to the direction of motion. Additionally, the classification accuracy depends on the number of features used.

Figure 12.12 depicts the classification accuracy as a function of the number of principal components, where the dashed lines refer to a reduced feature vector with PCA-based features only. The classification accuracy denotes the number of correct classifications divided by the total number of samples in the test set. In general, the classifier performs better when including the CV-based features. Moreover, toward radar gait measurements are classified with a higher accuracy than away from radar measurements. For classification, we thus choose a number of seven eigenimages for PCA-based features and include the CV-based features.

12.3.4.3 Classification

The available data sets, as listed in Table 12.4 for walks toward and away from radar, are each split into subsets of 70% and 30% of data samples for training and testing, respectively. We aim at distinguishing between an unassisted (NW) and an assisted walk. Here, the latter can be walk where the cane is synchronously moved with either leg (CW) or is out of sync with any leg (CW/oos). The nearest neighbor (NN) classifier is used to discriminate between unassisted and the two types of assisted walks, where the Euclidean distance is used as a distance measure. Results are averaged over 100 classification results, where training and test data are randomly chosen for each run.

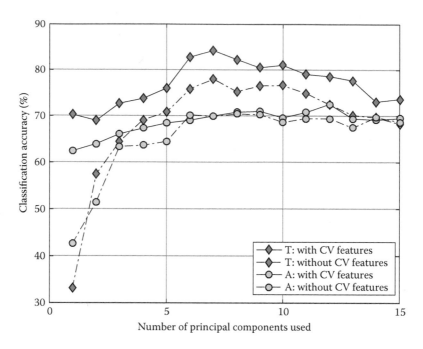

FIGURE 12.12

Classification accuracy versus the number of principal components used in the formation of the feature vector for toward (T) and away from (A) radar measurements.

TABLE 12.4

Data Set 2

Walking Style	Toward	Away
Normal walk	17	22
Walking with a cane—aligned with one leg	25	23
Walking with a cane—out of sync with any leg	13	12
Total	55	57

Source: Seifert, A.-K. et al., Radar-based human gait recognition in cane-assisted walks, in *Proceedings of the 2017 IEEE Radar Conference* © 2017 IEEE.

Table 12.5 presents the classification performance for toward radar measurements. Toward radar walks can be classified correctly in 84% of the cases using PCA-based features using $l = 7$ eigenimages and CV features. The FA rate and the MD rate are both 17%. Here, MD refers to the case in which a cane is present, but the gait is wrongly classified as unassisted. Without CV features, the classification accuracy decreases to 76% (FA = 24%, MD = 20%).

For measurements of a human walking away from the radar system, the classification results are presented in Table 12.6. Here, the classification accuracy is 71% (FA = 37%, MD = 33%), using $l = 7$ eigenimages for PCA-based features and CV features. Without the

TABLE 12.5

Classification Results for Walking toward the Radar

		Predicted Class		
		NW	CW	CW/oos
True Class	Normal walk (NW)	83	14	3
	Walking with a cane—aligned with one leg (CW)	12	86	2
	Walking with a cane—out of sync with any leg (CW/oos)	1	16	83

Source: Seifert, A.-K. et al., Radar-based human gait recognition in cane-assisted walks, in *Proceedings of the 2017 IEEE Radar Conference* © 2017 IEEE.

TABLE 12.6

Classification Results for Walking away from Radar

		Predicted Class		
		NW	CW	CW/oos
True Class	Normal walk (NW)	63	34	3
	Walking with a cane—aligned with one leg (CW)	27	71	2
	Walking with a cane—out of sync with any leg (CW/oos)	6	7	87

Source: Seifert, A.-K. et al., Radar-based human gait recognition in cane-assisted walks, in *Proceedings of the 2017 IEEE Radar Conference* © 2017 IEEE.

CV features, the classification accuracy is comparable to the previous case. For away from radar measurements, the use of CV features only improves the classification accuracy when using five or less eigenimages for the PCA-based features, as can be deduced from Figure 12.12.

12.4 Conclusions

This chapter focuses on normal and abnormal gait analysis using joint-variable time–frequency representations of the radar backscatters. It was shown that there are differences in the micro-Doppler signatures of a person walking toward and away from the radar. These differences are related to the nature of the leg, knee, and foot motions, and their impact on the radar signal returns from the front and back of a walking human. We established the distinction of these motion articulations in the time–frequency domain and used them to distinguish the cane's Doppler behavior from that of the leg under various conditions and alignments. Both spectrograms and CVDs were utilized in motion classifications. Physically interpreted features extracted from these representations were combined with those of PCA to successfully indicate the presence of a cane in human gait.

Acknowledgment

This work was supported by the Alexander von Humboldt Foundation, Bonn, Germany.

References

Amin, M. G., F. Ahmad, Y. D. Zhang, and B. Boashash, Micro-Doppler characteristics of elderly gait patterns with walking aids, in *Proceedings of SPIE Conference on Radar Sensor Technology XIX*, vol. 9461, 2015a.

Amin, M. G., F. Ahmad, Y. D. Zhang, and B. Boashash, Human gait recognition with cane assistive device using quadratic time-frequency distributions, *IET Radar, Sonar & Navigation*, 9 (9), 1224–1230, 2015b.

Amin, M. G., Y. D. Zhang, F. Ahmad, and K. C. D. Ho, Radar signal processing for elderly fall detection: The future for in-home monitoring, *IEEE Signal Processing Magazine*, 33 (2), 71–80, 2016.

Ancortek Inc., SDR-KIT 2500B, http://ancortek.com/sdr-kit-2500b, retrieved: November 9, 2016.

Barak, Y., R. C. Wagenaar, and K. G. Holt, Gait characteristics of elderly people with a history of falls: A dynamic approach, *Physical Therapy*, 86 (11), 1501–1510, 2006.

Bilik, I., and P. Khomchuk, Minimum divergence approaches for robust classification of ground moving targets, *IEEE Transactions on Aerospace and Electronic Systems*, 48 (1), 581–603, 2012.

Björklund, S., H. Petersson, and G. Hendeby, Features for micro-Doppler based activity classification, *IET Radar, Sonar & Navigation*, 9 (9), 1181–1187, 2015.

Björklund, S., T. Johansson, and H. Petersson, Evaluation of a micro-Doppler classification method on mm-wave data, in *Proceedings of the 2012 IEEE Radar Conference*, 2012.

Boashash, B., N. A. Khan, and T. Ben-Jabeur, Time-frequency features for pattern recognition using high-resolution TFDs: A tutorial review, *Digital Signal Processing*, 40, 1–30, 2015.

Boser, B. E., I. M. Guyon, and V. N. Vapnik, A training algorithm for optimal margin classifiers, in *Proceedings of the 5th Annual Workshop on Computational Learning Theory*, ACM, New York, pp. 144–152, 1992.

Chen, V. C., *The Micro-Doppler Effect in Radar*. London: Artech House, 2011.

Chen, V. C., and H. Ling, *Time-Frequency Transforms for Radar Imaging and Signal Analysis*. Boston, MA: Artech House, 2001.

Chen, V. C., D. Tahmoush, and W. J. Miceli, *Radar Micro-Doppler Signatures: Processing and Applications*. Stevenage: Institution of Engineering and Technology, 2014.

Clemente, C., A. Miller, and J. Soraghan, Robust principal component analysis for micro-Doppler based automatic target recognition, in *3rd IMA Conference on Mathematics in Defense*, Malvern, UK, October 2013.

Clemente, C., L. Pallotta, A. D. Maio, J. Soraghan, and A. Farina, A novel algorithm for radar classification based on Doppler characteristics exploiting orthogonal pseudo-Zernike polynomials, *IEEE Transactions on Aerospace and Electronic Systems*, 51 (1), 417–430, 2015.

Cohen, L., Time-frequency distributions—A review, *Proceedings of the IEEE*, 77 (7), 941–981, 1989.

Cohen, L., *Time-frequency Analysis*. Englewood Cliffs, NJ: Prentice Hall, 1995.

Cortes, C., and V. Vapnik, Support-vector networks, *Machine Learning*, 20 (3), 273–297, 1995.

Fairchild, D. P., and R. M. Narayanan, Classification of human motions using empirical mode decomposition of human micro-Doppler signatures, *IET Radar, Sonar & Navigation*, 8 (5), 425–434, 2014.

Gell, N. M., R. B. Wallace, A. Z. Lacroix, T. M. Mroz, and K. V. Patel, Mobility device use in older adults and incidence of falls and worry about falling: Findings from the 2011–2012 national health and aging trends study, *Journal of the American Geriatrics Society*, 63 (5), 853–859, 2015.

Gürbüz, S. Z., C. Clemente, A. Balleri, and J. Soraghan, Micro-Doppler based in-home aided and unaided walking recognition with multiple radar and sonar systems, *IET Radar Sonar & Navigation*, 11, 107–115, 2016.

Gürbüz, S. Z., B. Erol, B. Cagliyan, and B. Tekeli, Operational assessment and adaptive selection of micro-Doppler features, *IET Radar, Sonar & Navigation*, 9 (9), 1196–1204, 2015.

Gürbüz, S. Z., B. Tekeli, M. Yuksel, C. Karabacak, A. C. Gürbüz, and M. B. Guldogan, Importance ranking of features for human micro-Doppler classification with a radar network, in *Proceedings of the 16th International Conference on Information Fusion*, Instanbul, Turkey, pp. 610–616, 2013.

Javier, R. J., and Y. Kim, Application of linear predictive coding for human activity classification based on micro-Doppler signatures, *IEEE Geoscience and Remote Sensing Letters*, 11 (10), 1831–1834, 2014.

Jokanović, B., M. G. Amin, and F. Ahmad, Radar fall motion detection using deep learning, in *Proceedings of the 2016 IEEE Radar Conference*, 2016.

Jokanović, B., M. G. Amin, Y. D. Zhang, and F. Ahmad, Multi-window time-frequency signature reconstruction from undersampled continuous-wave radar measurements for fall detection, *IET Radar, Sonar & Navigation*, 9 (2), 173–183, 2015.

Kim, Y., and H. Ling, Human activity classification based on micro-Doppler signatures using a support vector machine, *IEEE Transactions on Geoscience and Remote Sensing*, 47 (5), 1328–1337, 2009.

Li, J., S. L. Phung, F. H. C. Tivive, and A. Bouzerdoum, Automatic classification of human motions using Doppler radar, in *Proceedings of the 2012 IEEE International Joint Conference on Neural Networks*, Brisbane, Australia, June 2012.

Lyonnet, B., C. Ioana, and M. G. Amin, Human gait classification using micro-Doppler time-frequency signal representations, in *Proceedings of the 2010 IEEE Radar Conference*, pp. 915–919, 2010.

Marek, K. D., L. Popejoy, G. Petroski, D. Mehr, M. Rantz, and W.-C. Lin, Clinical outcomes of aging in place, *Nursing Research*, 54 (3), 202–211, 2005.

Mobasseri, B. G., and M. G. Amin, A time-frequency classifier for human gait recognition, in *Proceedings of SPIE*, vol. 7306, 2009.

Molchanov, P., J. Astola, K. Egiazarian, and A. Totsky, Ground moving target classification by using DCT coefficients extracted from micro-Doppler radar signatures and artificial neuron network, *IEEE Microwaves, Radar and Remote Sensing Symposium*, 2011.

Muro-de-la-Herran, A., B. Garcia-Zapirain, and A. Mendez-Zorrilla, Gait analysis methods: An overview of wearable and non-wearable systems, highlighting clinical applications, *Sensors*, 14 (2), 3362–3394, 2014.

Orović, I., S. Stanković, and M. Amin, A new approach for classification of human gait based on time-frequency feature representations, *Signal Processing*, 91 (6), 1448–1456, 2011.

Otero, M., Application of a continuous wave radar for human gait recognition, *Proceedings of SPIE*, 5809, 538–548, 2005.

Phillips, C. E., J. Keller, M. Popescu, M. Skubic, M. J. Rantz, P. E. Cuddihy, and T. Yardibi, Radar walk detection in the apartments of elderly, in *Proceedings of the 2012 Annual International Conference of the IEEE Engineering in Medicine and Biology Society*, pp. 5863–5866, 2012.

Ricci, R., and A. Balleri, Recognition of humans based on radar micro-Doppler shape spectrum features, *IET Radar, Sonar & Navigation*, 9 (9), 1216–1223, 2015.

Seifert, A.-K., M. G. Amin, and A. M. Zoubir, New analysis of radar micro-Doppler gait signatures for rehabilitation and assisted living, in *Proceedings of the 2017 IEEE International Conference on Acoustics, Speech and Signal Processing*, 2017.

Seifert, A.-K., A. M. Zoubir, and M. G. Amin, Radar-based human gait recognition in cane-assisted walks, in *Proceedings of the 2017 IEEE Radar Conference*, 2017.

Stanković, L., M. Daković, and T. Thayaparan, *Time-Frequency Signal Analysis with Applications*. Boston, MA: Artech House, 2013.

Su, B. Y., K. Ho, M. J. Rantz, and M. Skubic, Doppler radar fall activity detection using the wavelet transform, *IEEE Transactions on Biomedical Engineering*, 62 (3), 865–875, 2015.

Tekeli, B., S. Z. Gürbüz, and M. Yuksel, Information-theoretic feature selection for human micro-Doppler signature classification, *IEEE Transactions on Geoscience and Remote Sensing*, 54 (5), 2749–2762, 2016.

Tivive, F. H. C., A. Bouzerdoum, and M. G. Amin, A human gait classification method based on radar Doppler spectrograms, *EURASIP Journal on Advances in Signal Processing*, 389716, 1–12, 2010.

Tivive, F. H. C., S. L. Phung, and A. Bouzerdoum, Classification of micro-Doppler signatures of human motions using log-Gabor filters, *IET Radar, Sonar & Navigation*, 9 (9), 1188–1195, 2015.

United Nations, World population prospects: The 2015 revision, Technical Report, Department of Economic and Social Affairs, New York, NY, 2015.

Wang, F., M. Skubic, M. Rantz, and P. E. Cuddihy, Quantitative gait measurement with pulse-Doppler radar for passive in-home gait assessment, *IEEE Transactions on Biomedical Engineering*, 61 (9), 2434–2443, 2014.

World Health Organization, *WHO Global Report on Falls Prevention in Older Age.* Geneva, Switzerland: World Health Organization, 2007.

Yardibi, T., P. Cuddihy, S. Genc, C. Bufi, M. Skubic, M. Rantz, L. Liu, and C. Phillips, Gait characterization via pulse-Doppler radar, in *Proceedings of the IEEE International Conference on Pervasive Computing and Communications Workshops*, pp. 662–667, 2011.

Yessad, D., A. Amrouche, M. Debyeche, and M. Djeddou, Micro-Doppler classification for ground surveillance radar using speech recognition tools, in *Proceedings of the 16th Iberoamerican Congress Conference on Progress in Pattern Recognition, Image Analysis, Computer Vision, and Applications*, pp. 280–287, 2011.

13

Radar for Disease Detection and Monitoring

Huiyuan Zhou, Ram M. Narayanan, Ilangko Balasingham, and Rohit Chandra

CONTENTS

13.1 Introduction

Microwave imaging of the human body, such as human breast, head, and intestine, for tumor and other disease detection has been a topic of interest for several decades. Its advantages include nonionizing and low-risk nature of microwave signals at low levels, low-cost implementation of practical systems, and the exploitation of high dielectric contrast between normal and abnormal human tissue (Rosen et al. 2002). Signals in the microwave frequency range are able to penetrate the human body and are able to collect useful information for detection and imaging of anomalies. Frequencies up to 4 GHz can penetrate skin, tissues, and clothing and can ease the requirement for the preferred half-wavelength spacing when architecting aperture antenna arrays (Zhuge et al. 2008). Good down-range resolution requires a wide operational bandwidth, whereas good cross-range resolution requires large physical or synthetic aperture.

Basically, there are two major directions to reconstruct the image of the human body using medical radar. The first approach is based on electromagnetic (EM) inverse scattering algorithms, whereas the other approach is based on ultra-wideband (UWB) radar principles. Inverse scattering methods for imaging are based on the estimation of the profile of the internal dielectric properties of human tissue, usually relative permittivity and conductivity, from the data collected outside human body. These types of problems are different from conventional direct problems. In direct problems, the EM properties and parameters of objects are known, and the scattered field is the output of the system. On the contrary, for inverse scattering problems, the inputs for the system are the measured scattered fields from receivers outside the human body and the incident field from transmitters wherein the dielectric profile is the unknown variable to be determined. Inverse problems are proven to be ill-posed problems in the sense of Hadamard's characterization (O'Sullivan 1986), which contain the existence, uniqueness, and stability of the solution. Generally, the methods for solving inverse problem can be grouped into stochastic and deterministic categories. The stochastic methods involve population-based or iterative methods. At each iteration, several trial solutions are considered. The elements of the population are processed iteratively in different methods with predefined cost function. The iteration steps will stop when one of the populations matches the stopping requirement. Researchers have developed various algorithms, such as simulated annealing (SA) (Garnero et al. 1991), evolutionary algorithms (EAs) as genetic algorithms (GA) (Rahmat-Samii and Michielssen 1999), differential evolution (DE) algorithm (Massa et al. 2004), particle swarm optimization algorithm (PSO) (Robinson and Rahmat-Samii 2004), and ant colony optimization algorithm (ACO) (Dorigo et al. 1996).

The advantages of stochastic methods are (1) avoiding being trapped in a local minimum, which leads to a false result; (2) having no requirement on *a priori* information for initial guess of the unknown parameters; and (3) the ability to reach the global minimum solution (Pastorino 2007). However, considering calculation efficiency and cost, and for mulation properties of inverse problem, some research groups are working on local optimization methods (deterministic methods), such as Newton-type methods and conjugate gradient method.

For deterministic methods, two approaches are widely used to solve the problem based on the formulation of scattering problem. The first one relies on electric field integral equations (EFIE) involving the electric field inside the investigation area and the contrast function as unknowns, which includes the information on the dielectric properties of the scatterers. These types of methods need forward solver one or more times to process during each iteration. The forward solver can also be categorized as two groups: The first is the partial differential equation (PDE)-based solver, and the second is the integral equation (IE)-based solver. Typical methods belonging to PDE-based solver include finite element (FE) method and finite difference (FD) method. IE-based solution is based on the method of moments (MoM).

A second formulation of the scattering problem has been introduced by Van den Berg and Kleinman (1997), which is based on the contrast source (CS) integral equations and assumes as problem unknowns the induced current (i.e., the contrast source) and the contrast function. Due to the structure of the involved equations, it solves inverse scattering problems without the need to call a forward solver. Contrast source inversion (CSI) technique belongs to this group. In each iteration of CSI, two variables (the contrast source and the contrast) are updated simultaneously using a conjugate gradient (CG) method. The variables are updated to minimize a given cost function.

Solving ill-posed problems with deterministic methods also requires regularization procedures in the algorithms to enhance the quality of the reconstruction, to increase the robustness to noise, and to speed up convergence rate. Most of the regularization methods have been investigated in the context of Hilbert spaces. Recently, regularization methods have been introduced in Banach spaces (Estatico et al. 2012), which is more accurate and efficient for nonsmoothing nature of biomedical objects (Daubechies et al. 2004).

UWB radars achieve excellent down-range resolution, because this is inversely proportional to the bandwidth. They can be implemented as pulsed radars (e.g., impulse) or frequency-modulated (FM) radars (e.g., chirp, linear FM, stepped FM, or UWB noise). UWB systems have been used for patient monitoring, vital signs detection, and in applications as diverse as cardiology, pneumology, obstetrics, and so on (Pan 2008).

Several interrelated design choices exist for UWB systems, with corresponding trade-offs in system performance (Paulson et al. 2005). A large bandwidth is beneficial to achieve good spatial resolution, whereas low frequencies are desirable to achieve good penetration through lossy tissue. To focus lower frequency signals into the human body, a larger antenna is required, which increases the overall size of the device. A large peak signal power is important for obtaining good penetration through materials and lossy tissue. In order to improve the signal-to-noise ratio (SNR), pulse averaging can be used, which increases the operating range and sensitivity but decreases the responsiveness to rapid motion.

Generally, different parts of the human body (e.g., head, breast) can be modeled, to a first approximation, as a multilayer structure with each layer corresponding to a particular tissue (Pancera 2010). Each tissue/layer is characterized by its relative permittivity and conductivity (both of which are frequency dependent), and by its thickness. Due to the difference of the dielectric profile between healthy and malignant tissue, it is possible to reconstruct a suitable image and detect tumors or other diseases inside the human body.

UWB technology has also been applied to wireless body area networks and capsule endoscopy (Chávez-Santiago et al. 2012). For the latter, the endoscopes are swallowed and travel through the body; hence, the circuitry must be very small and simple yet capable of high transmission rates to transmit real-time video. Significant signal processing is needed to process the low-level signals, which have traversed several tissue layers.

Most medical radar applications basically need two components: a sensor and a communication infrastructure (transceiver and protocols) to share the data gathered by the former (Bilich 2006). By combining sensing and communications in the same package, UWB radar could be used to measure vital signs information, and UWB communication standards could be used to transmit these measurements to a processor. Wireless networking is necessary to transmit and share the monitored data with a central repository.

It is important to ascertain which phenomenon impacts measured data. It was shown that body surface movements dominated remote radar measurements of heartbeat, compared to blood perfusion in the skin or internal body organ movements (Aardal et al. 2013).

A comprehensive list of relevant literature on medical radar is presented by Aardal and Hammerstad (2010), which includes references on general UWB radar, radar calibration, radar heartbeat and respiration measurements, medical radar systems, and medical radar imaging. In addition, a summary of important considerations in medical radar with an extensive list of references is provided in Narayanan (2013).

This chapter is organized as follows. In Section 13.2, the dielectric properties of human tissue are introduced in detail. In addition, the microwave penetration performance at different frequencies is briefly discussed. Section 13.3 briefly introduces forward solvers at first. Then, different medical radar techniques are introduced in two major categories. The first discusses inverse algorithms, including Gauss–Newton type methods, contrast source method, and distorted Born iterative method (DBIM). Second, reconstruction algorithms used in UWB radar imaging are described. Section 13.4 lists and discusses applications where medical radar techniques are primarily employed. Challenges and some future research directions are discussed in Section 13.5.

13.2 Dielectric Properties of Human Tissue

The differences between dielectric properties in normal tissue and malignant tissue are the foundation stone for medical radar imaging. Although dielectric property research started decades ago (starting in the 1950s), more recent publications include Joines et al. (1994) who performed experiments and summarized the measurements for a variety of normal and malignant human tissues, Foster and Schwan (1995) who summarized early work on human tissue dielectric properties, and Gabriel et al. (1996a) who supplied extensive data for various healthy tissues. Researchers continue to this date to expand the varieties of tissues and frequency ranges for dielectric property characterization.

In the range of microwave frequencies, the properties of human tissues show the feasibility of the imaging process because of the difference between normal human tissues and malignant tissues. When exposed to the EM field, dielectric properties of biological tissues show frequency dependence characteristics called the dispersion property (Joines et al. 1980). In recent studies, a variety of factors have been explored, which led to explanations of the distinction in dielectric properties in normal compared with malignant tissues (Sha et al. 2002), including necrosis and inflammation, charging of the cell membrane, relaxation times, sodium concentration, and water content.

Necrosis and inflammation cause breakdown of cell membranes, and thus increasing a larger fraction of the tissue can carry current at low frequencies (Sha et al. 2002). On the other hand, based on Pethig (1984), charging of the cell membrane is another reason because malignant cells can reduce membrane potentials and tend to have altered ability to absorb positive ions, and they have a higher negative surface charge on membranes. For relaxation times, the one found in malignant tissues is much different from those in normal tissue. This time is an important parameter for modeling tissue dielectric properties. According to Lazebnik et al. (2007a), the relaxation time difference between breast malignant tissue and normal tissue is up to 4 ps, whereas normal tissues' relaxation time varies from 9.2 to 14.1 ps. Sodium concentration in malignant cells is higher than normal cells (Pethig 1984). The high concentration of sodium affects the cell membrane potentials and makes malignant tissue to retain more fluid (Sha et al. 2002), which would lead to greater conductivity and permittivity values in malignant tissues compared to normal tissues. For water content, malignant tissues have higher water contents than normal tissues (Foster and Schepps 1981). The tissue dielectric properties correlate well with water contents, and the conductivity of the tissue increases with the volume fraction of water in the tissue (Schepps and Foster 1980). Water content is the main factor affecting the dielectric property differences between malignant tissues and normal tissues (Hagness et al. 2012).

13.2.1 Dielectric Property Modeling and Measurement

To model the frequency dispersive nature of the tissues, both the Debye model and the Cole–Cole model have been developed. Based on Gabriel et al. (1996b), for the first-order approximation, the expression for the complex relative permittivity is

$$\hat{\varepsilon} = \varepsilon_\infty + \frac{\varepsilon_s - \varepsilon_\infty}{1 + j\omega\tau} \tag{13.1}$$

where:
ε_∞ is the permittivity at field frequencies satisfying $\omega\tau \gg 1$
ε_s is the permittivity when $\omega\tau \ll 1$
ω is the angular frequency
τ is the time constant called the relaxation time

Subsequently, a summation of five Debye dispersions with additional conductivity terms were used to model the complex relative permittivity, which is given as (Gabriel et al. 1996b)

$$\hat{\varepsilon}(\omega) = \varepsilon_\infty + \sum_{n=1}^{5} \frac{\Delta\varepsilon_n}{1 + j\omega\tau_n} + \frac{\sigma_i}{j\omega\varepsilon_0} \tag{13.2}$$

where:
ε_0 is the permittivity of free space
σ_i is the static ionic conductivity
$\Delta\varepsilon = \varepsilon_s - \varepsilon_\infty$ represents the magnitude of the dispersion

The broadening of the dispersion could be empirically accounted for by introducing a distribution parameter; thus an alternative Debye expression is given and is known as the Cole–Cole equation:

$$\hat{\varepsilon} = \varepsilon_\infty + \frac{\Delta\varepsilon}{1 + \left(j\omega\tau\right)^{(1-\alpha)}} \tag{13.3}$$

where α is the distribution parameter, which is a measure of the broadening of the dispersion. For a more accurate representation of the complex relative permittivity, a multiple Cole–Cole dispersion expression is given as

$$\hat{\varepsilon}(\omega) = \varepsilon_\infty + \sum_{n} \frac{\Delta\varepsilon_n}{1 + \left(j\omega\tau_n\right)^{(1-\alpha_n)}} + \frac{\sigma_i}{j\omega\varepsilon_0} \tag{13.4}$$

which is based on Gabriel et al. (1996b) and can be used to model the dielectric properties over a wide frequency range.

In the microwave reconstruction process, typically employing inverse scattering algorithms, the results are obtained at a single frequency. For multiple frequencies, the process is implemented at different single-frequency values, and the results are then combined to obtain the final image. As the result, the complex relative permittivity is simplified as

$$\hat{\varepsilon}(r) = \varepsilon_r(r)\varepsilon_0(r) - j\frac{\sigma(r)}{\omega\varepsilon_b} \qquad (13.5)$$

where:

$\varepsilon_r(r)$ is the real relative permittivity at position $r(x,y)$
σ is the electric conductivity
ε_b is the dielectric permittivity of the background

Joines et al. (1994) provided a variety of conductivity and relative permittivity of malignant and normal human tissues at frequencies from 50 to 900 MHz.

Figure 13.1 shows a sample of the data showing changes in the properties of the malignant tissues with respect to the healthy tissue (Chandra et al. 2015b). It gives the percentage change in the dielectric property of the different malignant tissues from 50 to 900 MHz. It can be seen that the largest differences in the measured relative permittivity and conductivity occurred for breast tissue. Among all of the five tissues, kidneys have the least permittivity difference between the normal and malignant tissues. In the conductivity difference plot (Figure 13.1b), breast tissues still have the biggest difference, and kidneys have the least difference. All the aforementioned results indicated that breast tissues would be

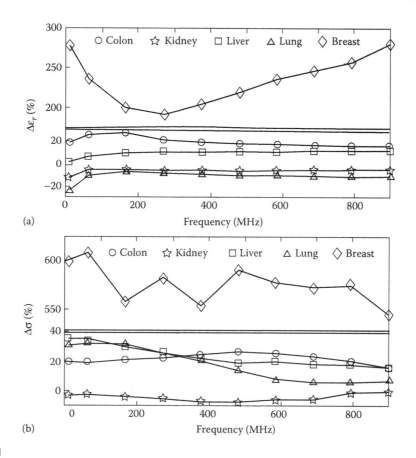

FIGURE 13.1
Percentage change in the electrical properties of malignant tissues with respect to the healthy tissue from 50 to 900 MHz. (a) Percentage change in the relative permittivity and (b) percentage change in the conductivity. (From Chandra, R. et al., *IEEE Trans. Biomed. Eng.*, 62(7), 1667–1682 © 2015 IEEE. With Permission.)

more suitable for the medical radar technique, compared to the other four tissue types. As a result, most of the research is focused in the mammary or breast tissue area.

In breast cancer research, extensive experiments have been performed to measure the dielectric property difference between the malignant breast tissues and normal breast tissues, as well as the dielectric properties of malignant breast tissues at different frequencies (Bindu and Mathew 2007; Lazebnik et al. 2007b, 2007c; Kim et al. 2008). It is pointed out that the microwave frequency dielectric-properties contrast between malignant breast tissues and normal adipose-dominated breast tissues is large, ranging up to a 10:1 contrast when considering almost entirely adipose breast tissue as the reference. In contrast, the dielectric-properties contrast between malignant and normal fibroconnective/glandular breast tissues is considerably lower, less than approximately 10% (O'Rourke et al. 2007).

Besides breast tissues, the dielectric properties of liver tissues have also been extensively characterized. O'Rourke et al. (2007) supplied measurement data at 915 MHz and 2.45 GHz, which indicates that the dielectric properties of *ex vivo* malignant liver tissue are 19%–30% higher than normal tissue, and the effective conductivity of malignant tissue properties are 16% higher than normal tissues at 915 MHz. So far, researchers have investigated and measured the dielectric properties of a variety of normal and malignant tissues, including lymph nodes (Deighton 2013), skin (Sunaga et al. 2001), and bone (Meaney et al. 2012).

Because *in situ* tissue measurements are not always possible and excised tissues show rapid changes in their dielectric properties with time, it is advantageous to develop simulated human biological materials. This was done by using appropriate chemical mixtures (Hartsgrove et al. 1987). Formulas were presented for simulating bone, lung, brain, and muscle tissue over the frequency range of 100 MHz to 1 GHz. A realistic equivalent to the human body could be constructed using these preparations. By characterizing the dielectric properties of normal liver and liver tumor over the 300 MHz–3 GHz frequency range, it was determined that the dielectric constant and conductivity of liver tumor were higher by 12% and 24%, respectively, when compared to normal liver (Stauffer et al. 2003). Furthermore, by comparing the dielectric data of human liver to measurements of homogeneous phantom mixtures, *in vitro* bovine liver, and *in vivo* canine and porcine liver tissues, it was determined that several animal tissues could be used to model the average dielectric properties of human liver reasonably well.

The age dependence of dielectric properties of biological tissues mostly relies on the fact that permittivity and conductivity may be expressed as a function of tissue water content, which decreases with age. Empirical formulas for estimating permittivity and conductivity of human biological tissues as a function of age have been developed (Ibrani et al. 2012).

13.2.2 Microwave Penetration in Human Tissue

Microwave penetration depth depends on the frequency of the incident electric field. Most published models use the industrial, scientific, and medical (ISM) frequency bands or UWB. Among these, the most commonly used frequency band is dedicated by International Telecommunication Union (ITU) for medical use. Frequencies in the range 9–600 kHz, 30–37.5 MHz, 401–406 MHz, 868 MHz, and 2.4 GHz are permitted for medical use. The Medical Implant Communication Service (MICS) is in the 401–406 MHz band (ITU 2015); the wireless Medical Telemetry Service is defined by the Federal Communications Commission (FCC) in the band 608–614 MHz, 1.395–1.4 GHz, and 1.429–1.432 GHz (FCC 2016); the ISM radio band is at 2.42–2.48 GHz and 5.3–5.9 GHz, which can be used without licenses (Garg et al. 1997).

For different imaging or detection applications, one needs to carefully choose the optimum frequency for the applications that will determine the penetration depth and path loss during propagation within the human body. Because the human skin is the first layer encountered by the EM wave, its dielectric properties are of special interest. The complex permittivity of living human skin was reported over the frequency range 8–18 GHz (Hey-Shipton et al. 1982), 28–57 GHz (Ghodgaonkar et al. 2000), 10–60 GHz (Chahat et al. 2011), 37–74 GHz (Alekseev and Ziskin 2007), and 60–100 GHz (Alabaster 2003, 2004). Differences between human skin *in vivo* and excised human skin tissue were attributed to variations of water content, blood content, and epidermal thickness. Multilayer models provided better fits to both forearm and palmar skin reflection data, especially for the latter with a thick stratum corneum. The skin depth of human skin at 10 GHz was about 2.7 mm, indicating that for frequencies above 10 GHz, the skin could be considered opaque to EM waves. Thus, it can be assumed that millimeter-wave radiation incident on the human body will be almost entirely absorbed (or reflected) by the skin layer.

The complex dielectric constants of blood and blood plasma were measured over the 1.7–2.4 GHz frequency range and compared to those of water (Cook 1951, 1952). Dispersion was found to occur, which was attributed entirely to dipolar orientation of part of the water present in blood and the ionic conductivity.

Based on Balanis (2012), the penetration depth δ is given as

$$\delta = \left[\frac{k^2}{2} \left[\sqrt{\varepsilon_r^2 + \left[\frac{\sigma}{\omega\varepsilon_0} \right]^2} - 1 \right] \right]^{-\frac{1}{2}} \tag{13.6}$$

where k is the wave number. Based on the values of relative permittivity and conductivity at frequency of 418 MHz, the penetration depth in muscle is around 30.4 mm, reducing to 25.5 mm at 916.5 MHz. Simulations of the radio frequency (RF) path loss over the frequency range of 200 MHz–3 GHz for abdominal fat and thigh muscle tissues were shown in Khaleghi and Balasingham (2015). In addition, depth effects have also been investigated. The frequency range 200 MHz–1.4 GHz has almost same loss characteristics, whereas the loss increases significantly when frequency goes up to 1.5 GHz. The superficial tissue has wider low loss characteristics up to 3 GHz.

Although it is feasible to obtain an image of human tissue in the microwave frequency range, there are still some significant challenges and difficulties linked to microwave imaging. The first problem is that different tissues have different properties, which are mainly related to the water content. At the same time, the dielectric properties suggest that tissues can be differentiated using microwave signals, and detection of disease or anomalies relies on differences between properties of healthy and diseased tissues or biological materials. However, the detection may be further complicated by the presence of multiple tissues with different properties, resulting in a complex scattering situation. Another problem is the variation of properties of tissues with frequency. Further, the conductivity of tissues typically increases with frequency, resulting in a trade-off between resolution and depth of penetration, as well as adding to the aforementioned dynamic range challenges.

All of these complex set of challenges implies that microwave imaging may achieve success with particular tissues or biological materials where significant differences in normal and diseased states exist at a specific frequency or over specific frequency bands.

13.2.3 Electromagnetic Field Interaction with Human Tissue

Accurate numerical and analytical techniques to predict the propagation of EM signals in biological tissue are essential for developing signal processing algorithms for medical diagnosis. A composite human body slice is modeled in terms of different tissue layers of appropriate complex permittivity and thicknesses. The reflection at the interfaces and the propagation loss through each layer determines the overall radar reflected signal. A technique to compute the reflected and transmitted fields from the first layer and to progressively include succeeding layers, while neglecting mutual influence between contiguous layers without error, was developed and applied to normal incidence (Taoufik et al. 2010) and subsequently extended to oblique incidence (Taoufik et al. 2011). Figure 13.2 shows an example of the multilayer modeling geometry and wave propagation considerations in each layer based upon its dielectric properties (Staderini 2002). The total reflected signal is assembled by taking into consideration the reflections at the layer interfaces and the transmissions through each layer.

Two techniques for modeling the propagation of UWB pulses in human tissue, namely the planar technique and finite-difference time-domain (FDTD) technique, were developed and applied to normal tissues as well as cancerous soft tissue (sarcoma). Both approaches were able to identify the presence of the soft tissue sarcoma quite easily (O'Halloran et al. 2006a, 2006b).

Moreover, for inverse scattering methods, the scattered field is the key for the reconstruction and detection. Usually, the human tissues are considered as inhomogeneous dielectric objects. Consequently, the scattered field can be expressed precisely based on the wave propagation equation.

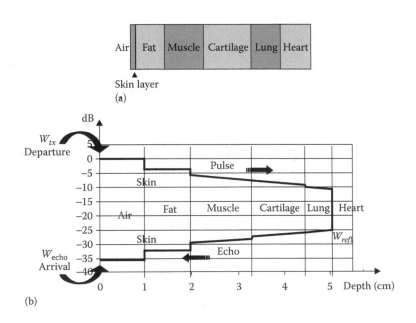

FIGURE 13.2
(a) Multilayered flat tissue model and (b) model predicted attenuation of pulse-echo intensity traveling from the transmitting antenna to the receiving antenna. Each step accounts for echo at the boundary. Decreasing of the curve accounts for linear attenuation in the tissue. (From Staderini, E. M., *IEEE Aeros. Electron. Syst. Mag.*, 17(1), 13–18 © 2002 IEEE. With Permission.)

13.3 Microwave Medical Radar Imaging Technique

Microwave imaging is generally based on the inverse scattering problem or the UWB radar approach. The inverse scattering problem is used to determine the internal dielectric profile of an inhomogeneous object from the difference between measurement data collected outside the object and calculated scattered field by the estimated dielectric profile.

Radar-based techniques are time-domain formulations and use computationally efficient synthetic aperture radar (SAR) algorithms to provide accurate object shape and location results. On the other hand, tomography-based techniques tend to be frequency domain approaches and use different forms of backprojection algorithms. They are based on nonlinear iterative inversion algorithms and give accurate information on the dielectric properties of the objects for moderate size-contrast products but suffer from a lack of accuracy (ill-conditioned matrices) or low computational efficiency (time-expensive inversion or iterative methods).

In this section, we discuss the forward problems and inverse algorithms to obtain the quantitative interior result of the object. The other part involves the UWB radar technique.

13.3.1 Forward Solvers

For inverse scattering methods, the scattered field is the key for the reconstruction and detection. Usually, the human tissues are considered as inhomogeneous dielectric objects. Consequently, the scattered field can be expressed precisely based on the wave propagation equation. In microwave imaging, forward solvers are applied to calculate the estimated scattered field at every iteration. All kinds of forward solvers are part of computational EM analyses. Basically, we can categorize the methods based on the equations to be solved. Some methods belong to IE solvers, such as MoM and boundary element method (BEM). The second kind is developed to solve the differential equation. FDTD method and FE method belong to this category. Generally, the most popular methods include the FDTD method, the FE method and the MoM.

The configuration discussed in this section is shown in Figure 13.3. In this arrangement, the antennas are placed around the object as a circular antenna array. At each time instant, one antenna acts as the transmitter, and all the others act as receivers. This system can also be accomplished by a two-antenna system using a fixed position antenna, which transmits the signal toward the object. The second antenna rotates around the object to collect the data at different locations.

For a lossy, inhomogeneous, dielectric body, the total field can be expressed as

$$E^t(r) = E^i(r) + E^s(r) \tag{13.7}$$

where:

r is the coordinate of the point or cell inside the image domain or object
E^t, E^i, and E^s represent the total electric field, the incident field, and the scattered field, respectively

For IE solvers, the integral equation in the imaging domain can be written as

$$E^s(r) = E^t(r) - E^i(r) = \omega^2 \mu \int_S E^t(r')\overline{\overline{G}}(r,r')C(r')dr' \tag{13.8}$$

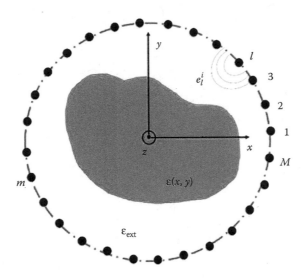

FIGURE 13.3
TM polarization (z-axis) system configuration. (Reproduced with permission of the International Society for Optics and Photonics; Zhou, H. et al., Microwave imaging of circular phantom using the Levenberg–Marquardt method, in *SPIE Conference on Radar Sensor Technology XIX and SPIE Conference on Active and Passive Signatures VI*, Baltimore, MD, pp. 946117-1–946117-12, 2015.)

where $C(r)$ is the dielectric contrast defined as the difference between the object complex relative permittivity and background permittivity, which is

$$C(r) = \varepsilon_r(r) - \varepsilon_b \tag{13.9}$$

where ε_b is the relative complex permittivity of the surrounding material. We set the imaging domain as S, then if $r \notin S$, the dielectric difference $C(r) = 0$.

The Green's function in Equation 13.8 is the dyadic Green's function for the background medium. The complete expression for the dyad Green's function is given by

$$\overline{\overline{G}}(r,r') = \left(\overline{\overline{I}} + \frac{1}{k_b^2} \nabla\nabla \right) G(r,r') = \begin{bmatrix} G_{xx} & G_{xy} & G_{xz} \\ G_{yx} & G_{yy} & G_{yz} \\ G_{zx} & G_{zy} & G_{zz} \end{bmatrix}$$

$$= \begin{bmatrix} k_b^2 + \dfrac{\partial^2}{\partial x^2} & \dfrac{\partial^2}{\partial x \partial y} & \dfrac{\partial^2}{\partial x \partial z} \\[2ex] \dfrac{\partial^2}{\partial y \partial x} & k_b^2 + \dfrac{\partial^2}{\partial y^2} & \dfrac{\partial^2}{\partial y \partial z} \\[2ex] \dfrac{\partial^2}{\partial z \partial x} & \dfrac{\partial^2}{\partial z \partial y} & k_b^2 + \dfrac{\partial^2}{\partial z^2} \end{bmatrix} \cdot \frac{1}{k_b^2} G(r,r') \tag{13.10}$$

and k_b is the wave number of the background medium, denoted by $k_b = \sqrt{\omega^2 \mu_0 \varepsilon_b}$.

The aforementioned formulations give the relationship between the incident field and dielectric property of the objects. In the inverse scattering problem, a forward solver needs to solve for the scattered field at each iteration. It is quite obvious that equation 13.8 is nonlinear; as a result, some approximation methods or iteration processes are introduced to make the calculation accurate and efficient.

For differential equation solvers, more details about FDTD method and FE method can be found in Taflove and Hagness (2005), Fang et al. (2004), Rekanos and Tsiboukis (1999), and so on.

13.3.2 Inverse Algorithms

Inverse problems are usually formulated as the minimization over the dielectric contrast of the data misfit cost function. The data misfit cost function is generally given as

$$F(c) = \min \left\| O(c) - e_s^{\mathrm{meas}} \right\|^2 \tag{13.11}$$

where:
e_s^{meas} is the measured scattered field at the observation domain
$O(c)$ is the complex valued nonlinear vector function of the scattered field calculated by estimated dielectric contrast

The difference is minimized iteratively by finding the contrast for the next iteration from the current iteration. The iteration is stopped either if the minimum acceptable error is achieved or the difference between the errors of the two iterations is below a certain threshold.

13.3.2.1 Gauss–Newton Type Algorithm (Levenberg–Marquardt Method)

The Gauss–Newton inversion approach is based on the Gauss–Newton optimization method (Chong and Zak 2013). It approximates the nonlinear cost function $F(c)$ as a quadratic form corresponding to the current iteration. The stationary point of the quadratic model is set to be the next iterate. As the result, the cost function is treated as data misfit cost function or augmented form. As the cost function is in term of the contrast dielectric profile, it can be considered equivalent to minimizing the cost function over vector Re{c} and Im{c}, which are the real and imaginary parts of the contrast dielectric vector, respectively. Then the Newton correction parameter Δc can be found using the following equation:

$$H \begin{pmatrix} \Delta c \\ \Delta c^* \end{pmatrix} = -G \tag{13.12}$$

where:
c is the complex contrast dielectric vector
c^* denotes its complex conjugate
H and G are the Hessian and the gradient of the cost function $F(c)$, respectively

These matrices are defined as

$$H = \begin{pmatrix} \dfrac{\partial^2 F}{\partial^2 c} & \dfrac{\partial^2 F}{\partial c \partial c^*} \\[2ex] \dfrac{\partial^2 F}{\partial c^* \partial c} & \dfrac{\partial^2 F}{\partial^2 c^*} \end{pmatrix} \tag{13.13}$$

$$G = \left(\frac{\partial F}{\partial c}, \frac{\partial F}{\partial c^*} \right)^T \tag{13.14}$$

In the Gauss–Newton method, the derivatives $\partial^2 F/\partial^2 c$ and $\partial^2 F/\partial^2 c^*$ are approximated to zero to reduce the computational cost. As the result, the Gauss–Newton correction at the nth iteration can be calculated by

$$\left. \frac{\partial^2 F}{\partial c^* \partial c} \Delta c \right|_{c=c_n} = \left. \frac{\partial F}{\partial c^*} \right|_{c=c_n} \tag{13.15}$$

Having the Gauss–Newton correction, the contrast is then updated as

$$c_{n+1} = c_n + v_n \Delta c_n \tag{13.16}$$

where v_n is the step length chosen to enforce the error reduction of the cost function. The equation to calculate the correction to the contrast dielectric vector is sometimes ill posed depending on the cost function. As the result, some regularization is applied to transform the ill-posed problem to a well-posed problem. Before the discussion of the regularization method, we can rewrite the derivative in Equation 13.12 as

$$\left. \frac{\partial^2 F}{\partial c^* \partial c} \right|_{c=c_n} = \frac{1}{\Pi} J_n^H J_n \tag{13.17}$$

$$\left. \frac{\partial F}{\partial c^*} \right|_{c=c_n} = \frac{1}{\Pi} J_n^H d_n \tag{13.18}$$

where:
  Π is the normalized constant which equals to $e_s^{\text{meas}2}$
  J_n is the Jacobian matrix, which is the derivative of $O(c)$ with respect to c evaluated at nth iteration
  J_n^H is the Hermitian transformation of J_n

The vector d_n is the discrepancy between the measured scattered field data and calculated scattered data, which is given by

$$d_n = e_s^{\text{meas}} - O(c_n) \tag{13.19}$$

Equations 13.20-13.26 below give a simple derivation of the Jacobian matrix J. First, the integral equation (13.8) can be rewritten in matrix form as

$$\left[I - G^r \text{diag} \{ \Delta \varepsilon \} \right] E^t = E^i \tag{13.20}$$

where:
  diag$\{\Delta\varepsilon\}$ denotes the diagonal matrix with diagonal elements having the value of $\Delta\varepsilon$, which is the dielectric profile difference between the object and background
  G^r is the Green's function in the form of an $N \times N$ matrix, which builds the relation between the incident field and total field in every cell of the object
  E^t is a vector supplying the total field in every cell
  E^i is a vector describing the incident field in every cell with size N

Equation 13.20 is also known as object equation.

For the scattered field, the matrix equation can be expressed as

$$E^s = G^s E^t [\Delta\varepsilon] \tag{13.21}$$

where:
G^s is the Green's function, which describes the relation between the scattered field at the receivers' position and total field at every cell
E^s is the calculated scattered field at every receivers' position, which is a M element vector $\Delta\varepsilon$ is the dielectric difference value with background material at every cell
E^t is an $N \times 1$ vector describing the total field in every cell

This equation is also called the data equation.
Then considering the first-order variations of Equations 13.20 and 13.21, we have

$$\delta[E^t] = G^r \delta[CE^t] \tag{13.22}$$

$$\delta[E^s] = G^s \delta[E^t C] \tag{13.23}$$

It is obvious that $CE^t = E^t C$, and based on Morozov (1984), there exists an approximation for $\delta[CE^t]$, which is

$$\delta[CE^t] = [\delta C]E^t + C[\delta E^t] \tag{13.24}$$

Substituting Equation 13.22 into Equation 13.24 gives

$$\delta[CE^t] = [I - CG^r]^{-1} [\delta C] E^t \tag{13.25}$$

Then, substituting Equation 13.25 into Equation 13.23 leads to the following formula:

$$J = G^s [I - CG^r]^{-1} E^t \tag{13.26}$$

The Jacobian matrix can be computed by the information supplied in forward process.
The next step for the Gauss–Newton method is the regularization process applied to the cost function. Basically, there are two general strategies for regularizing the inverse problem, which are distinguished by the type of the cost function to be minimized. The first kind regularizes the data misfit function at each iteration. The second kind strategy regularizes the nonlinear cost function first and then employs the Gauss–Newton method to the regularized cost function. The basic idea of the appropriate regularization weight for the Gauss–Newton method is that the regularization weight should be high in early iterations where the prediction of the dielectric profile is still far from the real solution, then it should gradually decrease when the algorithm gets closer to the true solution.
In this section, we introduce the Gauss–Newton-type inverse algorithm called the Levenberg–Marquardt method. The Levenberg–Marquardt algorithm neglects the second-order derivatives in the Hessian matrix of the cost function thereby making the computation simple. Another advantage of the algorithm is that the correction of the contrast

permittivity is never locally uphill. Based on the former discussion and Equation 13.16, the updated correction for contrast complex permittivity can be expressed as

$$J_n^H J_n \Delta c = J_n^H d_n \tag{13.27}$$

It is quite obvious that the correction Δc is never locally uphill because $J_n^H J_n$ is a positive semidefinite matrix. However, the approximation has some limitations to make sure that it is valid, such as the problem should not be highly nonlinear, and residuals d_n should not be very large. Furthermore, when the condition number of the matrix $J_n^H J_n$ is large, the correction is very sensitive to the noise in the data, which may lead to solution divergence and unstablility.

Therefore, the Levernberg–Marquardt algorithm sets the equation as

$$\left[J_n^H J_n + \alpha I \right] \Delta c_\alpha = J_n^H d_n \tag{13.28}$$

where:
α is a real positive number
I is the identity matrix

It is also equivalent to

$$\min \left\| J_n \Delta c_\alpha - d_n^{\,2} \right\| + \alpha \left\| \Delta c_\alpha^{\,2} \right\| \tag{13.29}$$

Based on Franchois and Pichot (1997), the advantages of the Levernberg–Marquardt method include (1) improvement in the convergence of the nonlinear problems or the initial guess of the contrast complex permittivity is far from the true solution and (2) possibility to control the condition number of the matrix $J_n^H J_n$, which corresponds to the sensitivity of the cost function to the data noise.

To solve Equation 13.29, the singular value decomposition (SVD) method can be applied because this facilitates the application of the generalized cross validation (GCV) method for choosing the regularization parameter. The SVD of Jacobian matrix J_n can be written as

$$J_n = U \Sigma V^* \tag{13.30}$$

where:
U and V are unitary matrices
Σ is the diagonal matrix with $[\Sigma]_{kk} = \lambda_k$, $k = 1 \cdots \mathrm{rank}(J_n)$, and λ_k are the singular values with descending order

Then, the Levenberg–Marquardt correction is given by

$$\Delta c_\alpha = V \mathrm{diag} \left[\frac{1}{\left| \lambda_k \right|^2 + \alpha} \right] V^* J_n^H d_n \tag{13.31}$$

which is equivalent to Gauss–Newton correction for $\alpha = 0$.

The next step is to set up the selection strategy for the parameter α. Two approaches are widely used. The first one is based on an empirical formula (Hugonin et al. 1990), which controls the step length with the perspective of avoiding overcorrections due to nonlinearity rather than of reducing ill-posed property, which is quite similar to trust-region method. The second approach focuses more on reducing the effects of the ill-posed property on the correction when the data are noisy. In this section, we provide a brief introduction to the empirical formula.

The empirical formula is given as

$$\alpha = \beta \frac{\mathrm{Trace}\left[J_n^H J_n\right]}{N} \frac{\left\|O(c_n) - e_s^{\mathrm{meas}}\right\|^p}{\left\|e_s^{\mathrm{meas}}\right\|^p} \tag{13.32}$$

where N is the total number of the discretization cells. Equation 13.32 is proportional to the normalized field error raised to the p power, which is usually set as 2 or 3. The normalized field error is defined as

$$\mathrm{err}_k = \frac{\left\|O(c_n) - e_s^{\mathrm{meas}}\right\|^p}{\left\|e_s^{\mathrm{meas}}\right\|^p} \tag{13.33}$$

In addition, the selection rule for the parameter β is given as

$$\beta_{k+1} = \begin{cases} \beta_k & \text{if } \mathrm{err}_k - \mathrm{err}_{k-1} < -0.1\,\mathrm{err}_k \\ \dfrac{1}{2}\beta_k & \text{if } -0.1\,\mathrm{err}_k \le \mathrm{err}_k - \mathrm{err}_{k-1} \le 0.1\,\mathrm{err}_k \\ 2\beta_k & \text{if } \mathrm{err}_k - \mathrm{err}_{k-1} > 0.1\,\mathrm{err}_k \end{cases} \tag{13.34}$$

13.3.2.2 Contrast Source Inversion Algorithm

In CSI algorithm, the contrast sources are defined different from the one we use in Gauss–Newton-type methods. It is defined as

$$\chi(r) = c(r)E^t(r) \tag{13.35}$$

Substituting Equation 13.35 into the data equation, we obtain the new data equation as

$$E^s = G^s \chi \tag{13.36}$$

whereas the object equation becomes

$$E^t = E^i + G^r \chi \tag{13.37}$$

Next, substituting Equation 13.35 into Equation 13.20, we obtain a new object formula in terms of the contrast sources as

$$cE^i = \chi - cG^r \chi \tag{13.38}$$

The sequences of contrast sources χ and contrast complex permittivity c can be iteratively found by minimizing the cost function as

$$F(\chi, c) = \frac{\sum_n \left\|e_s^{\mathrm{meas}} - G^s \chi\right\|_S^2}{\sum_n \left\|e_s^{\mathrm{meas}}\right\|_S^2} + \frac{\sum_n \left\|cE^i - \chi + cG^r \chi\right\|_D^2}{\sum_n \left\|cE^i\right\|_D^2} \tag{13.39}$$

where $\|\cdot\|_{S,D}$ denotes the norm on imaging domain D and measurement domain S. The CSI method starts with backpropagation as the initial guess of the contrast sources and contrast complex permittivity. In each iteration, CG directions are applied to update the contrast c and contrast sources χ.

The advantage of this method is that it does not require carrying out any matrix inversion calculation in each iterative step. Also, the computational complexity of the method is equivalent to the complexity of solving two forward problems using a CG method. Recently, some multiplicative regularization factors were applied to the CSI algorithm. The multiplicative technique allows the method to use a regularization factor without the necessity of determining an artificial weighting parameter. In addition, the regularization parameter is determined by the iterative process itself. The new cost function is

$$F_r(\chi,c) = F(\chi,c) F_k^R(c) \tag{13.40}$$

In Equation 13.40, F_k^R is the weighted $L^2(D)$-norm total variation regularization factor at kth iteration, which is given as

$$F_k^R(c) = \frac{1}{V} \int_D \frac{|\nabla c(r)|^2 + \delta_{k-1}^2}{|\nabla c_{k-1}(r)|^2 + \delta_{k-1}^2} \, dv(r) \tag{13.41}$$

where V is the volume of the imaging domain D. The regularization factor F_k^R is a multiplicative constraint, which makes the cost function determined by the inversion problem itself. The parameter δ_{k-1}^2 is chosen as

$$\delta_{k-1}^2 = F_{k-1}^D \Delta_s^2 \tag{13.42}$$

where Δ_s denotes the reciprocal mesh size of the discretized domain and F_{k-1}^D is the normalized error, given by

$$\frac{\sum_n \left\| cE^i - \chi + cG^r\chi \right\|_D^2}{\sum_n \left\| cE^i \right\|_D^2}$$

This regularized CSI algorithm builds the contrast sources and contrast permittivity by using CG steps. As the result, the computational complexity of the algorithm is approximately equal to solving two forward problems.

13.3.2.3 Distorted Born Iterative Method

The DBIM is designed to solve the conventional cost function in Equation 13.11. It first utilizes the Born approximation, which approximates the total field inside the object (in the image domain) to the incident field. In each iteration, we use the updated estimate of the contrast permittivity as the input to the forward solver to calculate the total field in the imaging domain. Then, we check the convergence of the algorithm based on the relative data equation error, which is given by

$$F_s = \frac{\sum \left\| \text{err}_k \right\|_S^2}{\sum \left\| e_s^{\text{meas}} \right\|_S^2} \tag{13.43}$$

where the data error is given by $\mathrm{err}_k = e_s^{\mathrm{meas}} - O(c_k)$. Next, we utilize the distorted Green's operator, and the fields estimated from updated contrast dielectric profile before, to calculate the contrast correction by solving the following minimization problem:

$$\Delta c = \arg\min \sum_m \left\| \mathrm{err}_k - \mathcal{G}\{\Delta c E^t\} \right\|_S^2 \tag{13.44}$$

where the distorted Green's operator $\mathcal{G}\{\delta c E^t\}$ is given as

$$\mathcal{G}\{\Delta c E^t\} = \int_S k_b^2 E^t(r') G_n(r,r') \delta c(r') dr' \tag{13.45}$$

where G_n is the distorted Green's function of the inhomogeneous object. Then, we update the contrast as

$$c_{n+1} = c_n + \Delta c_n \tag{13.46}$$

The total process will stop when the error in Equation 13.43 falls below a set threshold. The linearized inversion problem in Equation 13.43, while simpler than full nonlinear inverse problem, is still ill posed, as the integral operator is continuous and forms a Fredlholm integral equation of the first kind. There are some different approaches to stabilize the problem, such as

$$\Delta c = \arg\min \sum_m \left\| \mathrm{err}_k - \mathcal{G}\{\Delta c E^t\} \right\|_S^2 + \lambda^2 \left\| \nabla^2 \Delta c \right\|_D^2 \tag{13.47}$$

where λ is the regularization parameter determining the weight of the regularization.

There are also different approaches for finding a good regularization parameter in the framework of a Tikhonov function: for example, the discrepancy principle (Morozov 1984) and GCV (Golub et al. 1979). There is an equivalence of DBIM and the Gauss–Newton optimization method. The distorted Green's function operation is approximated by the Jacobian matrix operating on the update, which is given as

$$\mathcal{G}\{\Delta c E^t\} = J_n \Delta c \tag{13.48}$$

The minimization problem for Gauss–Newton method is then identical to that of DBIM.

13.3.3 Ultrawideband Radar Technique

UWB microwave imaging is another promising technique for medical detection and imaging. UWB radar illuminates the object with microwave pulses and collects reflected signals to detect and image the objects or features, which show dielectric contrast with the surrounding tissue. Beamformers are used to spatially focus the reflected signals and to compensate for path-dependent attenuation and phase effects. The most popular algorithms include the monostatic and multistatic delay-and-sum (DAS) method, the delay-multiply-and-sum (DMAS) method, and the improved delay-and-sum (IDAS) method. In this section, we will provide brief introductions to the DAS and the DMAS methods.

The original monostatic DAS beamformer is based on the confocal microwave imaging method (Hagness et al. 1998). In a monostatic beamformer, a UWB microwave signal illuminates the breast, and microwave energy scattered by potential tumor sites is recorded by the transmitting antenna array element. The DAS beamformer involves time shifting and summing the backscattered signals from the breast to create a synthetic focus. If a tumor exists at a specific focal point, then the returns from the tumor site will add coherently.

Assume that there are M monostatic antennas and set S_n as the nth backscattered signal, then the corresponding received energy can be expressed as (Byrne et al. 2010)

$$I(r) = \int_0^{T_t} \left[\sum_{n=1}^{M} S_n \left(t - \tau_n(r) \right) \right]^2 dt \qquad (13.49)$$

where r is the coordinate of the focal point. The nth discrete time delay is represented as $\tau_n(r) = \left(2d_n(r)\right)/\left(vT_s\right)$, where $d_n(r) = |r - r_n|$ is the discrete time distance between nth transmitter antenna r_n, and the focal point r, v denotes the average velocity of the signal propagation in the object, and T_t and T_s are the window length and the sampling interval, respectively.

The improved DAS method introduces a weighting factor called the quality factor $QF(r)$ and a weighing component w_n to the energy equation. It is a measure of the coherence of UWB backscattering at a particular focal point within the object. The improved DAS energy equation is given as (Klemm et al. 2008)

$$I(r) = QF(r) \cdot \int_0^{T_t} \left[\sum_{n=1}^{M} w_n S_n \left(t - \tau_n(r) \right) \right]^2 dt \qquad (13.50)$$

DMAS beamformer involves signals being time shifted, multiplied in pairs, and their products summed in order to calculate the energy at a focal point. In this case, the energy equation is written as (Lim et al. 2008)

$$I(r) = \int_0^{T_t} \left[\sum_{n=1}^{M-1} \sum_{j=n+1}^{M} S_n \left(t - \tau_n(r) \right) S_j \left(t - \tau_j(r) \right) \right]^2 dt \qquad (13.51)$$

13.4 Applications

Due to the advantages afforded by microwave medical radar, it is fast emerging as an important tool for visualizing the interior of the body for clinical purposes. Most of the recent research focuses on the all phases of cancer imaging and detection, identifying bone fracture and tumors, head stroke detection, localization of wireless capsule endoscopy, and intestinal tumor detection. In this section, several examples are introduced to demonstrate the application of microwave radar technique to medical utilization.

13.4.1 Breast Imaging

The University of Bristol team is working on multistatic UWB radar for breast cancer detection. In 2009/2010, their radar system was based on a real aperture antenna array, which consisted of 31 cavity-backed UWB wide-slot antennas arranged over a section of hemi-sphere (Klemm et al. 2009). In their measurement setup, the array is connected with coaxial cables to a custom-built network of electromechanical switches. The switches determine the pairs of antennas and connects all the antennas to vector network analyzer to measure the frequency domain parameter S_{21}. Subsequently, the measured data are transformed to time domain. In their system, 465 independent measurements are recorded, which takes about 80 s to acquire.

Figure 13.4a shows the 31-element array based on the wide-slot antenna. Figure 13.4b shows the complete imaging system. A modified DAS algorithm is used to reconstruct the images of scattered energy. Figure 13.5 shows the imaging results for the breast phantom

(a)   (b)

FIGURE 13.4
(a) Thirty-one-element hemispherical array with wide-slot antenna. (b) Complete imaging system. (From Klemm, M. et al., *IEEE Trans. Antennas Propag.*, 58(7), 2337–2344 © 2010b IEEE. With Permission.)

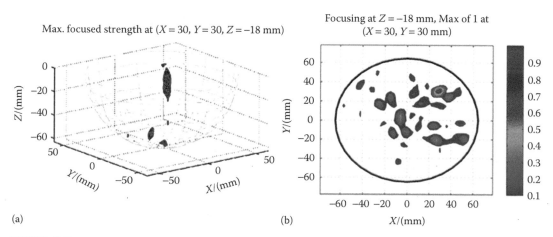

(a)   (b)

FIGURE 13.5
Microwave radar imaging results. (a) 3D radar image, patient in prone position and (b) 2D cross section image, view from behind the patient. (From Klemm, M. et al., *IEEE Trans. Antennas Propag.*, 58(7), 2337–2344 © 2010b IEEE. With Permission.)

from the 3D abreast phantom (Klemm et al. 2010b). The dielectric contrast between tumor and the surrounding tissue is 2:1. Later the work of Klemm et al. improves the ratio of peak clutter energy to a peak tumor energy from −0.6 dB to −3.7 dB (Klemm et al. 2010a). In early 2010, they developed a 60-slot element hemispherical antenna system, which is already installed at the Breast Care Center at Frenchay Hospital in North Bristol (Klemm et al. 2011).

Another group from Dartmouth College has developed an imaging system consisting of 16 monopole antennas placed around the container (Rubæk et al. 2007), as shown in Figure 13.6. The antennas are operated over the frequency range from 500 to 2300 MHz. The patient lies prone on top of the measurement tank with the breast to be examined suspended through an aperture in the top of the tank. The container is filled with coupling liquid, which has similar dielectric property as the average constitutive parameter of the breast.

The implemented image reconstruction algorithm is the iterative Gauss–Newton method combined with conjugate gradient least squares (CGLS) algorithm. The iterative Gauss–Newton method with Tikhonov regularization consists of five steps in each iteration:

1. Hybrid-element algorithm is applied as the forward solver to calculate the electric fields from the estimated complex wave number k_n^2 and compared to the stop condition.

2. Calculation of the Jacobian matrix J based on the current estimated complex wave number k_n^2.

3. Calculation of the Newton direction d_n by solving the linear equation

$$J_n(k_n^2)d_n = E^{\text{meas}} - E^{\text{est}}(k_n^2) \tag{13.52}$$

using the normal equation and the Tikhonov regularization algorithm.

FIGURE 13.6
The imaging system developed at Dartmouth College. The monopoles are positioned in a circular setup and during the measurements, the tank is filled with a coupling liquid. (From Rubæk, T. et al., *IEEE Trans. Antennas Propag.*, 55(8), 2320–2331 © 2007 IEEE. With Permission.)

4. Calculation of the Newton step length α_n using

$$\alpha_n = \arg\min\left\{E^{\text{meas}} - E^{\text{est}}\left(k_n^2 + \alpha_n d_n\right)_2^2\right\} \tag{13.53}$$

5. Updating of the complex wave number using

$$k_{n+1}^2 = k_n^2 + \alpha_n d_n \tag{13.54}$$

To improve the computation efficiency by reducing the iteration steps, a Gauss–Newton method combining with CGLS algorithm is presented (Rubæk et al. 2007). The difference is in the way of determining the updated values of complex wave number at each iteration step. It is given by

$$\Delta k_n^2 = \arg\min\left\{\left\|J\left(\Delta k_n^2\right)\Delta k_n^2 - (E^{\text{meas}} - E^{\text{est}}\left(\Delta k_n^2\right)\right\|_2^2\right\} \tag{13.55}$$

The solution to this linear equation after m CGLS iteration (m CGLS iterations per each Gauss–Newton iteration n) is given by

$$\left[\Delta k_n^2\right]^m = \arg\min\left\{\left\|J\left(\Delta k_n^2\right)^m - (E^{\text{meas}} - E^{\text{est}}\left(\Delta k_n^2\right)\right\|_2^2\right\} \tag{13.56}$$

$$\text{subject to } \left[\Delta k_n^2\right]^m \in \Re_m\left\{J^T J, \left[J^T\left[E^{\text{meas}} - E^{\text{est}}\right]\right]\right\}$$

where $\Re_m$ is the m-dimensional Krylov subspace defined by the Jacobian matrix and the vector of the difference between the measured and estimated fields.

Figure 13.7 shows the results for the right breast of the test patient, which is obtained at 1100 MHz by the Gauss–Newton method with Tikhonov regularization and Gauss–Newton method with CGLS algorithm. In the experiment, seven planes of the right breast are reconstructed, and Figure 13.7 shows three of them, namely plane 1, 4, and 7. Plane 1 is closest to the chest wall.

The CGLS method converges faster than the method using Tikhonov regularization. In addition, the tumor object inside breast is accentuated more in the CGLS approach than in the Tikhonov approach.

13.4.2 Head Stroke Detection

Brain stroke is another area which medical radar technique contributes. Brain stroke is the rapid loss of brain functions due to a disturbance in the blood supply. One system, shown in Figure 13.8, has been built with 16 corrugated tapered slot antenna (Mustafa et al. 2013). The beamformer forms a spatially filtered combination of time-delayed response of scattering points in the head with incident field excited by signal with bandwidth 1–4 GHz. The signal transmission and reception are performed in the frequency domain. Then, the frequency domain signal is transformed to positive value samples in the time domain. The background reflections are eliminated to construct the target response that shows up at different time-shifts with two different methods. To perform the stroke detection and to optimize the beamformer's performance, the background

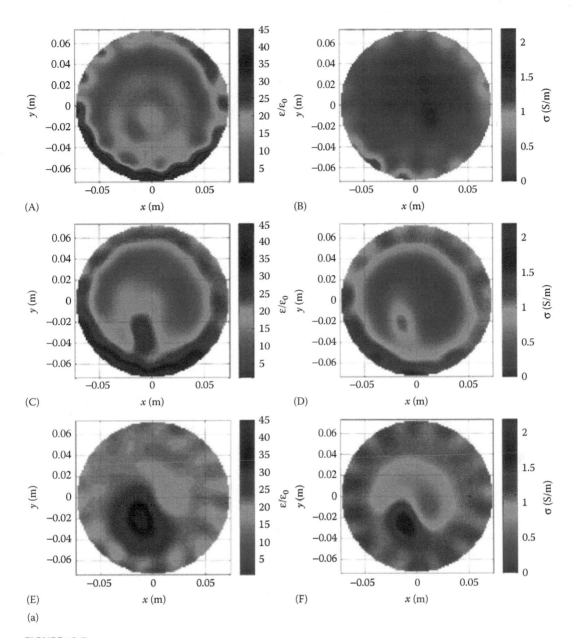

FIGURE 13.7
Results obtained using the Gauss–Newton with (a) Tikhonov regularization method: (A) Plane 1, perm.; (B) Plane 1, cond.; (C) Plane 4, perm.; (D) Plane 4, cond.; (E) Plane 7, perm.; (F) Plane 7, cond. *(Continued)*

reflections are removed by determining the difference in scattered signals. The first approach constructs the difference signal by the subtraction of backscattering signal pairs, whereas the second approach utilizes the symmetrical distribution property of brain tissue in both sides to remove the background reflection by using the difference backscattered signals. After applying the DAS beamforming algorithm, the reconstructed images are shown in Figure 13.9.

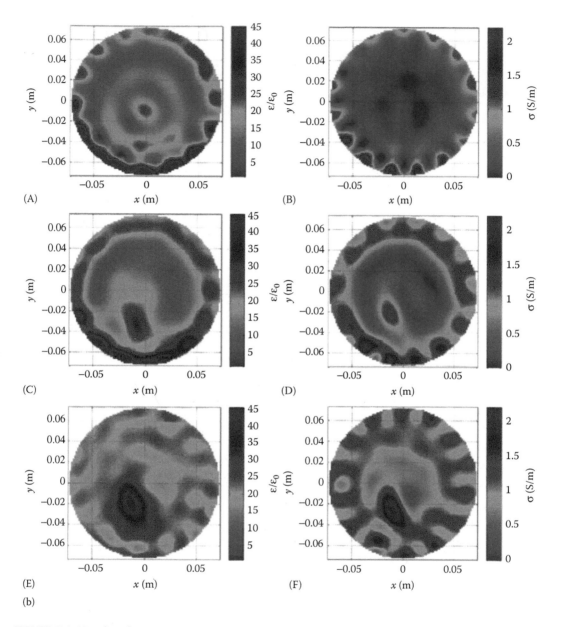

(b)

FIGURE 13.7 (Continued)

Results obtained using the Gauss–Newton with (b) conjugate gradient least squares (CGLS) method: (A) Plane 1, perm.; (B) Plane 1, cond.; (C) Plane 4, perm.; (D) Plane 4, cond.; (E) Plane 7, perm.; (F) Plane 7, cond. for the right breast of the test patient at three of the seven planes. (From Rubæk, T. et al., *IEEE Trans. Antennas Propag.*, 55(8), 2320–2331 © 2007 IEEE. With Permission.)

13.4.3 Intestinal Tumor Imaging

Another application for medical radar is the localization of an in-body RF source inside the gastrointestinal track. With the knowledge of the location, therapeutic operations can be performed precisely at the exact position. On the other hand, the cooperation between the inside RF source and surrounding antennas yields a great possibility for developing images or for

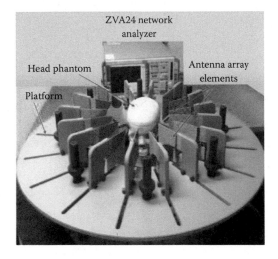

FIGURE 13.8
Imaging system with 16 corrugated tapered slot antenna. (From Mustafa, S. et al., *IEEE Antennas Wirel. Propag. Lett.*, 12, 460–463 © 2013 IEEE. With Permission.)

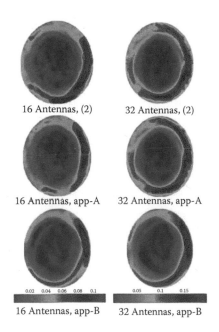

16 Antennas, (2)   32 Antennas, (2)

16 Antennas, app-A   32 Antennas, app-A

16 Antennas, app-B   32 Antennas, app-B

FIGURE 13.9
Reconstructed brain imaging using 16 and 32 antenna array. The ellipse with black color denotes the actual stroke position. (From Mustafa, S. et al., *IEEE Antennas Wirel. Propag. Lett.*, 12, 460–463 © 2013 IEEE. With Permission.)

detecting the small tumors underneath the inner surface of gastrointestinal track. In Chandra et al. (2015a), a localization method for the inside RF source is given. Based on the position predicted, Zhou et al. (2016) presented a potential method to detect and image the small tumor underneath the inner surface of the intestine, which is not detectable for the visual camera.

The localization method involves three parts as shown in Figure 13.10. Basically, the antenna around the body collects the measured scattered data. The Levenberg–Marquardt

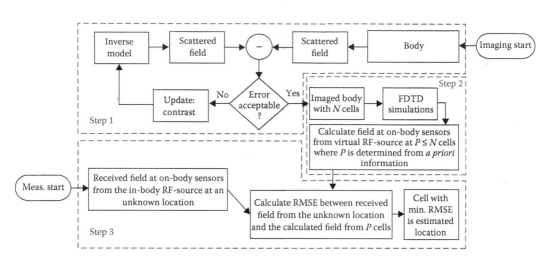

FIGURE 13.10
Schematic diagram of the three-step localization method. (From Chandra, R. et al., *IEEE Trans. Biomed. Eng.*, 62(5), 1231–1241 © 2015 IEEE. With Permission.)

algorithm is used to make the initial image of the object, which is an estimated dielectric profile at discrete points of the object (Zhou et al. 2015). Once the image of the body is obtained, *a priori* information is used to find the possible location of the internal RF source.

There are two major *a priori* information, which contribute to narrowing the range of the search points. The first is based on the electrical properties of the tissues. As the in-body RF source is traveling along the gastrointestinal track, the respective electrical properties are much different from those of fat, bone, or any other tissues. This *a priori* information is supplied by the good quality of the image reconstructed from the first step. The other *a priori* information is based on the human anatomy. According to basic human anatomy, only the front portion of the abdomen is occupied by the small intestine. From the image constructed and the *a priori* information available, one can successfully narrow the possible position of the in-body RF source. In the third step, using the same setup as the first step, the outside antenna collects the electric field. The root-mean-square error (RMSE) between these received field values and the field received by the outside antennas is calculated at all the possible positions. The estimated position of the in-body RF source is determined to be the position that gives the least value of RMSE. It is given mathematically as

$$e_i = \sqrt{\frac{1}{M} \sum_{k=1}^{M} \left\| E^{\text{meas}} - E_i^{\text{est}} \right\|^2} \tag{13.57}$$

where:
$i = 1, 2, \ldots, N$ and N is the total number of the possible position needed for calculation
M is the total number of the surrounding antenna outside human
E^{meas} is the actual measured field at the M antenna
E_i^{est} is the calculated electric field at outside antenna position when the RF source is assumed to be placed at ith position

Then, the position yielding the minimum RMSE is set to be the position of the RF source.

In the simulation, the incident field is set to 403.5 MHz. The algorithm is tested on FDTD simulated data of simple circular phantom and Billie phantom (Christ et al. 2010).

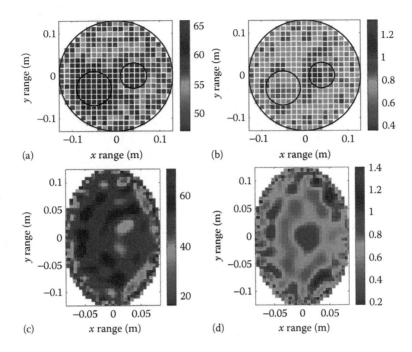

FIGURE 13.11
Imaged body showing: (a) estimated relative permittivity of the circular phantom; (b) estimated conductivity (in S/m) of the circular phantom; (c) estimated relative permittivity of Billie; (d) estimated conductivity (in S/m) of Billie. The image has been obtained at an SNR of 20 dB. (From Chandra, R. et al., *IEEE Trans. Biomed. Eng.,* 62(5), 1231–1241 © 2015 IEEE. With Permission.)

The imaging results for the first step are shown in Figure 13.11. The true path and the estimated path for the source inside the object are shown in Figure 13.12.

Combined with the estimated position of the in-body transmitter, Zhou et al. (2016) developed a possible method to combine the in-body transmitter and outside antenna to make a more accurate image around the inside RF source. The imaging process involves two

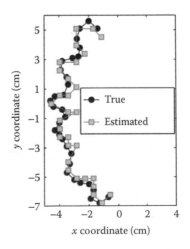

FIGURE 13.12
True path and the estimated path for the source inside the small intestine of Billie phantom. (From Chandra, R. et al., *IEEE Trans. Biomed. Eng.,* 62(5), 1231–1241 © 2015 IEEE. With Permission.)

FIGURE 13.13
Circular phantom (three cylinders inside) containing an inhomogeneous permittivity profile. (Reproduced with permission of the International Society for Optics and Photonics; Zhou, H. et al., Microwave reconstruction method using a circular antenna array cooperating with an internal transmitter, in *SPIE Conference on Radar Sensor Technology XX*, Baltimore, MD, pp. 98290F-1–98290F-12, 2016.)

parts: The first part is to use the aforementioned method to estimate the possible position of the in-body transmitter. The second step is to use the additional data transmitted from the inside RF source and collected by the antenna outside the object to reconstruct the image of the object. The image is discretized by dual mesh method, which has a denser mesh grid around the inside RF source and coarse mesh grid for the remaining part. The MoM with pulse-basis functions and point matching are modified to discretize and used as the forward solver. The Levenberg–Marquardt method is applied to the modified cost function:

$$F(c) = \left\| \left[O(c), \rho O_c(c) \right] - \left[e^s_{\text{meas}}, \rho e^s_{\text{meas},c} \right] \right\|^2 \qquad (13.58)$$

where:

$O_c(c)$ is the calculated scattered field from forward problem

$e^s_{\text{meas},c}$ represents the measured scattered data from antennas outside the object

Parameter ρ provides the weight of the additional data. Because the in-body transmitter is closer to the small (tumor) object compared to the outside antennas, the weight factor $\rho \geq 1$. The circular phantom is shown in Figure 13.13, containing three smaller objects inside with different dielectric values.

The simulation result is shown in Figure 13.14. The reconstructed image can detect the small object near the inside RF source, which is invisible to the conventional system. On the other hand, only the relative permittivity provides some useful information for the small object.

13.5 Challenges and Future Research Directions

Section 13.4 shows that microwave imaging technique has many promising medical applications. So far, the feasibility studies for many applications have been mostly done through numerical simulations. The ultimate objective of any microwave imaging system is to be

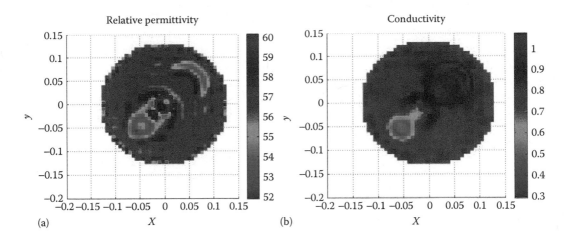

FIGURE 13.14
(a) Relative permittivity and (b) conductivity reconstruction result of modified algorithm. (Reproduced with permission of the International Society for Optics and Photonics; Zhou, H. et al., Microwave reconstruction method using a circular antenna array cooperating with an internal transmitter, in *SPIE Conference on Radar Sensor Technology XX*, Baltimore, MD, pp. 98290F-1–98290F-12, 2016.)

used in a clinical setting as a cost-effective alternative to the existing sensing and imaging modalities. To meet this objective, the microwave imaging system should be developed beyond theory and simulations into working prototypes that can be used for clinical trials. However, there are many technical obstacles and challenges that need to be solved before going for clinical trials.

Effectively, coupling of transmitted microwave signal is one of the major challenges. Due to a high difference in the electrical properties of the medium in which the antenna is placed and the body, strong reflection occurs at the tissue-medium boundary, leaving a weak signal to penetrate the body that is further attenuated by the lossy tissues. This leaves very weak scattered signal to be used for imaging and requires a system with very large dynamic range and great sensitivity. A clinical environment is another challenge for microwave imaging system, as it will be susceptible to EM interference and noise resulting in error in the measured data. The sources of such interference and noise in a clinical environment are RF emissions from wireless devices like mobile phones, medical telemetry, wireless local area network, and other devices as radiology equipments, electrocautery equipments, fluorescent lights, and computers.

Despite the aforementioned hardware and system challenges, contrast agents also provide challenges. A significant difference in the electrical properties of the malignant tissue and the healthy tissues of the same kind is used as a basis for detection of such malignant tissues. However, if any healthy tissue having a very small difference in the electrical properties with this malignant tissue is present in a close proximity, microwave imaging may fail to distinguish between these two kinds of tissues.

On the other hand, although various algorithms were developed recently, they still suffer from several limitations. To relax the nonuniqueness and the ill-posedness of nonlinear inverse problems to some extent, a large number of antennas have to be used so that the scattered field dataset from which the electrical properties are retrieved is large. However, the number of antennas that can be used is limited by their finite size. Moreover, placing the antennas too close to each other may result in high mutual coupling between the antennas introducing error in the measured scattered field. Some other challenges, such as medical frequency requirement, image resolution limitations, and commercial competition with other imaging methods, also need to be solved before microwave imaging system used in a clinical setting.

Some of the other future research work will naturally follow the ways and means to improve the signal excitation, detection, and the reconstruction techniques to obtain high quality, low noise signals, and images. In addition, novel approaches like using multiple-input multiple-output techniques can be considered for reducing the complexity of imaging systems or for improving the image by increasing the signal-to-clutter ratio when compared with a bistatic or a monostatic configuration. Moreover, 3D reconstruction with temporal information in real time will be useful for new applications like obtaining temporal images of a beating heart.

Another direction that can be explored for microwave imaging is compressive sensing. Compressive sensing is a method by which signals can be reconstructed by sampling them at a rate much lower than the Nyquist rate. However, for such a reconstruction, following criteria have to be met: *sparse* representation of the unknown signal in some domain and *incoherence* of the signal used for measurement with respect to the unknown signal. Compressed sensing has been shown to substantially improve the performance of ranging with UWB medical radar in comparison with more conventional methods in scenarios with low SNR.

13.6 Conclusions

Radar-based approaches are fast emerging as promising techniques for disease detection and monitoring. Their advantages, such as nonionizing and noninvasive nature and low cost, are attracting many research groups to contribute to their design, development, and implementations. In recent years, several simulation and modeling methods, algorithms, and hardware designs have been introduced, which can primarily be categorized in two approaches. The first approach is based on the inverse scattering method, which involves the EM theory calculations as the forward solver and numerical optimization methods as the inverse solver. Although this approach is able to provide high resolution for imaging, it also needs relative intensive computational resources and long processing time. The second approach is based on the UWB radar technique in which waveforms such as short pulses or wideband frequency modulated signals are used for probing.

To date, medical radar has been well developed for a variety of applications, such as heartbeat monitoring, brain stroke imaging, breast cancer detection, bone imaging, and capsule position monitoring. Several new applications are under investigation and rely on advancements in RF circuit technologies and signal and image processing algorithms. It is obvious that medical radar will soon be able to play an important role in clinical settings and gain acceptance by clinicians in diagnosis and assessment of diseases. At the same time, there are still many challenges and difficulties that need to be addressed and solved in the future.

Although medical radar is making rapid progress, there are several related issues that merit consideration in future designs. Multimodal approaches combine complementary or uncorrelated information from different sources and with different sensitivities or resolution for improving diagnostic capability. By combining magnetic resonance imaging (MRI) and UWB radar, improved functional diagnosis and imaging were found to be feasible (Thiel et al. 2009). EM compatibility is the most challenging issue when combining

MRI with other modalities. A medical facility is an environment in which EM interference abounds. Factors influencing the medical EM environment include radiated EM fields (due to other devices and cellular phones), noisy electrical power supplies and grounding (earth), magnetic fields (static and alternating), and surges (static discharge, lightning) (Hanada and Kudou 2009). Thus, the sensor electronics must be robust to EM interference. Also, the sensors themselves must not be a source of interference; thus the appropriate regulations and standards in different countries must be followed (Hanna 2009).

References

Aardal, Ø., and J. Hammerstad. 2010. Medical radar literature overview. Oslo, Norway: Norwegian Defence Research Establishment (FFI), Report 2010/00958.

Aardal, Ø., Y. Paichard, S. Brovoll, T. Berger, T. S. Lande, and S. E. Hamran. 2013. Physical working principles of medical radar. *IEEE Transactions on Biomedical Engineering* 60(4): 1142–1149.

Alabaster, C. M. 2003. Permittivity of human skin in millimetre wave band. *Electronics Letters* 39(21): 1521–1522.

Alabaster, C. M. 2004. The microwave properties of tissue and other lossy dielectrics. PhD thesis, Cranfield University.

Alekseev, S. I., and M. C. Ziskin. 2007. Human skin permittivity determined by millimeter wave reflection measurements. *Bioelectromagnetics* 28(5): 331–339.

Balanis, C. A. 2012. *Advanced Engineering Electromagnetics*. New York: Wiley.

Bilich, C. G. 2006. Bio-medical sensing using ultra wideband communications and radar technology: A feasibility study. In *2006 IEEE Pervasive Health Conference and Workshops*, Innsbruck, Austria. doi:10.1109/PCTHEALTH.2006.361671.

Bindu, G., and K. T. Mathew. 2007. Characterization of benign and malignant breast tissues using 2-D microwave tomographic imaging. *Microwave and Optical Technology Letters* 49(10): 2341–2345.

Byrne, D., M. O'Halloran, M. Glavin, and E. Jones. 2010. Data independent radar beamforming algorithms for breast cancer detection. *Progress in Electromagnetics Research* 107: 331–348.

Chahat, N., M. Zhadobov, R. Augustine, and R. Sauleau. 2011. Human skin permittivity models for millimetre-wave range. *Electronics Letters* 47(7): 427–428.

Chandra, R., A. J. Johansson, M. Gustafsson, and F. Tufvesson. 2015a. A microwave imaging-based technique to localize an in-body RF source for biomedical applications. *IEEE Transactions on Biomedical Engineering* 62(5): 1231–1241.

Chandra, R., H. Zhou, I. Balasingham, and R. M. Narayanan. 2015b. On the opportunities and challenges in microwave medical sensing and imaging. *IEEE Transactions on Biomedical Engineering* 62(7): 1667–1682.

Chávez-Santiago, R., I. Balasingham, and J. Bergsland. 2012. Ultrawideband technology in medicine: A survey. *Journal of Electrical and Computer Engineering* 2012: 716973. doi:10.1155/2012/716973.

Chong, E. K. P., and S. H. Zak. 2013. *An Introduction to Optimization*. Hoboken, NJ: Wiley.

Christ, A., W. Kainz, E. G. Hahn, K. Honegger, M. Zefferer, E. Neufeld, W. Rascher et al. 2010. The Virtual Family—development of surface-based anatomical models of two adults and two children for dosimetric simulations. *Physics in Medicine and Biology* 55(2): N23–N38.

Cook, H. F. 1951. Dielectric behaviour of human blood at microwave frequencies. *Nature* 168(4267): 247–248.

Cook, H. F. 1952. A comparison of the dielectric behaviour of pure water and human blood at microwave frequencies. *British Journal of Applied Physics* 3(8): 249–255.

Daubechies, I., M. Defrise, and C. De Mol. 2004. An iterative thresholding algorithm for linear inverse problems with a sparsity constraint. *Communications on Pure and Applied Mathematics* 57(11): 1413–1457.

Deighton, A. M. 2013. Differentiating between healthy and malignant lymph nodes at microwave frequencies. *Journal of Undergraduate Research in Alberta* 3(1): 6.

Dorigo, M., V. Maniezzo, and A. Colorni. 1996. Ant system: Optimization by a colony of cooperating agents. *IEEE Transactions on Systems, Man, and Cybernetics, Part B (Cybernetics)* 26(1): 29–41.

Estatico, C., M. Pastorino, and A. Randazzo. 2012. A novel microwave imaging approach based on regularization in Banach spaces. *IEEE Transactions on Antennas and Propagation* 60(7): 3373–3381.

Fang, Q., P. M. Meaney, S. D. Geimer, A. V. Streltsov, and K. D. Paulsen. 2004. Microwave image reconstruction from 3-D fields coupled to 2-D parameter estimation. *IEEE Transactions on Medical Imaging* 23(4): 475–484.

FCC. 2016. Wireless medical telemetry service (WMTS). https://www.fcc.gov/general/wireless-medical-telemetry-service-wmts (Accessed July 31, 2016).

Foster, K. R., and J. L. Schepps. 1981. Dielectric properties of tumor and normal tissues at radio through microwave frequencies. *Journal of Microwave Power* 16(2): 107–119.

Foster, K. R., and H. P. Schwan. 1995. Dielectric properties of tissues. In *Handbook of Biological Effects of Electromagnetic Fields*, C. Polk and E. Postow (Ed.), pp. 25–102. Boca Raton, FL: CRC Press.

Franchois, A., and C. Pichot. 1997. Microwave imaging-complex permittivity reconstruction with a Levenberg-Marquardt method. *IEEE Transactions on Antennas and Propagation* 45(2): 203–215.

Gabriel, C., S. Gabriel, and E. Corthout. 1996a. The dielectric properties of biological tissues: I. Literature survey. *Physics in Medicine and Biology* 41(11): 2231–2249.

Gabriel, S., R. W. Lau, and C. Gabriel. 1996b. The dielectric properties of biological tissues: III. Parametric models for the dielectric spectrum of tissues. *Physics in Medicine and Biology* 41(11): 2271–2293.

Garg, V. K., K. Smolik, and J. E. Wilkes. 1997. *Applications of CDMA in Wireless/Personal Communications*. Upper Saddle River, NJ: Prentice-Hall.

Garnero, L., A. Franchois, J.-P. Hugonin, C. Pichot, and N. Joachimowicz. 1991. Microwave imaging-complex permittivity reconstruction-by simulated annealing. *IEEE Transactions on Microwave Theory and Techniques* 39(11): 1801–1807.

Ghodgaonkar, D. K., O. P. Gandhi, and M. F. Iskander. 2000. Complex permittivity of human skin in vivo in the frequency band 26.5-60 GHz. In *IEEE Antennas and Propagation Society International Symposium*, Salt Lake City, UT, pp. 1100–1103.

Golub, G. H., M. Heath, and G. Wahba. 1979. Generalized cross-validation as a method for choosing a good ridge parameter. *Technometrics* 21(2): 215–223.

Hagness, S. C., E. C. Fear, and A. Massa. 2012. Guest editorial: Special cluster on microwave medical imaging. *IEEE Antennas and Wireless Propagation Letters* 11: 1592–1597.

Hagness, S. C., A. Taflove, A. and J. E. Bridges. 1998. Two-dimensional FDTD analysis of a pulsed microwave confocal system for breast cancer detection: Fixed-focus and antenna-array sensors. *IEEE Transactions on Biomedical Engineering* 45(12): 1470–1479.

Hanada, E., and T. Kudou. 2009. Electromagnetic noise in the clinical environment. In *3rd International Symposium on Medical Information & Communication Technology*, Montreal, PQ.

Hanna, S. A. 2009. Regulations and standards for wireless medical applications. In *3rd International Symposium on Medical Information & Communication Technology*, Montreal, PQ.

Hartsgrove, G., A. Kraszewski, and A. Surowiec. 1987. Simulated biological materials for electromagnetic radiation absorption studies. *Bioelectromagnetics* 8(1): 29–36.

Hey-Shipton, G. L., P. A. Matthews, and J. McStay. 1982. The complex permittivity of human tissue at microwave frequencies. *Physics in Medicine and Biology* 27(8), 1067–1071.

Hugonin, J. P., N. Joachimowicz, and C. Pichot. 1990. Quantitative reconstruction of complex permittivity distributions by means of microwave tomography. In *Inverse Methods in Action*, P. C. Sabatier (Ed.), pp. 302–310. Berlin, Germany: Springer-Verlag.

Ibrani, M., E. Hamiti, and L. Ahma. 2012. The age-dependence of microwave dielectric parameters of biological tissues. In *Microwave Materials Characterization*, S. Costanzo (Ed.). INTECH Open Access Publisher. doi:10.5772/51400.

ITU. 2015. Technical and operating parameters and spectrum use for short-range radiocommunication devices, Report ITU-R SM.2153-5. Geneva, Switzerland: International Telecommunications Union.

Joines, W. T., R. L. Jirtle, M. D. Rafal, and D. J. Schaefer. 1980. Microwave power absorption differences between normal and malignant tissue. *International Journal of Radiation Oncology • Biology • Physics* 6(6): 681–687.

Joines, W. T., Y. Zhang, C. Li, and R. L. Jirtle. 1994. The measured electrical properties of normal and malignant human tissues from 50 to 900 MHz. *Medical Physics* 21(4): 547–550.

Khaleghi, A., and I. Balasingham. 2015. On selecting the frequency for wireless implant communications. In *2015 Loughborough Antennas & Propagation Conference*, Loughborough, UK. doi:10.1109/LAPC.2015.7366023.

Kim, T., J. Oh, B. Kim, J. Lee, S. Jeon, S., and J. Pack. 2008. A study of dielectric properties of fatty, malignant and fibro-glandular tissues in female human breast. In *Asia-Pacific Symposium on Electromagnetic Compatibility and 19th International Zurich Symposium on Electromagnetic Compatibility*, Singapore: IEEE, pp. 216–219.

Klemm, M., I. J. Craddock, J. A. Leendertz, A. Preece, and R. Benjamin. 2008. Improved delay-and-sum beamforming algorithm for breast cancer detection. *International Journal of Antennas and Propagation* 2008: 761402. doi:10.1155/2008/761402.

Klemm, M., I. J. Craddock, J. A. Leendertz, A. Preece, D. R. Gibbins, M. Shere, and R. Benjamin. 2010a. Clinical trials of a UWB imaging radar for breast cancer. In *Fourth European Conference on Antennas and Propagation*, Barcelona, Spain: EuMA, pp. 1–4.

Klemm, M., D. Gibbins, J. Leendertz, T. Horseman, A. W. Preece, R. Benjamin, and I. J. Craddock. 2011. Development and testing of a 60-element UWB conformal array for breast cancer imaging. In *5th European Conference on Antennas and Propagation*, Rome, Italy: EuMA, pp. 3077–3079.

Klemm, M., J. A. Leendertz, D. Gibbins, I. J. Craddock, A. Preece, and R. Benjamin. 2009. Microwave radar-based breast cancer detection: Imaging in inhomogeneous breast phantoms. *IEEE Antennas and Wireless Propagation Letters* 8: 1349–1352.

Klemm, M., J. A. Leendertz, D. Gibbins, I. J. Craddock, A. Preece, and R. Benjamin. 2010b. Microwave radar-based differential breast cancer imaging: Imaging in homogeneous breast phantoms and low contrast scenarios. *IEEE Transactions on Antennas and Propagation* 58(7): 2337–2344.

Lazebnik, M., S. C. Hagness, J. H. Booske, D. Popovic, L. McCartney, M. Okoniewski, M. J. Lindstrom et al. 2007c. The dielectric properties of normal and malignant breast tissue at microwave frequencies: Analysis, conclusions, and implications from the Wisconsin/Calgary study. In *2007 IEEE Antennas and Propagation Society International Symposium*, Honolulu, HI, pp. 2172–2175.

Lazebnik, M., M. Okoniewski, J. H. Booske, and S. C. Hagness. 2007a. Highly accurate Debye models for normal and malignant breast tissue dielectric properties at microwave frequencies. *IEEE Microwave and Wireless Components Letters* 17(12): 822–824.

Lazebnik, M., D. Popovic, L. McCartney, C. B. Watkins, M. J. Lindstrom, J. Harter, S. Sewall et al. 2007b. A large-scale study of the ultrawideband microwave dielectric properties of normal, benign and malignant breast tissues obtained from cancer surgeries. *Physics in Medicine and Biology* 52(20): 6093–6115.

Lim, H. B., N. T. T. Nhung, E. P. Li, and N. D. Thang. 2008. Confocal microwave imaging for breast cancer detection: Delay-multiply-and-sum image reconstruction algorithm. *IEEE Transactions on Biomedical Engineering* 55(6): 1697–1704.

Massa, A., M. Pastorino, and A. Randazzo. 2004. Reconstruction of two-dimensional buried objects by a differential evolution method. *Inverse Problems* 20(6): S135–S150.

Meaney, P. M., T. Zhou, D. Goodwin, A. Golnabi, E. A. Attardo, and K. D. Paulsen. 2012. Bone dielectric property variation as a function of mineralization at microwave frequencies. *International Journal of Biomedical Imaging* 2012: 649612. doi:10.1155/2012/649612.

Morozov, V. A. 1984. *Methods for Solving Incorrectly Posed Problems*. New York: Springer Science & Business Media.

Mustafa, S., B. Mohammed, and A. Abbosh. 2013. Novel preprocessing techniques for accurate microwave imaging of human brain. *IEEE Antennas and Wireless Propagation Letters* 12: 460–463.

Narayanan, R. M. 2013. Technical considerations in medical radar. In *8th International Conference on Body Area Networks: Workshop on Perspectives and Future Trends for Body Area Networks (PFT-BAN)*, Boston, MA: ACM, pp. 526–535.

O'Halloran, M., M. Glavin, and E. Jones. 2006a. Comparison of a planar and finite difference time domain technique to simulate the propagation of electromagnetic waves in biological tissue. In *International Conference on Microwaves, Radar & Wireless Communications*, Krakow, Poland: IEEE, pp. 1037–1040.

O'Halloran, M., M. Glavin, and E. Jones. 2006b. Frequency-dependent modeling of ultra-wideband pulses in human tissue for biomedical applications. In *IET Irish Signals and Systems Conference*, Dublin, pp. 297–301.

O'Rourke, A. P., M. Lazebnik, J. M. Bertram, M. C. Converse, S. C. Hagness, J. G. Webster, and D. M. Mahvi. 2007. Dielectric properties of human normal, malignant and cirrhotic liver tissue: in vivo and ex vivo measurements from 0.5 to 20 GHz using a precision open-ended coaxial probe. *Physics in Medicine and Biology* 52(15): 4707–4719.

O'Sullivan, F. 1986. A statistical perspective on ill-posed inverse problems. *Statistical Science* 1(4): 502–518.

Pan, J. 2008. Medical applications of ultra-wideband (UWB). Survey paper. Washington University in St. Louis. http://www1.cse.wustl.edu/~jain/cse574-08/ftp/uwb.pdf (Accessed July 31, 2016).

Pancera, E. 2010. Medical applications of the ultra wideband technology. In *2010 Loughborough Antennas & Propagation Conference*, Loughborough: IEEE, pp. 52–56.

Pastorino, M. 2007. Stochastic optimization methods applied to microwave imaging: A review. *IEEE Transactions on Antennas and Propagation* 55(3): 538–548.

Paulson, C. N., J. T. Chang, C. E. Romero, J. Watson, F. J. Pearce, and N. Levin. 2005. Ultra-wideband radar methods and techniques of medical sensing and imaging. In *SPIE Conference on Smart Medical and Biomedical Sensor Technology III,* Boston, MA, pp. 60070L-1–60070L-12.

Pethig, R. 1984. Dielectric properties of biological materials: Biophysical and medical applications. *IEEE Transactions on Electrical Insulation* 19(5): 453–474.

Rahmat-Samii, Y., and E. Michielssen. 1999. *Electromagnetic Optimization by Genetic Algorithms*. New York: Wiley.

Rekanos, I. T., and T. D. Tsiboukis. 1999. A finite element based technique for microwave imaging of two-dimensional objects. In *IEEE Instrumentation and Measurement Technology Conference*, Venice, Italy, pp. 1576–1581.

Robinson, J., and Y. Rahmat-Samii. 2004. Particle swarm optimization in electromagnetics. *IEEE Transactions on Antennas and Propagation* 52(2): 397–407.

Rosen, A., M. A. Stuchly, and A. Vander Vorst. 2002. Applications of RF/microwaves in medicine. *IEEE Transactions on Microwave Theory and Techniques* 50(3): 963–974.

Rubæk, T., P. M. Meaney, P. Meincke, and K. D. Paulsen. 2007. Nonlinear microwave imaging for breast-cancer screening using Gauss–Newton's method and the CGLS inversion algorithm. *IEEE Transactions on Antennas and Propagation* 55(8): 2320–2331.

Schepps, J. L., and K. R. Foster. 1980. The UHF and microwave dielectric properties of normal and tumour tissues: variation in dielectric properties with tissue water content. *Physics in Medicine and Biology* 25(6): 1149–1159.

Sha, L., E. R. Ward, and B. Stroy. 2002. A review of dielectric properties of normal and malignant breast tissue. In *IEEE SoutheastCon*, Columbia, SC, pp. 457–462.

Staderini, E. M. 2002. UWB radars in medicine. *IEEE Aerospace and Electronic Systems Magazine* 17(1): 13–18.

Stauffer, P. R., F. Rossetto, M. Prakash, D. G. Neuman, and T. Lee. 2003. Phantom and animal tissues for modelling the electrical properties of human liver. *International Journal of Hyperthermia* 19(1): 89–101.

Sunaga, T., H. Ikehira, S. Furukawa, H. Shinkai, H. Kobayashi, Y. Matsumoto, E. Yoshitome et al. 2001. Measurement of the electrical properties of human skin and the variation among subjects with certain skin conditions. *Physics in Medicine and Biology* 47(1): N11–N15.

Taflove, A., and S. C. Hagness. 2005. *Computational Electrodynamics: The Finite-Difference Time-Domain Method*. Norwood, MA: Artech House.

Taoufik, E., S. Nabila, and B. Ridha. 2010. The reflection of electromagnetic field by body tissue in the UWB frequency range. In *2010 IEEE Radar Conference*, Washington, DC, pp. 1403–1407.

Taoufik, E., S. Nabila, and B. Ridha. 2011. The interaction between the human body and the ultra wide band radar pulse. In *Microwaves, Radar and Remote Sensing Symposium (MRRS)*, Kiev, Ukraine: IEEE, pp. 105–109.

Thiel, F., M. Hein, U. Schwarz, J. Sachs, and F. Seifert. 2009. Combining magnetic resonance imaging and ultrawideband radar: A new concept for multimodal biomedical imaging. *Review of Scientific Instruments* 80(1): 014302. doi: 10.1063/1.3065095.

Van Den Berg, P. M., and R. E. Kleinman. 1997. A contrast source inversion method. *Inverse Problems* 13(6): 1607–1620.

Zhou, H., R. M. Narayanan, and I. Balasingham. 2016. Microwave reconstruction method using a circular antenna array cooperating with an internal transmitter. In *SPIE Conference on Radar Sensor Technology XX*, Baltimore, MD, pp. 98290F-1–98290F-12.

Zhou, H., R. M. Narayanan, R. Chandra, and I. Balasingham. 2015. Microwave imaging of circular phantom using the Levenberg-Marquardt method. In *SPIE Conference on Radar Sensor Technology XIX and Proceedings of SPIE Conference on Active and Passive Signatures VI*, Baltimore, MD, pp. 946117-1–946117-12.

Zhuge, X., T. G. Savelyev, A. G. Yarovoy, L. P. Ligthart, J. Matuzas, and B. Levitas. 2008. Human body imaging by microwave UWB radar. In *EuMA 5th European Radar Conference*, Amsterdam, The Netherlands, pp. 148–151.

14

Wireless Sensing for Device-Free Recognition of Human Motion

Stefano Savazzi, Stephan Sigg, Monica Nicoli, Vittorio Rampa,
Sanaz Kianoush, and Umberto Spagnolini

CONTENTS

14.1 Introduction: Using Radio Devices as Sensors

This chapter focuses on the use of radio signals for human body motion recognition and sensing purposes. The technology relies on the use of the same electromagnetic (EM) fields adopted for wireless transmission. Radio signals are attenuated and reflected by human bodies due to diffraction, reflection, and scattering phenomena. Therefore, the presence and even the small, that is, half/full-body, movements result in a change of the propagation environment that affects the EM field in a predictable way. Body-induced perturbations of the EM field can be measured and processed in real time to extract shape information about the subject (position, activity, or health status) or to compute an image of the environment that originated the EM perturbation.

This passive (or *device-free*) wireless sensing technology is designed to augment pre-existing radio transceivers that monitor the fluctuations of the EM field across the space. Personal sensing is the current scale at which these technologies are being studied by the research community: they are designed for sensing a single (or a limited number of) individual(s).

As depicted in Figure 14.1, monitoring devices may be deployed at arbitrary (or optimized) locations for communication purposes in the area of interest and exchange digital information by any wireless communication protocol. The presence/activity or location of the body could be recognized by considering a multitude of radio links and by analyzing several received wireless signal features, that is, the link channel quality

Device-free wireless sensing

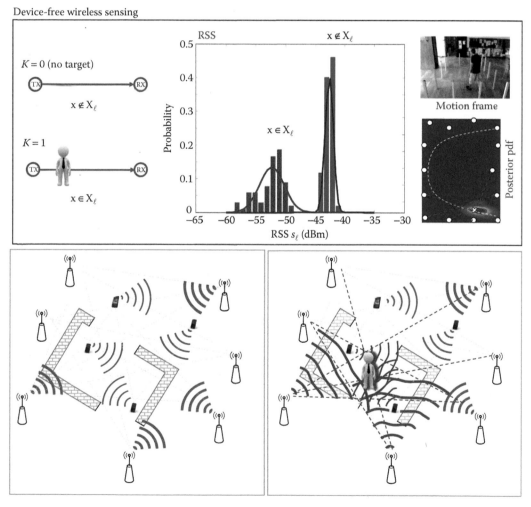

FIGURE 14.1
Wireless sensing for body recognition.

information (CQI), such as power, Doppler/phase shifts, in-phase and quadrature (IQ) raw signals, and so on. Each wireless link can thus double its functionality by acting also as a sensor, and the combination of multiple heterogeneous links enables powerful distributed sensing for accurate human motion understanding.

Radio signals adopted for recognition can be either narrowband or wideband, in licensed or unlicensed frequency bands, with carrier frequencies ranging from MHz to GHz, and above. In order to support human sensing and imaging functions, three key distinctive technological features are incorporated; these are summarized in the following:

Sensorless interaction and anonymous tracking: Simple and intuitive gesture-based interactions of the user with the environment are detected without the need of instrumenting the human body (device-free) or deploying sensors to be set up and calibrated for each user (sensorless). Subjects are thus anonymously tracked and localized without the use of privacy intrusive video cameras, nor image processing (Buehrer et al., 2014), but simply by inferring the EM perturbations from devices' CQI.

Ubiquitous monitoring: Unlike existing infrared (IR) recognition platforms (Lloret et al., 2015), device-free wireless sensing systems support ubiquitous user detection in complex nonline-of-sight (NLOS) indoor space (Seifeldin et al., 2013), using both fixed (e.g., WiFi access points, ZigBee/Bluetooth devices) and nomadic (e.g., smartphones, tablets) radio devices that are interacting over mixed line-of-sight (LOS) and NLOS, or through-the-wall links (Patwari and Wilson, 2010). As a matter of fact, unlike vision- and ultrasonic-based systems, radio frequency (RF) signals with wavelengths that are long enough to penetrate dense objects, such as doors or walls, can be exploited to recognize human motion and gestures even if these gestures are visually in shadow or in a different room adjacent to the one where the RF device itself is operating.

Scalable CQI analytics: The technology typically requires information aggregation, processing, and computation of massive amounts of CQI data generated from, and delivered to, highly distributed heterogeneous wireless devices. Learning and running analytics from large volumes of CQI data require the use of signal processing tools designed to efficiently work on high-dimensional and often incomplete datasets (Slavakis et al., 2014), that is, due to random power cycling of devices or communication failures.

Over the last decade, remarkable progresses have been made toward the recognition of human activities and motions through opportunistic radio signal analytics. Device-free sensing approach has been experimented using heterogeneous networks, but it is also appropriate for the emerging low-power wireless standards, and for personal and device-to-device (D2D) communications (Buehrer et al., 2014), including WiFi, Bluetooth low energy (BLE), ZigBee, and D2D enabled machine-type communication (Laya et al., 2014). The technology also targeted several applications for both home and industrial automation, such as presence, crowd counting, motion detection, localization (Savazzi et al., 2016b), gesture/activity recognition (Sigg et al., 2014b), human–machine cooperation, fall/impact detection, and so on.

Sections 14.2 through 14.7 discuss relevant radio signal features that can be employed to provide device-free sensing, as well as realistic models to describe human body-induced RF fluctuations. Detection and tracking capabilities are investigated in different application contexts, with special focus on location tracking and body activity recognition. Potentials and current limitations related to this technology are also discussed from the feasibility perspective and practical applications. In particular, a specific case study is discussed that focuses on device-free localization (DF-L) and fall detection inside a human–robot-shared industrial workplace.

14.2 Radio Signals for Human-Scale Sensing: Channel Quality Information

Device-free sensing systems consist of wireless receivers that are exposed to modulated EM fields, carrying digital/analog information, and are configured to extract, process, and share RF data in the form of noisy estimations of the time-varying channel response. Carrier frequency can range from lower frequency bands (e.g., 100 MHz in case of FM radio) to microwaves (e.g., 2.4 GHz in case of WiFi/ZigBee or Bluetooth) to millimeter waves (e.g., 60 GHz or above for 5G-and-beyond networks). Let us consider a human body, located at position $\mathbf{x}$ inside the wireless link area, performing an activity $\Theta = \Theta(\mathbf{x})$ defined as an ensemble of nonrigid body motions in the surroundings of location $\mathbf{x}$. The effects of the user behavior (e.g., state, location, motion, activities, and intentions) on the channel response are typically observed over sets of consecutive received symbols (or frames) from which a sequence of CQI footprints can be extracted and processed (typically in real time). Any recognition system should be able to get the best performance, taking into account the specific characteristics of the CQI signals to be processed. In particular, CQI used as EM measurements for recognition can be of three forms as detailed in Sections 14.2.1–14.2.3 and summarized in Figure 14.2.

14.2.1 Upper-Layer Received Signal Strength

Received signal strength (RSS) comprises any aggregated medium access control (MAC) sublayer link quality metric, such as packet/frame error rate, and link quality information (LQI). The RSS metric is thus used to assess CQI at frame level, and it is commonly adopted for transmitter (TX)—receiver (RX) link adaptation and MAC layer transmission scheduling tasks. RSS is especially attractive for device-free wireless sensing due to its accessibility with commercial/commodity devices. Low-cost power estimators, or peak detectors, are typically used to acquire information about signal strength, and these are embedded in devices for many purposes (e.g., automatic gain control [AGC] loop or to adapt the dynamic range before A/D conversion). Existing RSS-based recognition schemes usually correlate the RSS profile or the variations with respect to the human-free state with the environmental changes caused by human motion. For example, simple presence detection systems are based on the real-time monitoring of RSS deviations as large fluctuations indicate a moving target in the surrounding of the link, whereas small variations infer absence of people. At MAC layer frame time t, the RSS s_t can be modeled in logarithmic (dB) scale as

$$s_t(\Theta) = s(\varnothing) + \Delta s_t(\Theta) \tag{14.1}$$

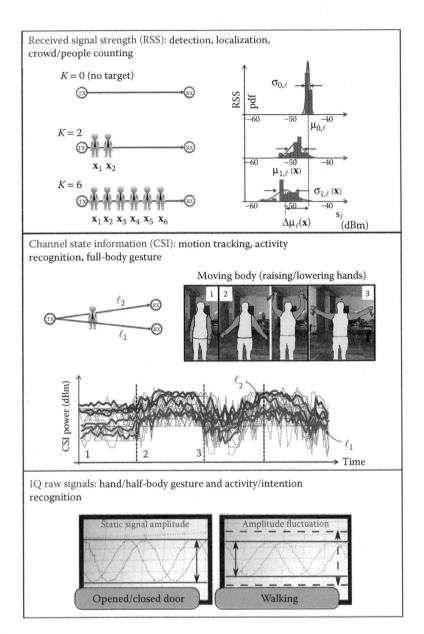

FIGURE 14.2
CQI signals: received signal strength (RSS), channel state information (CSI), and IQ raw signals.

where the additive deviation $\Delta s_t(\Theta)$ from $\Theta = \varnothing$ models the body-induced perturbation, and $s(\varnothing) = \mathbb{E}[s_t(\Theta = \varnothing)]$ is the (average) RSS observed in the human-free state. The sequence $\mathbf{s}(\Theta) = [s_t(\Theta)]_{t \in T} \in \mathbb{R}^{T \times 1}$ collects the human-induced RSS footprint over T frames. Similarly, the *RSS profile* is the deviation with respect to $\Theta = \varnothing : \Delta \mathbf{s}(\Theta) = [s_t(\Theta) - s(\varnothing)]_{t \in T} \in \mathbb{R}^{T \times 1}$. For IEEE 802.15.4 devices, the digital RSS indicator (RSSI) $\hat{s}_t$ can be used as estimator of the RSS with 8-bit resolution. Similarly, in WiFi standards, RSS estimation can be obtained from the received channel power indicator (RCPI).

14.2.2 Baseband Channel State Information

Channel state information (CSI) measures the channel response at symbol level: CSI estimation is typically obtained from training/reference signals multiplexed with information symbols and periodically placed in physical-layer data frames. Therefore, processing of CSI information for the purpose of sensing can leverage on multiple independent measurements at frame level and can be used to capture fast human body movements and gestures. CSI can be extracted by using either software defined radio (SDR) devices or also commercial off-the-shelf (COTS) network interface cards (NIC) (Halperin et al., 2011). Under the frequency-flat channel assumption, the received reference signal (RS) at symbol time $t \in T$, $r_t = h_t(\Theta)\omega_t + n_t$ (with ω_t and n_t the transmitted RS and the noise term, respectively), captures the moving body in state Θ through the corresponding complex channel envelope:

$$h_t(\Theta) = h(\varnothing) + \Delta h_t(\Theta) \tag{14.2}$$

Human body effects on channel response are now embedded into a characteristic footprint of channel variations over T received symbols $\mathbf{h}(\Theta) \in \mathbb{C}^{T \times 1} = [h_t(\Theta)]_{t \in T}$. The *CSI profile* set is $\Delta \mathbf{h}(\Theta) \in \mathbb{C}^{T \times 1} = [\Delta h_t(\Theta) = h_t(\Theta) - h(\varnothing)]_{t \in T}$, with $h(\varnothing) = \mathbb{E}[h_t(\Theta = \varnothing)]$ being the average response for the human-free state. Noisy profiles $\Delta \hat{\mathbf{h}}(\Theta) = [\hat{h}_t(\Theta) - \hat{h}(\varnothing)]_{t \in \Omega}$ with estimated channels $\hat{h}_t(\Theta)$ and human-free response $\hat{h}(\varnothing)$ are typically observed over a subset of times (or symbol indexes) $\Omega \subseteq T$ accounting for the training/data multiplexing, and missing symbols. CSI-related signals can also provide an indirect measurement on the frequency selectivity of the channel, which is related to multipath propagation. For example, in multicarrier orthogonal frequency-division multiplexing (OFDM) modulation as in any specification of the recent WiFi standards, multidimensional processing of CSI can be carried out over the time–frequency grid. The received reference signals $\mathbf{r}_t$ over the K pilot subcarriers $\{f_1, \ldots, f_K\}$ inside an OFDM symbol t can thus be written as $\mathbf{r}_t = \operatorname{diag}[\omega_t] \cdot \mathbf{h}_t(\Theta) + \mathbf{n}_t$ with vector ω_t collecting the transmitted training symbols and baseband channel vector $\mathbf{h}_t(\Theta) = [H_{f,t}(\Theta)]_{f=f_1}^{f_K}$ containing the Fourier Transform F ($\cdot$) of channel $h_t(\tau \mid \Theta)$

$$H_{f,t}(\Theta) = H_f(\varnothing) + \underbrace{\mathrm{F}\left(\Delta h_t(\tau \mid \Theta)\right)|_f}_{\Delta H_{f,t}(\Theta)} \tag{14.3}$$

The CSI footprint is now the matrix $\mathbf{H}(\Theta) \in \mathbb{C}^{K \times T} = [\mathbf{h}_1(\Theta), \ldots, \mathbf{h}_T(\Theta)]_{t \in T}$ with human-induced profile $\Delta \mathbf{H}(\Theta) = [\mathbf{H}(\Theta) - \mathbf{H}(\varnothing)]$.

14.2.3 I/Q Raw Signals Values

Other IQ raw signal features can be processed in addition to baseband CSI, such as *micro-Doppler, dynamic phase shifts*, and IQ *channel envelope*. In this case, the receiver is designed to collect the physical (PHY) layer raw signal samples (either continuously or over consecutive physical layer data frames) and to extract relevant signal features. This approach can be, in principle, applied to any RF system because it does not require the use of a wireless communication interface providing a standardized measurement of CSI or RSS. Processing of IQ values requires in general the use of SDR devices that implement the receiver in order to get access to the signal at baseband. For example, micro-Doppler shifts can be extracted for activity/gesture recognition (Youngwook and Hao, 2009). The challenge is that human activities (e.g., hand/body gestures) result in very small Doppler shifts. For instance, in

WiFi networks, Doppler shifts of few Hertz must be detected from typical signals of 20 MHz band: considering a 5 GHz WiFi transmission, a maximum Doppler shift of 17 Hz was observed as caused by a 0.5 m/s full-body gesture (Adib and Katabi, 2013).

14.3 Device-Free Sensing Approaches and Relevant Applications

Wireless human tracking via RF devices encompasses different research areas such as signal processing, computer vision, communication networks, and human–machine interfaces. In what follows we report the most recent research results for selected relevant applications.

14.3.1 Presence Detection and People Counting

The first experimental activities showed that body motions leave a characteristic footprint on RSS patterns (Woyach et al., 2006); therefore, it was possible to detect the presence of noncooperating subjects, enabling a large set of applications including intrusion detection, border protection, and home automation (Kosba et al., 2012). For example, anomaly or change-point detection techniques can be adopted to identify human presence. In addition to detection, people counting allows the estimation of the occupancy, as well as the number of people (or spatial density), in open or closed environments (Depatla et al., 2015). A model that correlates the statistical feature of RSS as a function of the number of subjects in the area has also been analyzed therein.

14.3.2 Localization and Tracking

Real-time processing of RSS fluctuations can be effectively used for body localization (Patwari and Wilson, 2010). One approach, the radio tomographic imaging (RTI) (Patwari and Wilson, 2010, Wilson and Patwari, 2011) adopts computed tomography methods to reconstruct an image of static objects. The methods introduced in Seifeldin et al. (2013), Savazzi et al. (2014), Saeed et al. (2014) allow to explicitly track the position of targets using a Bayesian approach that jointly processes the RSS mean and standard deviations.

14.3.3 Activity Recognition

Human gesture recognition and body motion detection have been addressed in recent research projects such as SenseWaves, WiSee, and Wi-Vi (Adib and Katabi, 2013, Pu et al., 2013, Sigg et al., 2014b), targeting both RSS (Sigg et al., 2014b) and baseband CSI analytics using SDR devices operating at 900 MHz (Shi et al., 2014), 2.4 GHz with 20 MHz band WiFi-compliant RF frontends, (Adib and Katabi, 2013, Wang et al., 2014a) and 5.8 GHz (Pu, 2014). Activities of a single subject can be either simple actions such as walking, standing, or sitting or more complex combinations. More recently, device-free systems based on tracking of RSS profiles have also been designed for obstacle/object 2D mapping (Mostofi, 2013), detection of human breathing (Patwari et al., 2014), and fall detection for assisted living domain (Mager et al., 2013, Kianoush et al., 2016).

14.4 Physical Modeling of Body-Induced Fading

A crucial problem for quantitative evaluation of device-free wireless sensing system performance is the availability of simple but realistic models to describe human body-induced shadowing. Ray tracing (Aly and Youssef, 2013), stochastic (Smith et al., 2013), empirical (Bultitude, 1987), and geometric-based models (Ghaddar et al., 2007, Kouitatis, 2010) have been investigated to predict the correlation between the human body position **x** and the corresponding channel perturbations. Usually, EM methods can exploit geometric/uniform theory of diffraction (GTD/UTD), see (McNamara et al., 1990) for a short review, for their ability to accurately evaluate the EM field at the receiver; however, they are very complex, time consuming, and, above all, require perfect knowledge of the shape, composition, and properties of the obstacle.

In what follows, a simplified model based on the Fresnel–Kirchhoff scalar diffraction theory is introduced and depicted in the scenario shown in Figure 14.3. Here, we consider the framework summarized according to the single-link configuration of Figure 14.3a where the link is horizontally placed at distance h from the ground. The human body is sketched as a 3D perfectly absorbing homogeneous cylinder (Ghaddar et al., 2007), having a height $2a_z$ and an elliptical base with semiaxes a_{yu} and a_{yv}. To account for a human body standing in a specific location but possibly changing its posture, the 3D cylinder rotates along the vertical axis x_φ with azimuth φ. As a trade-off between model simplicity and accuracy of the EM results, the 3D cylinder is reduced to an equivalent 2D rectangular knife-edge surface (Rampa et al., 2015) that is modeled according to the Fresnel–Kirchhoff scalar diffraction theory. This 2D surface has the same height $2a_z$ but variable traversal semisize $a_y = a_y(\varphi)$ with $a_{yv} \leq a_y(\varphi) \leq a_{yu}$ and $-\pi < \varphi \leq \pi$, whereas its barycenter G is placed in position $\mathbf{x} = \left[x_1 = x, x_2 = y \right]$ defined in the 2D horizontal space (x_1, x_2). We assume that the LOS path length is equal to d, and the carrier wavelength $\lambda = c/f_c$ is related to the carrier frequency f_c. If the obstacle is located near the LOS path with (first) Fresnel's ellipsoid radius $R = \sqrt{\lambda x(d-x)/d} < h$ (that defines the first Fresnel's zone) and $y, a_{yu}, a_{yv}, a_z \ll d$, then, neglecting multipath body-induced effects and reflections from the ground, ceiling, and other colocated objects, the additional attenuation term expressed in logarithmic (dB) scale (*i.e.*, with respect to the free-space case) is $\Delta E^2_{dB}(\mathbf{x}; \varphi) = -\left(\left| E(\mathbf{x}; \varphi)/E_0 \right|^2 \right)_{dB}$, with $(\cdot)_{dB}$ indicating dB-scale conversion. It depends only on the received electric field $E(\mathbf{x}; \varphi)$ for the obstacle in **x** and azimuth φ, and the free-space field E_0:

$$\frac{E(\mathbf{x}; \varphi)}{E_0} = 1 - j\left(\Gamma_C - j\Gamma_S\right)\left(\Psi_C - j\Psi_S\right) \qquad (14.4)$$

where $\Psi_C = C\left(\sqrt{2}a_z/R\right)$, $\Psi_S = S\left(\sqrt{2}a_z/R\right)$, $\Gamma_C = C\left(\sqrt{2}(y+a_y)/R\right) - C\left(\sqrt{2}(y-a_y)/R\right)$, and $\Gamma_S = S\left(\sqrt{2}(y+a_y)/R\right) - S\left(\sqrt{2}(y-a_y)/R\right)$, depend only on the Fresnel cosine $C(\cdot)$ and sine $S(\cdot)$ integrals (Abramowitz and Stegun, 1972), the geometrical constants of the obstacle and the length of the link path.

Focusing on body localization, because the human body moves around the current position not only rotating but also slightly changing position or posture, its location **x** (or state $\Theta = [\mathbf{x}]$) is known with an uncertainty $\Delta\mathbf{x} = \left[\Delta x, \Delta y\right]$. Therefore, the body-induced channel power perturbation $\Delta E^2_{dB}(\mathbf{x} + \Delta\mathbf{x}; \varphi)$ with respect to the free-space case, for variable azimuth φ and position uncertainty $\Delta\mathbf{x}$, is due to:

$$\Delta E^2_{dB}(\mathbf{x} + \Delta\mathbf{x}; \varphi) = -10\log_{10}\left(1 + \left(\Psi^2_C + \Psi^2_S\right)\left(\Gamma^2_S + \Gamma^2_C\right) - 2\Psi_S\Gamma_C - 2\Psi_C\Gamma_S\right) \qquad (14.5)$$

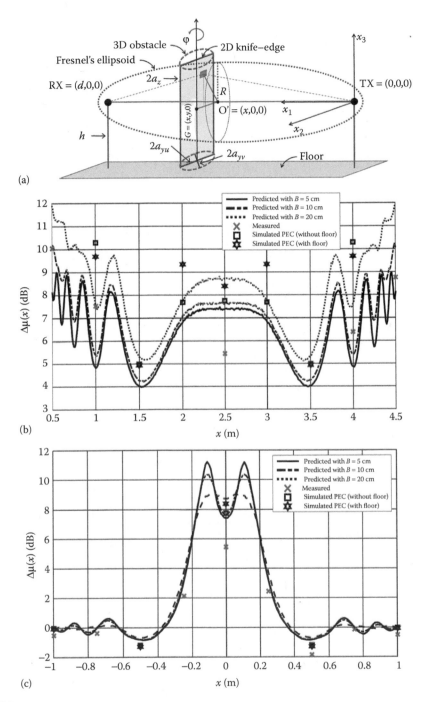

FIGURE 14.3
(a) Single-link layout, (b) predicted versus measured and simulated additional attenuation along, and (c) across the LOS path.

It is worth noticing that the effects of the positional uncertainty directly influence the Fresnel's radius $R(\Delta x) = \sqrt{\lambda(x + \Delta x)(d - x - \Delta x)/d}$ and, indirectly, also impact on all terms $\Psi_C = C(\sqrt{2}a_z/R)$, $\Psi_S = S(\sqrt{2}a_z/R)$, $\Gamma_C = C(\sqrt{2}(y + \Delta y + a_y)/R) - C(\sqrt{2}(y + \Delta y - a_y)/R)$, and $\Gamma_S = S(\sqrt{2}(y + \Delta y + a_y)/R) - S(\sqrt{2}(y + \Delta y - a_y)/R)$ used to compute Equation 14.5.

The attenuation term as shown in Equation 14.5 can be considered as random due to the varying orientations φ and the uncertainty Δx of the obstruction body. In fact, it is assumed here that the positional uncertainty Δx is represented by a 2D uniformly distributed random variable with $|\Delta x|, |\Delta y| \leq B/2$. Therefore, the body is allowed to randomly make small movements in a squared area of size B centered around the nominal position $\mathbf{x}$.

The simplified model as shown in Equation 14.4 neglects the true shape, surface conditions, complex composition, and EM properties of the obstacle (e.g., polarization, permittivity, and conductivity), but it is accurate enough to model human-induced attenuation effects. Figure 14.3b and c shows the comparison of the average channel power perturbation $\mathbb{E}_{\varphi, \Delta x}\left[\Delta E_{dB}^2(\mathbf{x} + \Delta \mathbf{x}; \varphi)\right]$ induced by a person against the values predicted by Equation 14.5 and the ones obtained by simulating the obstacle as a 2D perfect electric conductor (PEC), having the same size of the person, placed over a concrete floor with EM parameters $\epsilon_r = 5$ and $\tan \delta = 0.14$. For comparison, simulations without the floor are considered, too. The predicted terms account for small but unknown movements Δx and rotations φ of the target, herein simulated as uniformly distributed inside a squared bins of size $B = 5, 10$, and 20 cm, and inside the azimuth interval $[-\pi + \pi]$, respectively.

Figure 14.4 shows the mean $\Delta\mu(\mathbf{x}) = \mathbb{E}_{\varphi, \Delta x}[\Delta E_{dB}^2(\mathbf{x} + \Delta \mathbf{x}; \varphi)]$ (Figure 14.4a) and the standard deviation $\Delta\sigma(\mathbf{x}) = \sqrt{\mathbb{E}_{\varphi, \Delta x}[(\Delta E_{dB}^2(\mathbf{x} + \Delta \mathbf{x}; \varphi) - \Delta\mu(\mathbf{x}))^2]}$ (Figure 14.4b) perturbation maps of

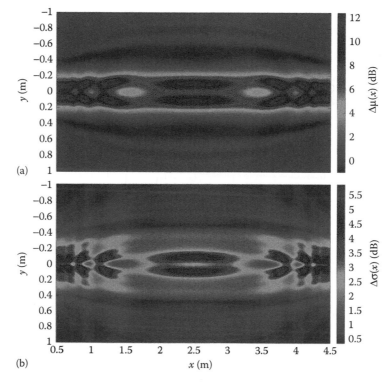

FIGURE 14.4

(a) Predicted mean $\Delta\mu(x)$ and (b) standard deviation $\Delta\sigma(x)$ attenuation maps for $B = 10$ cm. The simulation employs the same model parameters used for Figure 14.3.

the RF attenuation as a function of the obstacle position **x** after averaging out all variable azimuth and positional uncertainty effects. The simulation parameters are $d = 5$ m, $f_c = 2.48$ GHz, $a_z = 90$ cm, $a_{yu} = 27.5$ cm, $a_{yv} = 12$ cm, $B = 5,10,20$ cm, whereas additional multipath effects are precalibrated from measurements as shown in Rampa et al. (2015).

According to Figure 14.4, the *sensitivity area* X that is influenced by the presence of the obstacle has roughly a width of about 1 m around the LOS path. As expected, the impact of the target presence is higher along the LOS path and close to the transmitting/receiving devices, highlighting a significant sensitivity to the target location.

14.5 Localization and Motion Tracking

User location is needed in assisted living applications (Savazzi et al., [2016b]), (Seifeldin et al., 2013) for monitoring daily activities, detecting anomalies, or controlling appliances (e.g., lights, doors, windows, air conditioners). DF-L (Youssef et al., 2007, Wilson and Patwari, 2010, Savazzi et al., 2014) is a new technology that has recently emerged for passive detection and tracking of user movements in areas covered by preexisting wireless networks. It exploits RSS observations (Equation 14.1) periodically collected over the network links, without requiring the user to carry any radio equipment nor to take active part in the positioning process. The power $s_{\ell,t}$ is measured over each link $\ell = 1,\ldots,L$, at time instants $t = 1,2,3,\ldots$, with sampling interval Δt. Measurements are then gathered, either by a fusion center or in a distributed manner, to form a multilink dataset $\mathbf{s}_t = \left[s_{1,t},\ldots,s_{L,t}\right]^T$ and perform multiangulation. The goal is the estimation of the user position $\mathbf{x}_t$ within the network area $\mathcal{X}$ from the dataset $\mathbf{s}_{1:t} = \left[\mathbf{s}_1 \ldots \mathbf{s}_t\right]$ aggregated over all links and time instants. The following analysis focuses on single-user applications; for extension to a multiuser scenario refer to Nicoli et al., 2016.

As illustrated in Figure 14.5, each RSS $s_{\ell,t}$ (in logarithmic units) can be reasonably modeled as Gaussian distributed (Savazzi et al., 2014), i.e. lognormal (Coulson et al., 1998), with moments depending on the target location relative to the ℓth link and particularly on whether $\mathbf{x}_t$ falls or not within the link sensitivity area X_ℓ:

$$s_{\ell,t} = \begin{cases} \mu_{0,\ell} + w_{0,\ell,t}, & \mathbf{x}_t \notin X_\ell \\ \mu_{1,\ell}(\mathbf{x}_t) + w_{1,\ell,t}, & \mathbf{x}_t \in X_\ell \end{cases} \tag{14.6}$$

If the target is outside the link area ($\mathbf{x}_t \notin X_\ell$), the RSS has a deterministic mean μ_0 that accounts for static propagation effects (path-loss, shadowing and multipath) due to fixed obstructions or scattering objects. The random term $w_{0,\ell,t} \sim N(0,\sigma_{0,\ell}^2)$ accounts for measurement errors due to hardware limits as well as for small power fluctuations caused by variations in the surrounding environment. When the target is closer to the link (i.e., $\mathbf{x}_t \in X_\ell$), on the other hand, the received signal is typically subject to an increased attenuation due to the obstruction generated by the target itself and an amplified fluctuation due to its movements around $\mathbf{x}_t$ (i.e., turning, change of posture, arm movement). The mean RSS is thus $\mu_{1,\ell}(\mathbf{x}_t) = \mu_{0,\ell} + \Delta\mu_\ell(\mathbf{x}_t)$ with $\Delta\mu_\ell(\mathbf{x}_t) \leq 0$, whereas the random shadowing is $w_{1,\ell,t} \sim N\left(0,\sigma_{1,\ell}^2(\mathbf{x}_t)\right)$ with $\sigma_{1,\ell}(\mathbf{x}_t) = \sigma_{0,\ell} + \Delta\sigma_\ell(\mathbf{x}_t)$ and $\Delta\sigma_\ell(\mathbf{x}_t) \geq 0$. An experimental evidence of this is in Figure 14.5, where the RSS distribution built with real measurements is compared to the Gaussian model for a link of length 4 m, in the two cases of target outside (right) or inside (left) the sensitivity area.

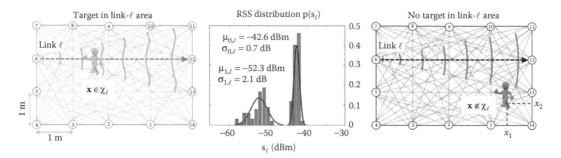

FIGURE 14.5
DF-L: RSS modeling for target outside (left) or inside (right) the link area. The Gaussian model is compared to the distribution obtained from real measurements over the link between nodes 6 and 12, in the two cases (left and right distributions in the central plot).

The perturbation maps $\{\Delta\mu_\ell(\mathbf{x}_t), \Delta\sigma_\ell(\mathbf{x}_t)\}$ are strongly related to the target position (Patwari and Wilson, 2010, Wilson and Patwari, 2011, Savazzi et al., 2014): they are, in fact, most relevant when the target is obstructing the LOS, especially if close to the transmitter or the receiver. This can be appreciated in the examples of Figure 14.6, where 2D experimental profiles $\{\Delta\mu_\ell(\mathbf{x}_t), \Delta\sigma_\ell(\mathbf{x}_t)\}$ are shown for the links of a network of 14 devices, regularly deployed along the boundary of a $4\text{m} \times 3\text{m}$ area, and for a human target position $\mathbf{x}_t$ ranging within the rectangular area, in both indoor and outdoor environments. The maps are shown on the bottom for a selected subset of the $L = (14 \cdot 13)/2 = 91$ links. As expected, the sensitivity area X_ℓ is centered on the LOS path (highlighted in green), and it is larger in indoor than in outdoor due to multipath effects. On the top right, the perturbation measured along the LOS path of a 4 m link is compared against the predicted one obtained by diffraction-based modeling (Rampa et al., 2015).

Results in Figure 14.6 confirm that both the RSS mean and variance provide significant information on the target location and should be jointly exploited for localization. Positioning, however, requires the knowledge of the reference parameters $\{\mu_{0,\ell}, \sigma_{0,\ell}\}$ as well as the perturbation maps $\{\Delta\mu_\ell(\mathbf{x}_t), \Delta\sigma_\ell(\mathbf{x}_t)\}$ for all position values $\mathbf{x}_t \in X_\ell$ and for all links $\ell = 1, \ldots, L$. Although $\{\mu_{0,\ell}, \sigma_{0,\ell}\}$ can be easily precalibrated when no target is in the network area, evaluation of $\{\Delta\mu_\ell(\mathbf{x}_t), \Delta\sigma_\ell(\mathbf{x}_t)\}$ is more critical as it requires extensive fingerprinting campaigns (Wilson and Patwari, 2010, Saeed et al., 2014, Savazzi et al., 2014) or ray-tracing simulations (Scholz et al., 2015). Analytical modeling, when viable, should be preferred to simplify the calibration and to enable predeployment performance assessment. In Wilson and Patwari (2010, 2011), a simple single-parameter model is considered where $\Delta\mu_\ell(\mathbf{x}_t)$ and $\Delta\sigma_\ell(\mathbf{x}_t)$ are assumed to be constant and inversely proportional to the square root of the link distance for $\mathbf{x}_t \in X_\ell$ with X_ℓ modeled as an ellipsoid with foci at the transmitter and receiver locations. Parametric shadowing models can also be found in Buehrer et al. (2014), whereas in Rampa et al. (2015) a diffraction-based model is proposed.

Once the digital maps have been built, the target location can be derived by combining the RSS measurements of the different links according to a selected estimation criterion and model equation (14.6). Assuming the measurements as independent over the links, the maximum likelihood (ML) criterion applies as $\hat{\mathbf{x}}_t = \arg\max_{\mathbf{x}_t} \mathcal{L}(\mathbf{s}_t \mid \mathbf{x}_t)$, with log-likelihood function:

$$\mathcal{L}(\mathbf{s}_t \mid \mathbf{x}_t) = -\ln\left(\det\left[\mathbf{C}(\mathbf{x}_t)\right]\right) - \left\|\mathbf{s}_t - \mu(\mathbf{x}_t)\right\|^2_{\mathbf{C}^{-1}(\mathbf{x}_t)} \qquad (14.7)$$

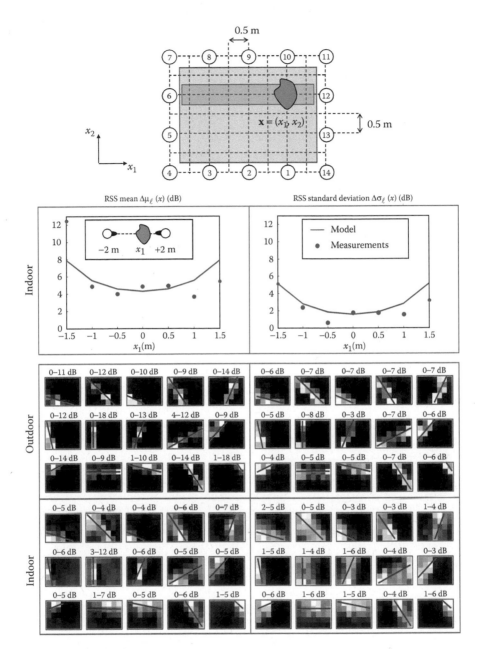

FIGURE 14.6
Digital maps of mean and standard deviation of the RSS perturbation $\{\Delta h_\ell(\mathbf{x}), \Delta\sigma_\ell(\mathbf{x})\}$ versus the target location $\mathbf{x}$, for the network layout in the top-center plot. Perturbations are shown along the LOS path of link (top-left and top-right plots) and over the entire network area for some selected links (bottom plots) in both indoor and outdoor scenarios. The figure is based on Savazzi et al. (2014) , but it shows different subsets of data and examples.

with det[·] denoting the determinant operator, $\left\|\mathbf{s}_t - \mu(\mathbf{x}_t)\right\|^2_{\mathbf{C}^{-1}(\mathbf{x}_t)} = [\mathbf{s}_t - \mu(\mathbf{x}_t)]^\mathsf{T}\mathbf{C}^{-1}(\mathbf{x}_t)[\mathbf{s}_t - \mu(\mathbf{x}_t)]$, $\mu(\mathbf{x}_t) = [\mu_1(\mathbf{x}_t) \cdots \mu_L(\mathbf{x}_t)]^\mathsf{T}$, and $\mathbf{C}(\mathbf{x}_t) = \mathrm{diag}[\sigma_1^2(\mathbf{x}_t), \ldots, \sigma_L^2(\mathbf{x}_t)]$. Using the weighted least squares (WLS) criterion, the estimate simplifies to $\hat{\mathbf{x}}_t = \arg\min_{\mathbf{x}_t} \left\|\mathbf{s}_t - \mu(\mathbf{x}_t)\right\|^2_{\mathbf{C}^{-1}(\mathbf{x}_t)}$.

Information extracted from the instantaneous measurement $\mathbf{s}_t$ can also be augmented with prior information about the target motion, using sequential Bayesian filtering. The a-posteriori pdf is derived recursively as (Arulampalam et al., 2002):

$$p(\mathbf{x}_t \mid \mathbf{s}_{1:t}) \propto p(\mathbf{s}_t \mid \mathbf{x}_t) \cdot \int_X p(\mathbf{x}_t \mid \mathbf{x}_{t-1}) p(\mathbf{x}_{t-1} \mid \mathbf{s}_{1:t-1}) d\mathbf{x}_{t-1} \qquad (14.8)$$

with transition probabilities $p(\mathbf{x}_t \mid \mathbf{x}_{t-1})$ computed according to a target motion model and initialization $p(\mathbf{x}_t \mid \mathbf{s}_{1:t-1}) = p(\mathbf{x}_1)$ for $t = 1$. Analytical solutions of Equation 14.8 are provided by Kalman filter (KF), grid-based filter (GF), or particle filter (PF) (Arulampalam et al., 2002). The location estimate is then obtained from the a-posteriori pdf using either the maximum-a-posteriori (MAP) criterion, $\hat{\mathbf{x}}_t = \arg\max_{\mathbf{x}_t \in X} p(\mathbf{x}_t \mid \mathbf{s}_{1:t})$, or the minimum mean-square error (MMSE) one, $\hat{\mathbf{x}}_t = \int_X \mathbf{x}_t p(\mathbf{x}_t \mid \mathbf{s}_{1:t}) d\mathbf{x}_t$.

An alternative approach for passive localization is the RTI, as originally proposed in Wilson and Patwari (2010), where DF-L is formulated as the estimation of a motion image of the area X capturing any variation with respect to the empty-space scenario observed during calibration. Assuming X divided into M voxels, $m = 1, \ldots, M$, the image to be estimated is $\mathbf{v}_t = [v_{1,t} \cdots v_{M,t}]^\mathsf{T}$ where $v_{m,t} = 0$ or $v_{m,t} = 1$ indicates whether a target is observed, or not, in the voxel m. For sparse motion, the received strength can be approximated as the sum of the contributions generated by all occupied voxels:

$$s_{\ell,t} = \sum_{m=1}^{M} \Delta\mu_{\ell,m} v_{m,t} + \mu_{0,\ell} + w_{\ell,t} \qquad (14.9)$$

where $\Delta\mu_{\ell,m}$ is the attenuation contribution due to target in voxel m and $w_{\ell,t} \sim N(0, \sigma_\ell^2)$ the fluctuation. For all links, we get $\mathbf{s}_t = \Delta\mu \cdot \mathbf{v}_t + \mu_0 + \mathbf{w}_t$, with $L \times M$ matrix $\Delta\mu = [\Delta\mu_{\ell,m}]$ collecting the average perturbations for all links and voxels, $\mu_0 = [\mu_{0,1} \cdots \mu_{0,L}]^\mathsf{T}$ the empty-space reference values, and $\mathbf{w}_t = [w_{1,t} \cdots w_{L,t}]^\mathsf{T} \sim N(\mathbf{0}, \mathbf{Q})$ the shadowing terms. The least squares (LS) approach has been proposed for imaging solution as $\hat{\mathbf{v}}_t = \mathbf{A}(\mathbf{s}_t - \mu_0)$ with $\mathbf{A} = \Delta\mu^\dagger$ or modified according to Tikhonov regularization (Wilson and Patwari, 2010). The target position is then estimated as the voxel with the maximum image value: $\hat{m}_t = \arg\max_m \hat{v}_{m,t}$. RTI based on RSS variances has also been investigated (Wilson and Patwari, 2011), modeling the variance as linearly increasing with the number of occupied voxels, as $\sigma_\ell^2 = \sigma_\ell^2(\mathbf{v}_t) = \sigma_{0,\ell}^2 + \sum_{m=1}^{M} \Delta\sigma_{\ell,m}^2 v_{m,t}$, and using the LS approach for image evaluation. Finally, the extension of RTI to jointly account for the target impact on *both* RSS average *and* fluctuation can be found in Nicoli et al. (2016), where the ML criterion is applied according to the stochastic model equation (14.6): $\hat{\mathbf{v}}_t = \arg\max_{\mathbf{v}_t} \mathcal{L}(\mathbf{s}_t \mid \mathbf{v}_t)$ with $\mathcal{L}(\mathbf{s}_t \mid \mathbf{v}_t) = -\ln|\mathbf{Q}(\mathbf{v}_t)| - \left\|\mathbf{s}_t - \Delta\mu \cdot \mathbf{v}_t - \mu_0\right\|^2_{\mathbf{Q}^{-1}(\mathbf{v}_t)}$ and $\mathbf{Q}(\mathbf{v}_t) = \mathrm{diag}\left[\sigma_1^2(\mathbf{v}_t), \cdots, \sigma_\ell^2(\mathbf{v}_t)\right]$.

Examples of DF-L results based on both mean and variance are in Figure 14.7, for indoor (left) and outdoor (right) scenarios, with a human target moving according to the highlighted trajectory. Three frames of the user trajectory are shown together with the location belief images evaluated as snapshot likelihood $\mathcal{L}(\mathbf{s}_t \mid \mathbf{x}_t)$ or GF a-posteriori pdf $p(\mathbf{x}_t \mid \mathbf{s}_{1:t})$. The indoor images clearly highlight the benefit provided by the Bayesian filtering in solving the ambiguities due to multipath.

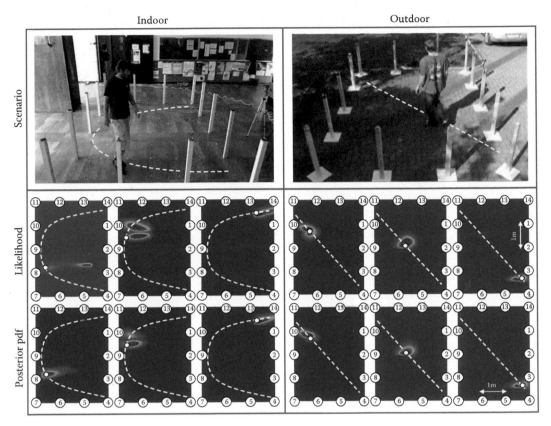

FIGURE 14.7
DF-L: Belief images of the target location, evaluated as instantaneous likelihood or a-posteriori pdf, for three frames of the user trajectory in a 4m × 3m area, in indoor (left) and outdoor (right) scenarios. The figure is based on Savazzi et al. (2014) but changing datasets and examples.

14.6 Activity and Gesture Recognition

Recognition of activity requires a higher sampling frequency as compared to localization systems. Typical sampling frequencies utilized are between 6 and 40 Hz where higher sampling potentially fosters good recognition accuracy (Sigg et al., 2014a, 2014b). Therefore, methods such as RTI are too slow to achieve reasonable accuracy. Also, fingerprinting is not useful for activity recognition as it covers static situations instead of actions in motion.

Device-free activity recognition (DF-AR) systems can be distinguished into active and passive (Sigg et al., 2014b); in addition, they can further be differentiated by the feature classes utilized for recognition. In particular, we distinguish between time-domain features, frequency domain (or phase) features as well as features exploiting multipath effects.

The achievable recognition accuracy for DF-AR is limited by the system class (active or passive), the signal utilized (baseband CSI vs. RSS), the receive nodes employed, the sampling rate (6+Hz), and features utilized (e.g., time- or frequency domain features).

The recognized classes range from basic whole-body activities such as walking or walking speed (Sigg et al., 2014a), over crowd size (Depatla et al., 2015) or crowd flow monitoring (Zhang et al., 2012), hand gestures (Adib and Katabi, 2013), and gait (Wang et al., 2016b) to breathing (Wang et al., 2016a) or recognition of lip movement (Sun et al., 2015), and (Wang et al., 2014a). Recently, researchers have also applied the same technology to detect changes in attention level (Shi et al., 2014) as well as emotion (Raja and Sigg, 2016) by means of sentiment-indicating gestures, movement patterns, or pose.

14.6.1 Using Off-the-Shelf Consumer Devices for Device-Free Activity Recognition

Apart from expensive real-time measurement equipment or SDR devices, recent consumer products also enable the exploitation of CSI for analysis and activity recognition. In contrast to traditional RSS measurements, which are a single quantity per packet representing signal-to-interference-plus-noise ratio (SINR), CSI contains amplitude and phase measurements separately (i.e., for each OFDM subcarrier). Therefore, different subcarriers experience different channel fading profiles.

For example, with IEEE 802.11n multiple-input multiple output (MIMO) 20 MHz channel, radios measure amplitude and phase for each of the 52 OFDM subcarriers. With 40 MHz channels, measurements are available over 128 subcarriers. The emerging IEEE 802.11ac standard supports even wider bandwidths. These measurements essentially allow estimating the channel frequency response.

Current freely available tools for CSI analysis include the Linux 802.11n CSI Tool (Halperin et al., 2011) as well as the newer Atheros CSI Tool, which can cover wider channel bandwidth and can allow complete access to subchannel phases (Xie et al., 2015).

14.6.2 Time-Domain Features

The first systems demonstrating DF-AR concepts have utilized simple time-domain features (Scholz et al., 2011). Typical features are root-mean-square (RMS), average magnitude squared (AMS), signal-to-noise ratio (SNR), signal amplitude, signal peaks, and the number of large delta in successive signal peaks. For instance, the E-eyes system (Wang et al., 2014b) is based on a DF-L system: it defines walking as the transition between two locations. The system combines WiFi 2.4 GHz OFDM links from Intel 5300 cards exploiting the diverse energy pattern over 30 CSI subcarriers. Although CSI provides significantly more information due to its multiple subcarriers, monitoring of RSSI information over a single link can also be utilized for activity recognition (Sigg et al., 2014b). The typical approach is to train a machine learning classifier on recorded samples (for instance, k-nearest neighbor, decision trees, support vector machines). In particular, for time-domain features, the recognizable classes of activities need to have a characteristic pattern over time so that generally whole-body movement or strong gestures can be recognized. For instance, walking and walking speeds have been identified from RF-enabled sensors and FM radio signals (Shi, 2013), which can also extend to the simultaneous recognition of activities from multiple subjects (Sigg et al., 2013). Utilizing off-the-shelf laptop and smartphone devices as wireless receivers, the environmental infrastructure can be exploited for ubiquitous recognition of presence, walking speed (Sigg et al., 2014a), and also gestures (Abdelnasser et al., 2015). In such settings, a precision and recall rate in the order of 90% for simple activities and 70% for gestures is possible. Further improvement is achieved by filtering noise and by focusing on the falling and rising edges of the composition of the signal features.

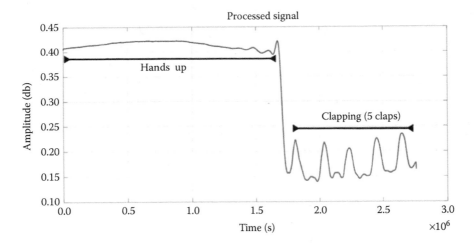

FIGURE 14.8
Gestures recognized from time-domain signal for intention/sentiment recognition. The figure depicts the processed and wavelet-filtered RSS. The gestures were performed by a subject in the LOS between a USRP transmit and receive pair, separated by 2 m in an indoor environment. Devices with omnidirectional antennas (2.4GHz centre frequency) are placed 1 m above ground.

In an indoor environment, such modalities can also help to recognize changing attention levels (e.g., from changes in walking speed, [Shi, 2013]) or emotion (e.g., joy or fear, see Figure 14.8) from sentiment-indicating gestures, motion patterns, or pose (Raja and Sigg, 2016).

14.6.3 Frequency Domain and Phase-Shift Features

By exploiting time domain only, the recognizable activities are restricted to those that cause a significant change on the signal strength. By exploiting frequency domain features such as Doppler variation and phase changes, fine-grained movements can be recognized. However, this approach requires precise orientation of the subject such that the movement is performed in the direction of the receive antenna. The accuracy of the recognition is then conditioned on the wavelength of the signal, where smaller wavelength enables the detection of smaller, more fine-grained movement. In addition, simultaneous detection of gestures from multiple individuals can in principle be obtained by utilizing multiantenna MIMO nodes. It has been demonstrated by leveraging micro-Doppler fluctuations in the WiSee system (Pu et al., 2013), which can distinguish nine predefined gestures from different people simultaneously with a detection probability of 94%. A main challenge is that Doppler shifts caused by human motion are several magnitudes smaller than the signal bandwidth. Focusing on OFDM modulation with multiple antenna transceivers, an analysis of such Doppler fluctuation is, however, possible after transforming the received symbols into narrowband pulses. Then, patterns from falling and rising signal edges of the Doppler fluctuation can be exploited for gesture and activity recognition.

For single-antenna nodes, other approaches can be adopted to emulate an antenna array at the receiver side. For example, in the WiVi system (Adib and Katabi, 2013), a single antenna receiver is used, whereas a preamble transmission stage is designed to isolate the time-varying reflections induced by the human body and null direct and wall-reflected disturbance. The system tracks the direction of the moving object with respect to the

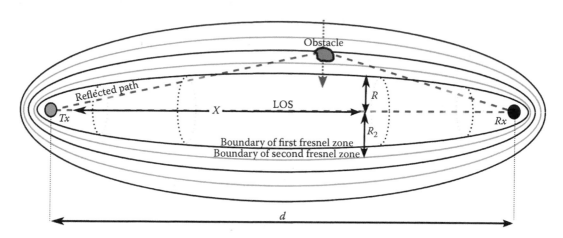

FIGURE 14.9

Schematic illustration of the use of the Fresnel zone concept for DF-AR first Fresnel radius R (Section 14.4) and high-order radii $R_n = \sqrt{n\lambda x(d-x)/d}$, for $n > 1$ superimposed.

receiver device using inverse synthetic aperture radar (ISAR): consecutive CSI measurements are collected over time to emulate an antenna array at the receiver.

Another example in this domain is contact-free RF respiration monitoring systems. Detection of breathing can be based on monitoring the RSS fluctuations from a preexisting IEEE 802.15.4 network surrounding the subject (Patwari et al., 2014). Using ML estimation, an error of 0.3 breaths per minute is shown as achievable.

14.6.4 Features Exploiting Fresnel Zones

A relatively new concept for DF-AR is the exploitation of Fresnel zones (Section 14.4) between a transmitter–receiver pair. In particular, assuming an in-phase transmission of a wave signal at both transceivers, the Fresnel ellipse defines the points in the 2D space where the two signals perfectly superimpose either constructively or destructively (Figure 14.9). Consequently, the Fresnel zones also define the reflection points in space from where the reflected signal would superimpose with the LOS path either constructively or destructively to result in largest or smallest signal amplitude with this effect decreasing gradually as the reflecting object is moving outward. Fresnel zones have been demonstrated useful in DF-AR scenarios for the detection of respiration rate (Wang et al., 2016b), localization, and tracking of individuals (Wu et al., 2016). In particular, the crossing of individual Fresnel zones by the human body or the lifting and falling chest are distinguishable from the received signal.

14.7 Case Study: Monitoring of Human Well-Being for Robotic-Assisted Manufacturing

This section highlights a case study where device-free sensing technologies are employed for joint wireless DF-L and device-free fall detection (DF-FD). This study focuses on worker protection in hybrid production systems characterized by human–robot cooperative tasks. Such tasks require advanced sensing capabilities and flexible solutions to track the movements of the operator in real time.

Systems and solutions developed for human tracking can be divided into two main families, namely active and passive, based on the cooperation from the monitored subject. The first approach is based on the use of wearable devices and sensors, such as accelerometers, wireless, and posture sensors (Hui et al., 2007, Kim et al., 2013). However, most of the research on human sensing has focused on active solutions (Yuwono et al., 2012, Cheng et al., 2013); looking to the industrial domain, the trend toward these solutions pose several problems (Patwari, 2010, Labrador and Lara Yejas, 2013). For example, personal protective equipment in the workplace involves wearable devices that are designed to protect only the single individual, whereas in the context of functional safety, not all plant operators are supposed to wear a personal radio tag. Wearable devices also strongly limit the operator mobility, visibility, and communication during working tasks.

The passive (device-free) approach is based on camera/video devices, laser-based devices, acoustic sources and/or other event sensors deployed in the monitoring area (Igual et al., 2013). It may exploit the properties of reflectivity with time-of-flight (ToF) cameras or radars, attenuation with tomographic sensors, IR sensors (Popescu et al., 2012), or emissivity with thermal images to extract human position and gestures. The main advantage of using passive approaches is that the operator does not need to wear any device. However, these approaches are typically characterized by high installation costs, limited detection range, and long calibration setup time as well as privacy constraints. Finally, they cannot be used in case of fire or in the presence of smoke.

The case study proposed in this section highlights the potential integration of a device-free wireless sensing technology into an industry-compliant architecture, by verifying its ability to jointly support both human localization (DF-L) and fall detection (DF-FD), in the surroundings of the body location, and in turn to provide augmented information about worker safety in robotic-assisted manufacturing (Figure 14.10).

Existing solutions to the human sensing, particularly fall detection problem (Rimminen et al., 2010, Ashfak et al., 2014), can gain only partial confidence by users about their

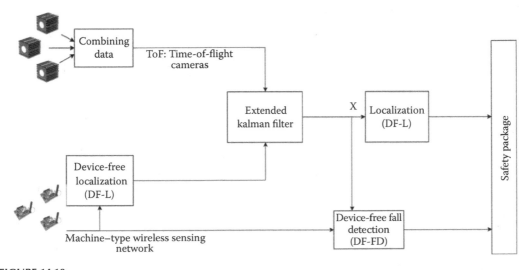

FIGURE 14.10
Integrated sensor fusion scheme for localization and fall detection modules, as multiple inputs to a safety package for worker monitoring.

reliability for the deployment in real industrial environments. Provided that no single technology can solve the problem of continuous worker monitoring (e.g., fall detection, localization, status of proximity to machines, onset of hazardous events), a solid evidence in detection performance, that is, by combining multiple sensor sources and technologies, is the major driver for system design choices.

Device-free operator monitoring makes use of an existing distributed wireless network infrastructure deployed for standard M2M communication tasks integrated with other sources (e.g., camera/video). System validation in this case study is based on IEEE 802.15.4 industry-standard wireless nodes, which are adopted for real-time process control and maintenance systems. The design and the experimental validation of a sensor fusion framework (Savazzi et al., [2016a]) is also tackled to integrate the device-free RF-based human sensing technology with image sensors (i.e., 3D ToF cameras). The proposed RF system thus *augments* the existing ToF-based vision platform at the cost of an additional deployment of wireless nodes.

14.7.1 Methodology

Analysis of human-induced RSSI fluctuations in a given position $\mathbf{x}$ is herein carried out to discriminate operator falling (described as state Θ or *fall* state) from a safe condition (indicated as $\varnothing$ or non*fall* state) corresponding to a human located in a known position $\mathbf{x}$ (or estimated by DF-L system, Section 14.5) inside the detection zone $\mathcal{X}$ and in safe conditions (i.e., sitting, standing, or walking inside the detection area). Fall detection is based on hard decisions over an optimized subset of links $\ell \in \mathbb{L}_\mathbf{x}$ selected during a calibration procedure, whereas noninformative links are purged. Calibration of both DF-L and DF-FD systems is implemented by an automatic software tool, collecting RSSI training measurements $\hat{\mathbf{s}}_\ell$ as described in Section 14.5. In particular, the calibration procedures should now acquire and process RSSI sequences $\hat{\mathbf{s}}_\ell(\Theta, \mathbf{x})$ and $\hat{\mathbf{s}}_\ell(\varnothing, \mathbf{x})$ corresponding to human body in safe and nonsafe state, respectively, for each position $\mathbf{x}$ of interest.

For fall state Θ, the observed sequences $\mathbf{s}_\ell$ are modeled by a hidden Markov model (HMM) with state space $\mathbf{Q}_\ell(\Theta) \in \mathbb{R}^{Q \times 1} = [q_j]_{j=1}^Q$ containing Q selected values from the CQI profiles $\hat{\mathbf{s}}_\ell(\Theta, \mathbf{x})$ learned during training for falling-state estimation. HMM parameters $\lambda_\ell(\Theta | \mathbf{x}) = [\mathbf{A}_\ell, \mathbf{B}_{\ell,\ell}]$, correspond to state transition $[\mathbf{A}_\ell]_{i,j} = p[q_t = q_i | q_{t-1} = q_j]$, observation probability $[\mathbf{B}_\ell]_{i,j} = p[s_{\ell,t} = s_i | q_t = q_j]$, and initial state $[\pi_\ell]_i = p[q_0 = q_i]$. The calibration of the HMM parameters can be carried out by the expectation-maximization, Baum–Welch algorithm (Rabiner, 1989) using the corresponding reference sequences $\hat{\mathbf{s}}_\ell(\Theta, \mathbf{x})$ and $\hat{\mathbf{s}}_\ell(\varnothing, \mathbf{x})$ for body falling in position $\mathbf{x}$.

The detection system iteratively computes the log-likelihood functions $\Lambda_\ell[\mathbf{s}_\ell | \Theta, \mathbf{x}] = \ln(\mathcal{L}[\mathbf{s}_\ell | \lambda_\ell(\Theta | \mathbf{x})])$ of the observation sequence and using the HMM parameters corresponding to the estimated body position $\mathbf{x}$. Each log-likelihood function $\Lambda_\ell[\mathbf{s}_\ell | \Theta, \mathbf{x}]$ is obtained iteratively (i.e., during the evaluation phase) as

$$\mathcal{L}[\mathbf{s}_\ell | \lambda_\ell(\Theta)] = \sum_{\forall \mathbf{q} \in \mathbf{Q}_\ell(\Theta)} p[\mathbf{s}_\ell, \mathbf{q} | \lambda_\ell(\Theta)] \tag{14.10}$$

with state sequence $\mathbf{q} = [q_1, \ldots, q_T]$ and joint probability $p[\mathbf{s}_\ell, \mathbf{q} | \lambda_\ell(\Theta)] = \prod_{t=1}^T p[s_{\ell,t} | q_t] \cdot p[q_t | q_{t-1}]$. The likelihood functions $\mathcal{L}[\mathbf{s}_\ell | \lambda_\ell(\Theta)]$ are continuously evaluated for new observed sequences over the selected links. Fall detection over input

RSSI sequence $\mathbf{s}_\ell$ can be based on hard decision with respect to precalibrated threshold τ_ℓ such that

$$\mathrm{LLR}_\ell(\mathbf{s}_\ell) = \Lambda_\ell[\mathbf{s}_\ell \mid \Theta, \mathbf{x}] - \Lambda_\ell[\mathbf{s}_\ell \mid \varnothing, \mathbf{x}] \geq \tau_\ell \tag{14.11}$$

where LLR is the log-likelihood ratio corresponding to the $\Lambda_\ell[\mathbf{s}_\ell \mid \Theta, \mathbf{x}] = \ln(\mathcal{L}[\mathbf{s}_\ell \mid \lambda_\ell(\Theta)])$, $\Lambda_\ell[\mathbf{s}_\ell \mid \varnothing, \mathbf{x}] = \ln(\mathcal{L}[\mathbf{s}_\ell \mid \lambda_\ell(\varnothing)])$. Likelihood $\mathcal{L}[\mathbf{s}_\ell \mid \lambda_\ell(\varnothing)]$ is obtained for HMM $\lambda_\ell(\varnothing)$ that considers arbitrary (but safe) body movements in the same position. Selection of the link subset $\mathbb{L}_\mathbf{x}$ for each monitored position $\mathbf{x}$ is crucial for maximize the accuracy: this can be obtained as shown in the following section.

14.7.2 Link Selection

Optimization of the link subset $\mathbb{L}_\mathbf{x}$ is based on the log-likelihood terms (Equation 14.10): for each link, LLR metrics are evaluated from reference measurement $\hat{\mathbf{s}}_\ell$, whereas fall ratio (FR) and nonfall ratio (NR) values are computed. The FR metric for Θ versus $\varnothing$ hypothesis evaluation compares fall $\Lambda_\ell[\hat{\mathbf{s}}_\ell \mid \Theta, \mathbf{x}] = \ln(\mathcal{L}[\hat{\mathbf{s}}_\ell \mid \lambda_\ell(\Theta)])$ versus nonfall $\Lambda_\ell[\hat{\mathbf{s}}_\ell \mid \varnothing, \mathbf{x}] = \ln(\mathcal{L}[\hat{\mathbf{s}}_\ell \mid \lambda_\ell(\varnothing)])$ likelihood functions:

$$\mathrm{FR}_\ell(\hat{\mathbf{s}}_\ell) = \Lambda_\ell[\hat{\mathbf{s}}_\ell \mid \Theta, \mathbf{x}] - \Lambda_\ell[\hat{\mathbf{s}}_\ell \mid \varnothing, \mathbf{x}] \tag{14.12}$$

with $\hat{\mathbf{s}}_\ell = \hat{\mathbf{s}}_\ell(\Theta, \mathbf{x})$. Similarly, the NR metric for Θ versus $\varnothing$ evaluation compares nonfall and fall probability using nonfall measurements as reference observations:

$$\mathrm{NR}_\ell(\hat{\mathbf{s}}_\ell) = \Lambda_\ell[\hat{\mathbf{s}}_\ell \mid \varnothing, \mathbf{x}] - \Lambda_\ell[\hat{\mathbf{s}}_\ell \mid \Theta, \mathbf{x}] \tag{14.13}$$

with $\hat{\mathbf{s}}_\ell = \hat{\mathbf{s}}_\ell(\varnothing, \mathbf{x})$. Optimal link subset is thus obtained as

$$\mathbb{L}_\mathbf{x} \triangleq \{\ell \in L : 1_{\mathrm{FR}_\ell > \tau_\ell} \times 1_{\mathrm{NR}_\ell > \tau_\ell} = 1\} \tag{14.14}$$

for each link ℓ, with $1_{x>y}(x)$ being the indicator function: $1_{x>y}(x) = 1$ if $x > y$ and $1_{x>y}(x) = 0$ otherwise. The optimal threshold value τ_ℓ for each link corresponds to the value that gives detection accuracy equal or better than a specified value (e.g., usually the target accuracy during calibration is set to 100%). In practice, a range of optimal threshold values that maximizes the detection accuracy is typically observed for each link. The optimal value τ_ℓ can then be selected as the mean point of each interval. Finally, τ_ℓ values are exploited for fall detection in real time.

14.7.2.1 Fall Detection inside the Shared Workspace

Detection of body fall at position $\mathbf{x}$ is based on a hard decision metric on each link defined in Equation 14.11. A fall event can be detected if the majority of the links votes for the fall event as shown in the decision variable:

$$\sum_{\ell \in L_\mathbf{x}} 1_{\mathrm{LLR}_\ell(\mathbf{s}_\ell) \geq \tau_\ell} > \sum_{\ell \in L_\mathbf{x}} 1_{\mathrm{LLR}_\ell(\mathbf{s}_\ell) < \tau_\ell} \tag{14.15}$$

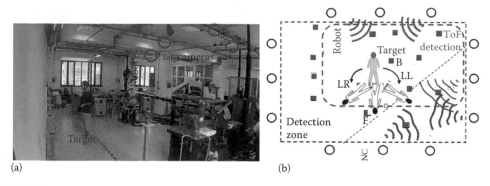

FIGURE 14.11

(a) Picture of shared human–robot test environment with ToF cameras (blue circles) and RF nodes and (b) deployment of the wireless network composed by RF nodes (circle markers) and the corresponding detection zones for wireless network and ToF cameras that cover the monitoring positions (green rectangular markers). LL, Lateral left; LR, lateral right; F, Forward; B, Backward.

The shared workspace, depicted in Figure 14.11a, is also monitored by five ToF cameras. In such scheme, the ToF channel acts as one of the sources to be fused, whereas the proposed RF system augments the existing sensing platform.

As far as localization of operator is concerned, most of errors occur in proximity of the robot body (see the position of the robot in Figure 14.11). Recall that the robot is not moving, but still it represents a massive disturbance contribution in direct and nearby RF links. DF-L observed accuracy is around 0.3 m, and it is expressed in root-mean-square error (RMSE). In these tests, the average computing latency is 360 ms.

Fall detection performance is evaluated in terms of sensitivity, specificity, and accuracy. Extensive experimental measurements with artificial and real human subjects, different body structure, age, and gender have been conducted (Figure 14.11b). Fall events that happen at different directions, that is, backward, forward, lateral left, and right, have also been considered. Finally, nonfall measurements are obtained for an operator/worker that is sitting or standing while losing his balance. Two scenarios are implemented with one artificial and two real human subjects. In the first scenario, experiments are conducted using a cylindrical anisotropic object whose surface is covered with adhesive aluminum foils that mimics a person located in a given position that falls toward different directions. In the second scenario, two real human subjects with different body structures have been involved as detailed in Table 14.1. In all the tests, the operator maintains the preimpact posture (sitting or standing in the workspace) for 4 s and then falls in different directions (fall gesture is repeated 10 times for each direction). RF measurements are also affected by WiFi interference originated by other preexisting networks deployed in the surrounding workspace area.

TABLE 14.1

Fall Scenario Profile

Subject	Height	Weight/Diameter	Age	Gender
1	1.60	60 kg/–	34	Male
2	1.81	95 kg/–	58	Male
3	1.80	1 kg/30 cm	–	Artificial object

TABLE 14.2

Fall Detection Performances Using 50 Selected
Links Over 4488 Tests in Different Directions

Sensitivity	Specificity	Accuracy	F-1	FPR
0.97	0.99	0.98	0.98	0.007

Distributed network configuration, used for experimental validation, is based on a deployment of 12 IEEE 802.15.4-compliant devices communicating over a full mesh topology. In particular, the network includes 11 field devices, one network coordinator (NC) device (see Figure 14.11b), and a processing unit (that might be physically separated from the NC). For all tests, omnidirectional, vertically polarized antennas have been employed with a gain of 2dBi. Decision about the fall/nonfall event is based on major voting (Equation 14.15) over the optimized link subset. Validation of detection results, after the link optimization phase, is based on a second dataset of fall and nonfall RSSI measurements having the same size of the training set used for calibration.

Table 14.2 summarizes the fall/nonfall detection capabilities in terms of the following metrics: sensitivity, specificity, accuracy, false positive rate, and F1 score (Abbate et al., 2012, Kianoush et al., 2016), using the optimized thresholds and link subset. These metrics refer to the true positive (TP), true negative (TN), false positive (FP), and false negative (FN) quantities computed for each trained detector during the verification phase. The results confirm the good performances of the proposed fall detection system for different falling directions.

The receiver operating characteristic (ROC) curve of the complete fall detection system, including the nonfall measurements and fall measurements in four directions, is depicted in Figure 14.12. The curve is obtained by comparing the sensitivity versus the false positive rate, after multilink selection (i.e., 50 links) and according to the majority voting scheme Equation 14.15. To plot this curve, a single threshold value τ_ℓ is now used for all link detectors: $\forall \ell = 1, \ldots, 50$, $\tau_\ell = \tau$. ROC curve is also based on the same selected subset of links, which is chosen during calibration. Sensitivity corresponds to the fall event measurements (toward all the predefined directions) and is computed by comparing the falling likelihood ratio in Equation 14.12 with respect to the corresponding threshold value τ_ℓ for a

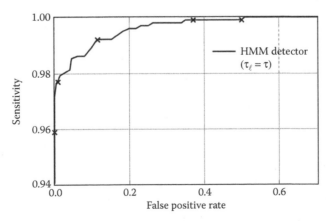

FIGURE 14.12
Receiver operating characteristic (ROC) curve for multilink fall detection over the optimized links.

given link ℓ. False positive rate is obtained from nonfall measurements by comparing the nonfalling likelihood ratio in Equation 14.13 with respect to the corresponding threshold value τ_ℓ for a given link.

14.8 Conclusions

This chapter is focused on the most recent developments on wireless sensing for device-free human body localization and motion recognition. Different methods have been discussed that exploit preexisting radio transceivers as sensing devices, and continuously monitor and process the radio CQI for the purpose of both human body localization (DF-L) and activity recognition (DF-AR). A case study has been presented that focused on the design and the implementation of a wireless sensing platform for fall detection and localization of an operator in the context of safe industrial workspaces.

References

Abbate, S., M. Avvenuti, F. Bonatesta, G. Cola, P. Corsini, and A. Vecchio, 2012. A smartphone-based fall detection system. *Pervasive and Mobile Computing*, 8, 883–899.

Abdelnasser, H., M. Youssef, and K.A. Harras, 2015. Wigest: A ubiquitous wifi-based gesture recognition system. *IEEE Conference on Computer Communication*, Hong Kong, China.

Abramowitz, M. and I. Stegun, 1972. *Handbook of Mathematical Functions*. National Bureau of Standards. Applied Mathematics Series, Washington, D.C., 55.

Adib, F. and D. Katabi, 2013. See through walls with Wi-Fi! *Proceedings of ACM SIGCOMM*.

Aly, H. and M. Youssef, 2013. New insights into Wi-Fi based device-free localization. *Proceedings of UbiComp'13*, September 8–12, Zurich, Switzerland, pp. 541–548.

Arulampalam, M.S., S. Maskell, N. Gordon, and T. Clapp, 2002. A tutorial on particle filters for online nonlinear/non-Gaussian Bayesian tracking. *IEEE Transactions on Signal Processing*, 50 (2), 174–188.

Ashfak, H.M., M.S. Mohktar, S.B. Kamaruzzaman, K.S. Lim, T.M. Pin, and F. Ibrahim, 2014. Smartphone-based solutions for fall detection and prevention: Challenges and open issues. *Sensors Journal*, 14 (4), 7181–7208.

Buehrer, R.M., C.R. Anderson, R.K. Martin, N. Patwari, and M.G. Rabbat, 2014. Introduction to the special issue on non-cooperative localization networks. *IEEE Journal of Selected Topics in Signal Processing*, 8 (1), 2–4.

Bultitude, R., 1987. Measurement, characterization and modeling of indoor 800/900 MHz radio channels for digital communications. *IEEE Communication Magazine*, 25 (6), 5–12.

Cheng, J., C. Xiang, and S. Minfen, 2013. A framework for daily activity monitoring and fall detection based on surface electromyography and accelerometer signals. *IEEE Journal of Biomedical and Health Informatics*, 17 (1), 38–45.

Coulson, A.J., A.G. Williamson, and R.G. Vaughan, 1998. A statistical basis for lognormal shadowing effects in multipath fading channels. *IEEE Transactions on Communications*, 46 (4), 494–502.

Depatla. S., A. Muralidharan, and Y. Mostofi, 2015. Occupancy estimation using only wifi power measurements. *IEEE Selected Areas in Communications*, 33 (7), 1381–1393.

Ghaddar, M., L. Talbi, T.A. Denidni, and A. Sebak, 2007. A conducting cylinder for modeling human body presence in indoor propagation channel. *IEEE Transactions on Antennas and Propagation*, 55 (11), 3099–3103.

Halperin D., W. Hu, A. Sheth, and D. Wetherall, 2011. Tool release: Gathering 802.11n traces with channel state information. *ACM SIGCOMM CCR*.

Hui, L., H. Darabi, P. Banerjee, and J. Liu, 2007. Survey of wireless indoor positioning techniques and systems. *IEEE Transactions on Systems, Man, and Cybernetics, Part C (Applications and Reviews)*, 37 (6), 1067–1080.

Igual, R., C. Medrano, and I. Plaza, 2013. Challenges, issues and trends in fall detection systems. *Journal of Biomedical Engineering Online*, 12 (66), 1–66.

Kianoush, S., S. Savazzi, F. Vicentini, V. Rampa, and M. Giussani, 2016. Device-free RF human body fall detection and localization in industrial workplaces. *IEEE Internet of Things Journal*, in pre-print, 2 (4), 351-362.

Kim, S., C. Mariotti, F. Alimenti, P. Mezzanotte, A. Georgiadis, A. Collado, L. Roselli, and M.M. Tentzeris, 2013. No battery required: Perpetual RFID-enabled wireless sensors for cognitive intelligence applications. *IEEE Microwave Magazine*, 14 (5), 66–77.

Kosba, A., A. Saeed, and M. Youssef, 2012. Robust wlan device-free passive motion detection. *Proceedings of Wireless Communications and Networking Conference (WCNC)*, pp. 3284–3289.

Kouitatis, G., 2010. Multiple human effects in body area networks. *IEEE Antennas and Wireless Propagation Letters*, 9, 938–941.

Labrador, M.A. and O.D. Lara Yejas, 2013. *Human Activity Recognition: Using Wearable Sensors and Smartphones*. Chapman and Hall/CRC, Boca Raton, FL.

Laya, A., W. Kun, A.A. Widaa, J. Alonso-Zarate, J. Markendahl, and L. Alonso, 2014. Device-to-device communications and small cells: Enabling spectrum reuse for dense networks. *IEEE Wireless Communications*, 21 (4), 98–105.

Lloret, J., A. Canovas, S. Sendra, and L. Parra, 2015. A smart communication architecture for ambient assisted living. *IEEE Communications Magazine*, 53, 26–33.

Mager, B., N. Patwari, and M. Bocca, 2013. Fall detection using RF sensor networks. *Proceedings of IEEE 24th International Symposium on Personal Indoor and Mobile Radio Communications (PIMRC)*, pp. 3472–3476.

McNamara, D.A., C.W. I. Pistorius, and J.A.G. Malherbe, 1990. *Introduction to the Geometrical Theory of Diffraction*. Artech House, Boston, MA.

Mostofi, Y., 2013. Cooperative wireless-based obstacle/object mapping and see-through capabilities in robotic networks. *IEEE Transactions on Mobile Computing*, 12 (5), 817–829.

Nicoli, M., V. Rampa, S. Savazzi, and S. Schiaroli, 2016. Device-free localization of multiple targets. *EURASIP European Signal Processing Conference (EUSIPCO'16)*, Budapest, Hungary.

Patwari, N., L. Brewer, Q. Tate, O. Kaltiokallio, and M. Bocca, 2014. Breathfinding: A wireless network that monitors and locates breathing in a home. *IEEE Journal of Selected Topics in Signal Processing*, 8 (1), 30–42.

Patwari, N. and J. Wilson, 2010. RF sensor networks for device-free localization: Measurements, models and algorithms. *Proceedings of the IEEE*, 98 (11), 1961–1973.

Popescu, M., B. Hotrabhavananda, M. Moore, and M. Skubic, 2012. VAMPIR- an automatic fall detection system using a vertical PIR sensor array. *6th International Conference on Pervasive Computing Technologies for Healthcare*, San Diego, CA, pp. 163–166.

Pu, Q., S. Gupta, S. Gollakota, and S. Patel, 2013. Whole-home gesture recognition using wireless signals. *Proceedings of ACM MobiCom*, ACM, Miami, FL.

Rabiner, L., 1989, A tutorial on Hidden Markov Models and selected applications in speech recognition. *Proceedings of the IEEE*, 77 (2), 257–286.

Raja, M. and S. Sigg, 2016. Applicability of RF-based methods for emotion recognition: A survey. *2016 IEEE International Conference on Pervasive Computing and Communication Workshops (PerCom Workshops)*, pp. 1–6, March.

Rampa, V., S. Savazzi, M. Nicoli, and M. D'Amico, 2015. Physical modeling and performance bounds for device-free localization systems. *IEEE Signal Processing Letters*, 22 (11), 1864–1868.

Rimminen, H., J. Lindstrom, M. Linnavuo, and R. Sepponen, 2010. Detection of falls among the elderly by a floor sensor using the electric near field. *IEEE Transactions on Information Technology in Biomedicine*, 14 (6), 1475–1476.

Saeed, A., A.E. Kosba, and M. Youssef, 2014. Ichnaea: A low-overhead robust WLAN device-free passive localization system.*IEEE Journal of Selected Topics in Signal Processing*, 8 (1), 5–15.

Savazzi, S., M. Nicoli, F. Carminati, and M. Riva, 2014. A Bayesian approach to device-free localization: Modelling and experimental assessment. *IEEE Journal of Selected Topics in Signal Processing*, 8 (1), 16–29.

Savazzi, S., V. Rampa, F. Vicentini, and M. Giussani, 2016a. Device-free human sensing and localization in collaborative human-robot workspaces: A case study. *IEEE Sensors Journal*, 16 (5), 1253–1264.

Savazzi, S., S. Sigg, M. Nicoli, V. Rampa, S. Kianoush, and U. Spagnolini, 2016b. Device-free radio vision for assisted living. *IEEE Signal Processing Magazine*, 33 (2), 45–58.

Scholz, M., L. Kohout, M. Horne, M. Budde, M. Beigl, and M.A. Youssef, 2015. Device-free radio-based low overhead identification of subject classes. *Proceedings of the 2nd Workshop on Physical Analytics WPA'15*, ACM, pp. 1–6.

Scholz, M., S. Sigg, D. Shihskova, G. von Zengen, G. Bagshik, T. Guenther, M. Beigl, and Y. Ji, 2011. Sensewaves: Radiowaves for context recognition, *Video Proceedings of the 9th International Conference on Pervasive Computing (Pervasive 2011)*, June, San Francisco, CA.

Seifeldin, M., A. Saeed, A. Kosba, A. El-Keyi, and M. Youssef, 2013. Nuzzer: A large-scale device-free passive localization system for wireless environments. *IEEE Transactions on Mobile Computing*, 12 (7), 1321–1334.

Shi, S., S. Sigg, and Y. Ji, 2014. Monitoring of attention from ambient FM-radio signals. *IEEE Pervasive Computing, Special Issue Managing Attention in Pervasive Environments*, Budapest, Hungary.

Sigg, S., S.U. Blanke, and G. Troster, 2014a. The telepathic phone: Frictionless activity recognition from WiFi-RSSI. *Proceedings of IEEE International Conference on Pervasive Computing*, pp. 148–155.

Sigg, S., M. Scholz, S. Shi, Y. Ji, and M. Beigl. 2014b. RF-sensing of activities from non-cooperative subjects in device-free recognition systems using ambient and local signals. *IEEE Transactions on Mobile Computing*, 13 (4), 907–920.

Sigg, S., S. Shi, and Y. Ji, 2013. Rf-based device-free recognition of simultaneously conducted activities, *Adjunct Proceedings of the 2013 ACM International Joint Conference on Pervasive and Ubiquitous Computing (UbiComp 2013), ser. UbiComp'13*.

Slavakis, K., G.B. Giannakis, and G. Mateos, 2014. Modeling and optimization for big data analytics. *IEEE Signal Processing Magazine*, 31 (5), 18–31.

Smith, D.B., D. Miniutti, T.A. Lamahewa, and L.W. Hanlen, 2013. Propagation models for body-area networks: A survey and new outlook. *IEEE Antennas and Propagation Magazine*, 55 (5), 97–117.

Sun, L., S. Sen, D. Koutsonikolas, and K.H. Kim, 2015. Widraw: Enabling hands-free drawing in the air on commodity WiFi devices. *Proceedings of the 21st Annual International Conference on Mobile Computing and Networking (MobiCom)*, New York, pp. 77–89.

Wang, G., Y. Zou, Z. Zhou, K. Wu, and L. Ni. 2014a. We can hear you with Wi-Fi! *Proceedings of the 20th Annual International Conference on Mobile Computing and Networking (MobiCom '14)*. ACM, New York, NY, pp. 593–604.

Wang, H., D. Zhang, J. Ma, Y. Wang, Y. Wang, D. Wu, and T. Gu, 2016a. Human respiration detection with commodity WiFi devices: Do user location and body orientation matter? *Proceedings of UbiComp'16*, September 12–16, Heidelberg, Germany.

Wang, W., A. Liu, and M. Shahzad, 2016b. Gait recognition using WiFi signals. *Proceedings of UbiComp'16*, September 12–16, Heidelberg, Germany.

Wang, Y., J. Liu, Y. Chen, M. Gruteser, J. Yang, and H. Liu, 2014b. E-eyes: Device-free location-oriented activity identification using fine-grained WiFi signatures. *Proceedings of the 20th Annual International Conference on Mobile Computing and Networking*, ACM, pp. 617–628.

Wilson, J. and N. Patwari, 2010. Radio tomographic imaging with wireless networks. *IEEE Transactions on Mobile Computing*, 9 (5), 621–632.

Wilson J. and N. Patwari, 2011. See-through walls: Motion tracking using variance-based radio tomography networks. *IEEE Transactions on Mobile Computing*, 10 (5), 612–621.

Woyach, K., D. Puccinelli, and M. Haenggi, 2006. Sensorless sensing in wireless networks: Implementation and measurements. *Proceedings of 4th International Symposium on Modeling and Optimization in Mobile, Ad Hoc and Wireless Networks*, IEEE, pp. 1–8.

Wu, D., D. Zhang, C. Xu, Y. Wang, and H. Wang, 2016. WiDir: Walking direction estimation using wireless signals. *Proceedings of UbiComp'16*, September 12–16, Heidelberg, Germany.

Xie, Y., Z. Li, and M. Li, 2015. Precise power delay profiling with commodity WiFi. *ACM MobiCom*.

Youngwook, K. and L. Hao, 2009. Human activity classification based on micro-Doppler signatures using a support vector machine. *IEEE Transactions on Geoscience and Remote Sensing*, 47 (5), 1328–1337.

Youssef, M., M. Mah, and A. Agrawala, 2007. Challenges: Device-free passive localization for wireless environments. *Proceedings of the 13th Annual ACM International Conference on Mobile Computing and Networking, ACM MobiCom*.

Yuwono, M., B. Moulton, S. Su, B. Celler, and H. Nguyen, 2012. Unsupervised machine-learning method for improving the performance of ambulatory fall-detection systems. *Journal of Biomedical Engineering Online*, 11, 9.

Zhang, D., Y. Liu, X. Guo, M. Gao, and L.M. Ni, 2012. On distinguishing the multiple radio paths in RSS-based ranging, *Proceedings of the 31st IEEE International Conference on Computer Communications*.

15

IR Sensors for Indoor Monitoring

Fatih Erden and A. Enis Cetin

CONTENTS

15.1 Introduction

There are many other sensing modalities, besides radars, utilized for indoor monitoring. The sensors deployed by those modalities can be categorized into two groups due to their mobility: (1) ambient sensors (at fixed locations), for example, infrared (IR) sensors, vibration sensors, pressure sensors, cameras, and microphones, and (2) mobile and wearable sensors, for example, accelerometers, thermal sensors, electrocardiogram, electromyography, and pulse oximeters. In addition to the ability to solve a particular problem, cost, accuracy, user acceptance, and privacy issues decide which sensor is selected (Chernbumroong et al. 2013). This is to say that the modalities to be developed should be affordable by an average user; as the aim is to somehow enhance the wellness and life quality of people, a reasonable error rate should be achieved; it should be easy to interact with those modalities; the modalities should gather minimum private information about the daily livings of the individuals.

Each sensor type generates different data formats, for example, numeric, audio, and image. The data are processed using appropriate techniques for monitoring purposes such as assisted living (AL) applications, home automation, health, safety, and so on. The data from two or more sensors can be fused and analyzed to extract rich context information. Some of the popular sensors and what they measure are listed in Table 15.1. If user convenience and cost are the priority, ambient sensors are deemed more advantageous. On the other hand, if accuracy is the priority, then the choice should be wearables.

This chapter will focus on IR sensors that complement and have similar architectures to radars (both have transmitters/receivers). However, the discussion here can be extended to cover the sensors that generate data on a time-series basis such as vibration and acoustic sensors. That is, similar techniques and algorithms can also be deployed for processing

TABLE 15.1

Some of the Popular Sensors Used for Indoor Monitoring and Their Measurements

	Sensor	Measurement
Ambient sensors	IR	Motion
	Vibration	Motion
	Microphone	Activity
	Pressure	Pressure on bed, chair, etc.
	Camera	Activity
	Magnetic switches	Door opening/closing
Wearable sensors	Accelerometer	Acceleration
	Gyroscope	Orientation
	ECG	Heart activity
	EEG	Brain activity
	Thermal	Body temperature
	Pulse oximeter	Blood oxygen

the outputs of these sensors. The feature extraction methods and machine learning algorithms tailored for each specific application will be presented. Radars will be compared to IR sensors for similar mode of operation highlighting the pros and cons of each.

15.2 IR Sensors: Types and Operating Principles

IR sensors are electronic devices designed to sense specific characteristics of subjects that are within their frame of view. This is achieved by either active illumination of a target with IR light, then receiving the response or by passive detection of IR emissions. IR radiation, which lies between the visible and microwave regions of the electromagnetic spectrum, is not visible to the human eye but can be detected and comparatively measured by an IR sensor. IR sensors typically use IR light emitting diodes (LEDs) as the emitter. IR waves emitted by the LEDs are reflected by the target then detected using IR photodiodes sensitive to the same wavelength of light emitted by the LEDs. When the photodiodes are exposed to IR radiation, an output voltage proportional to the IR flux is generated by the sensor circuitry. Based on this measure, it is possible to achieve various tasks such as temperature measurement, flame monitoring, moisture analysis, and gas analysis.

IR sensors described earlier may be referred to as *active* IR sensors because of actively emitting IR radiation. There is another type of IR sensor, which has a more widespread use in indoor monitoring applications and hence will be the focus of attention in this chapter, which is called as *passive* IR (PIR) sensor. A PIR sensor only has a collector to sense the motion of the subjects in its field of view (FOV). Indeed, rather than sensing the motion, the PIR sensor detects the change in the amount of IR radiation impinging upon it. A question arises herein: what is it that generates the IR radiation to be detected by the collector of the PIR sensor? All objects with a temperature above absolute zero (0 K) emit thermal energy in the form of IR radiation, and PIR sensors view those objects as the source of IR radiation. A PIR sensor is sensitive in a range of 0–14 μm. However, the sensitive range can be narrowed by using optics between the sensor and the target. For example, PIR sensors used with a Fresnel lens are sensitive to 8–14 μm where the aim is to detect human motion, because the human body radiates at roughly 10 μm.

There are two prevalent types of PIR sensors: single-element and dual-element. Higher pixel count IR sensors are also available but much more expensive in comparison. However, they will be more affordable in the near future as it was the case with single- and dual-element PIR sensors. A single-element PIR sensor responds to any changes in IR radiation in its field of view. Temperature changes in the environment can cause a bias in the output of this type; thus, they should be compensated to reduce the sensitivity to ambient temperature. Other PIR sensors have two pyroelectric elements, and the output of those elements can be denoted by (s_1) and (s_2) as in Figure 15.1 where (g) is ground. When a hot body moves into the PIR sensor's FOV, the two pyroelectric elements are exposed to different amounts of IR radiation. The PIR sensor gain circuit then produces an output voltage (typically between 0 and 5 V) at pin (d) corresponding to this difference. If the elements are subject to the same level of IR radiation, the effect in each element cancels the other out, and the sensor circuit produces a zero-output signal. For the rest of this chapter, *PIR sensors* will be used to refer to dual-element type PIR sensors unless otherwise stated.

PIR sensors provide low-cost, simple, and effective solutions for indoor monitoring. Commercially available PIR sensors generate binary outputs, that is, a digital "1" to indicate the presence of a hot body and a digital "0," otherwise, and are mostly used in automatically activated lighting systems and burglar alarms. However, it is possible to capture a continuous-time analog signal from the PIR sensor through slight modification of the sensor circuitry. Leveraging this technique, the PIR sensor can provide further definition of the monitored area and individual(s), for example, classification of the hot bodies and recognition of various human activities or even the rejection of unwanted objects. The main steps of the intelligent PIR sensor signal processing are summarized in Figure 15.2. Continuous-time analog signal $x(t)$ received from the modified PIR sensor circuitry is first digitized using an A/D converter. Then feature vectors v_n are extracted from the digital signal $x[n]$. A feature vector may represent a single sample as well as a frame of data. The choice depends on the constraints imposed by computational efficiency requirements. Finally, the feature vectors are classified using an appropriate classifier to detect the events of interest such as falls, fire, localization and tracking of a subject, distinguishing human

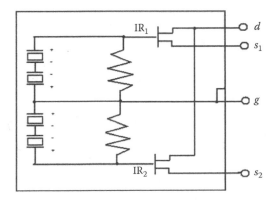

FIGURE 15.1
Model of a PIR sensor circuit. (From Erden, F. et al., *Infrared Phys. Technol.*, 73, 88–94, 2015. With Permission.)

FIGURE 15.2
Main steps for intelligent processing of PIR sensor signal.

motion from animal motion, and rejection of other noise sources. The classifier is usually trained using past and/or simulated data.

The output of a PIR sensor circuit is just a one-dimensional (1D) signal rather than an audio record or an image; thus, PIR sensor-based systems protect privacy. This simplicity is an advantage of PIR sensors when the application requires a stand-alone system with minimal energy and computational resources. As the computational cost of processing 1D data (of course with the proper methods) is low, PIR sensor-based system can run on a field-programmable gate array (FPGA) or on a low-cost digital signal processor. In addition, PIR sensors can function in complete darkness unlike ordinary vision-based systems because they image IR light. These advantages in turn feature the PIR sensors as one of the key players in indoor monitoring.

PIR sensor-based indoor monitoring applications can be categorized into the following fields: home security and safety, home automation, and health status monitoring. Besides being used alone, PIR sensors can also be integrated with other sensors to improve overall accuracy. The discussion in Sections 15.3 through 15.5 will also cover those instances.

15.3 Home Security/Safety

PIR sensors are primarily used for intrusion detection for home and commercial security. A number of PIR sensors are placed at fixed points, usually near ceilings to monitor a wider portion of the area. The system is activated manually by the residents before they leave, and an alarm is fired when at least one of the PIR sensors produces a peak at a time instant, that is, when they detect a motion. The output of the system may be an audible alarm, or the system may directly send a notification to the client via mobile, SMS text, e-mail, or the client can monitor his or her residence in real time by using a personal computer as suggested by Moghavvemi and Seng (2004).

Although the PIR sensor is effective in detecting motion of warm bodies, and has some capability to filter small objects, there is still room to reduce false alarms and improve detection quality. This being said, the discussion on intruder detection can be taken one step further by *identifying* a human using a PIR sensor array. Body parts such as arms and legs create a characteristic signature while a human is walking in front of PIR sensors. Each person, even each body part, acts as an IR source and by investigating the motion and patterns of these elements with respect to each other, it is possible to classify and in some instances to identify the subject. The identification process basically consists of two phases: training and testing. In the training phase, a model is generated to represent each registered person based on the features extracted from previous ground truth recordings. Then, a frame of test data is compared to each model in terms of a similarity metric, and the frame is assigned to the model with the highest score.

Fang et al. (2006) illustrate this concept using a sensor module, which includes four vertically aligned PIR sensors with modulated visibilities and is located at a predetermined height from the ground to best capture the different parts of a walker. Binary event index sequences of fixed length are used as features. These sequences consist of binary "1"s and "0"s and are formed as follows. First, matched filtering is applied to the raw output signal to maximize the signal-to-noise ratio. Second, a threshold test follows and if the absolute value of a sample is greater than the threshold, then the value of that sample is assigned to "1," otherwise to "0." Finally, a low-pass filter is applied. The event index sequences are generated on an embedded microcontroller and transmitted to the host computer via a wireless channel. The rest of the process is carried out in the computer. In the training phase, a hidden Markov model (HMM) is built for each walker. The maximum likelihood (ML) estimate of the parameters λ of an HMM, given a set of observed feature sequences O, is determined using the expectation-maximization (EM) algorithm. This is also known as the Baum–Welch algorithm and ensures maximizing the probability of a feature observation given the model, $P(O \mid \lambda)$. In the test phase, the probability of an unknown sequence X, given the model λ_i, $P(X \mid \lambda_i)$ is computed, where $1 \le i \le N$ and N is the number of registered walkers, and X is affiliated to the model producing the highest likelihood. The proposed system, which contains 8 detector units, is tested in a 9 m × 9 m room for 10 people walking randomly either on a definite path or a random path. The identification rates are reported to be 91% and 78.5% for definite-path and random-path cases, respectively.

It is important to note that the identification rates may differ depending on the number of states in HMM training, the length of the training sequences, and also the number of the registered people. The number of states should be set to an optimum value to avoid overfitting and high computational costs. Changing the length of the training sequences brings a trade-off between the identification rate and the training time, that is, increasing the length of the training sequence, while extending the training time, improves the identification rate and vice versa.

Yun and Lee (2014) introduce another setup and a set of techniques for human motion detection and identification using PIR sensors. Three modules, each has two pairs of orthogonally aligned PIR sensors with modified Fresnel lenses, are placed in a hallway: one on the ceiling and the other two on opposite walls facing each other. The setup is shown in Figure 15.3. Two types of feature sets are used: a raw dataset and a reduced feature set. The latter is composed of amplitude and time to peaks, and passage duration obtained from each PIR sensor. The goal is to decide the moving direction (back or forth), the distance interval (close to one or the other sensor on the wall or in the middle of them), and the speed level (slow, moderate, or fast) of the subjects. Various machine learning methods such as decision tree, multilayer perceptron, and support vector machine (SVM) are tested for the classification of the data of interest, and more than 92% and 94% of recognition accuracies are achieved for the raw and the reduced feature sets, respectively, for eight subjects. SVMs are reported to get the edge on the other machine learning methods.

It is also possible to make use of PIR sensors in the context of home safety. Erden et al. (2012) introduce a flame detector, which is capable of distinguishing flickering flame from human motions, with a single PIR sensor. The idea here is that a flickering flame has a significantly different IR emission pattern than human motion because of its uncontrolled nature. To make a distinction between these two event classes, first the output of the PIR sensor is captured in the form of a continuous-time signal. This is achieved by directly reading the analog signal rather than the binary output of the comparator unit. The analog signal should be sampled with a sampling frequency above the Nyquist rate, in this case, the double of the highest flame flicker frequency, which is 13 Hz (Albers and Agrawal 1999). The output signal is then digitized using an embedded microcontroller and transferred to

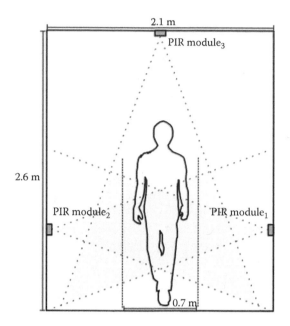

FIGURE 15.3
The setup for human motion detection and identification with three PIR sensor modules. (From Yun, J., and S.-S. Lee, *Sensors*, 14, 8057–8081, 2014.)

the host computer using a serial communication via RS232 port. The digitized PIR sensor signals due to walking motion of a human and flickering flame are shown in Figure 15.4a and b, respectively. As it is clear from the figure, the oscillation patterns of the two events are significantly different.

Wavelet coefficients $w[k]$, obtained after a single-stage subband decomposition, are used as features. Wavelet transform, while reducing the sensitivity to sudden ambient changes, halves the computational load due to downsampling by a factor of 2. In the training phase, the data collected for each event class are concatenated separately, and the wavelet

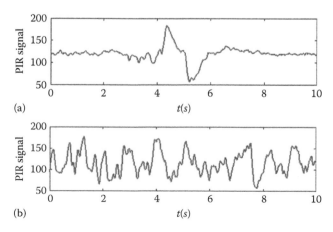

FIGURE 15.4
PIR sensor signal at 8-bit quantization for (a) a human walking and (b) a flame of an uncontrolled fire at a 5 m distance with a sampling frequency of 100 Hz.

sequences are computed. Based on the training wavelet sequences and the two thresholds T_1 and T_2 that are determined experimentally, a three-state (S_0, S_1, S_2) Markov model is generated for each class. The state of a wavelet coefficient is determined as follows:

$$S\{w[k]\} = \begin{cases} S_0, & w[k] < T_1 \\ S_1, & T_1 < w[k] < T_2 \\ S_2, & w[k] > T_2 \end{cases} \tag{15.1}$$

Then, the probability of each transition to occur is computed for each event class by using

$$(p_{a,b})_{ij} = \frac{(a,b)_{ij}}{L} \tag{15.2}$$

where:
  a_{ij} and b_{ij} are the number of transitions from S_i to S_j in the walking motion and flickering flame training state sequences
  L is the length of the training state sequence of the corresponding class

During testing, wavelet coefficients of the current data frame are inputted to each model and the probability that the frame belongs to each class is calculated by

$$P_{A,B}(F) = \prod_{i=0}^{2} \prod_{j=0}^{2} (p_{a,b})_{ij}^{f_{ij}} \tag{15.3}$$

where:
  F is the wavelet sequence of the test frame
  f_{ij} is the number of transitions from S_i to S_j in F

The frame is said to belong to the walking motion class if $P_A > P_B$, and flickering flame class if otherwise.

Note that it is critical to set the thresholds in such a way that the state transition probability vector of one class differs from that of another as much as possible. The thresholds T_1 and T_2, alternatively, can be determined by using a genetic algorithm rather than being experimentally assessed (Erden et al. 2015a). The algorithm finds the optimum values of T_1 and T_2 by maximizing the dissimilarity function:

$$D(T_1, T_2) = \left\| \overline{p_a} - \overline{p_b} \right\|^2 \tag{15.4}$$

where $\overline{p_a}$ and $\overline{p_b}$, each of which is a function of T_1 and T_2, are the state transition probability vectors of the two Markov models, and $\|x - y\|^2$ is the L^2 distance between the vectors x and y.

The flame detection system, unlike conventional smoke detectors, is capable of detecting flame in open spaces or large rooms in a few seconds after the flame is visible. Response time of the conventional detectors depends on the fire area, the distance to the fire, and where they are used, that is, indoor or outdoor and are much higher in comparison to that of the PIR sensor-based system. The detection range of the PIR sensor is 5 m, which enables covering a 10 m × 10 m room using a single sensor.

Yet another, and last, example of utilization of PIR sensors for home safety is the volatile organic compounds (VOC) gas leak detection (Erden et al. 2010). PIR sensor's spectral range intersects with the absorption bands of VOC gases. By using this fact, a similar procedure is followed to the one in flame detection. The main difference is seen in the feature extraction phase. This time, a four-stage subband decomposition is carried out to find the wavelet coefficients, because VOC gas leaks cannot be visualized before that stage. A bottled gas containing a mixture of butane and propane gases is used for validation. A true positive rate of above 90% is achieved within a distance of 1 m.

15.4 Home Automation

Another area where PIR sensors find applications in is home automation. The most known applications in this context are the smart lighting controls, which employ PIR sensors as presence detectors. However, there are a number of studies on the control of electrical home appliances using PIR sensors.

Bai and Ku (2008) propose a home light control module (HLCM) to adjust the room light intensity. The HLCM is made up of four blocks: a PIR sensor circuit, a light sensor circuit, a microprocessor, and an RF module. The PIR sensor is used in binary on/off mode to detect whether a human enters the surveillance area or not. If a human is present in the area, then the PIR sensor circuit triggers an alarm, so that the light sensor circuit measures the light intensity, and the HLCM maintains sufficient light. The RF module is responsible for providing the communication between different HLCMs. If no presence is detected, the controlled lights are turned off. The resulting system provides a substantial reduction in power consumption. Tsai et al. (2011) introduce a module including a circuit integrated with a PIR sensor. Although the PIR sensor controls the lights, the circuit provides a way to reducing the standby power consumption of the PIR sensor. Thus, the design is called an *ultralow standby power PIR-sensor-based lighting device.*

A similar approach can be followed for smart control of electrical devices in a home environment as in smart lighting control systems. Mrazovac et al. (2011) present a design and implementation of a software-based platform to automatically control audio/video devices. The controller uses audio and video technology in addition to PIR technology and interprets information about a user's position as a command issued to a list of devices such as PC, TV, or audio system. The command may enable activating, aborting, or changing video presentation; pausing and continuing playback; amplifying sound or silencing it.

Contact-free interaction with electrical appliances is an increasingly desired feature. This is primarily driven by enhanced user experience (entertainment) purposes, providing comfort in daily life, and to improve hygiene in public areas. Most of these enhancements are achieved using PIR sensors in mass-market devices. Although solutions offered by wearable sensors and cameras provide a set of advanced control options, they are more expensive and require a heavy computational load. Wojtczuk et al. (2011) present a prototype consisting of a 4×4 PIR sensor array capable of recognizing hand movements such as horizontal, vertical, and diagonal swipes, and in/out movements. The output of the system, which is a matrix of binary numbers, can be thought of as a 4 pixel × 4 pixel image. Hand movements in PIR sensors' FOV are tracked using a blob detection algorithm, which determines the type of movement.

PIR sensors respond to any hot body movements, which mean that the response may be due to a part or the whole of human body. If the aim is to develop a hand gesture-based remote control system, then either of the following two conditions must be met: the only motion in the field of view is a hand movement, or the other parts of the body other than hands are somehow eliminated. The latter can be satisfied by introducing a camera to a PIR sensor-based hand gesture recognition system (Erden and Cetin 2014). The visible light camera distinguishes a hand from other sources of motion, such as a face, by performing a convex hull-defect analysis. Once a hand is detected, PIR sensors are activated to recognize simple gestures such as left-to-right, right-to-left, clockwise, and counter-clockwise hand movements. Three PIR sensors are used, and each is located at a corner of a triangle. For this problem, the distinctive property of the output signals due to different hand gestures is not the oscillation characteristics, but rather the order in which signal peaks appear. Therefore, an ordinal measure should be used for feature extraction. The raw data obtained from the PIR sensors are concatenated first with respect to an order and transformed into the wavelet domain. Then, winner-take-all (WTA) hashing is used as a further feature extractor by converting the corresponding wavelet coefficients to binary codes. Finally, the Jaccard distances between the training and test binary codes are calculated to decide the class of the current test frame. If C_1 and C_2 are any two binary code sequences of the same length, the Jaccard distance between the sequences is defined as

$$d_{C_1, C_2} = \frac{(q+r)}{(p+q+r)} \tag{15.5}$$

where p, q, and r are the number of encounters "11," "10," and "01," respectively, when the code sequences are written one under the other. Note that other distance metrics can also be used to find the likelihood of a model such as Hamming distance, but additional operations may be required. The system described by Erden et al. provides higher accuracy rates when compared to that in Wojtczuk et al. (2011) because it processes the continuous-time analog sensor signals rather than using the raw binary outputs.

15.5 Health Status Monitoring

The increase in the elderly population has initiated intense studies focused on developing AL technologies. AL systems aim to improve the wellness of the elderly while allowing them to live independently. There are currently many PIR sensor-based AL systems that offer unobtrusive, privacy preserving, and practical solutions capable of assisting elderly health and emergency situations. These systems, in general, are designed to detect a predefined abnormal situation (such as falls, a significant deviation from the daily behavioral routines) by analyzing the information extracted from the PIR sensors then firing an alarm in case of occurrence of such events in order to inform a relative or a caregiver. Erden et al. (2016) PIR sensor applications for health status monitoring will be investigated in three sections: activity monitoring, fall detection, and vital sign monitoring.

15.5.1 Activity Monitoring

Elderly in-home activities can be monitored to a certain extent by a PIR sensor network that implements different localization and tracking algorithms. Ha et al. (2006) present a setup shown in Figure 15.5 for the precise localization of an individual in a room. Three PIR sensors are placed on the ceiling, and the location of the resident is determined by

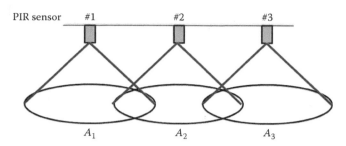

FIGURE 15.5
Sensor arrangement for localization of an individual in a room.

analyzing the sensor array states. If only sensor #1 is active, then the resident is somewhere in region A_1/A_2. If both the sensors #1 and #2 are on, then the resident is in the region $A_1 \cap A_2$, and so on for the other sensing areas. It is obvious that the precision can be increased by using a larger number of sensors. Interested readers can find the application of the same idea with different methods in Song et al. (2008), Luo et al. (2009), and Nam et al. (2015).

Zappi et al. (2010) introduce a wireless PIR sensor network to determine the direction of movement and the distance to the sensors. When a body passes in front of a dual-element PIR sensor, it produces an output signal consisting of two peaks: one positive and one negative. The direction of the movement can easily be determined by looking at the sign of the first peak. The distance of the movement to two sensors is classified as *close to 1*, *close to 2*, and *middle*. Peak duration and the amplitude of the signals are used as features for distance measurement. The features are classified using a Bayes classifier, SVMs, and *k*-nearest neighbor algorithms.

Individual behavior patterns or information regarding activity levels extracted from tracking and localization algorithms can be further examined to detect possible health problems. Banerjee et al. (2003) have developed a surveillance system to report the daily activities of an individual in a hospital room with a bathroom inside of it. Eight PIR sensors, each of which surveys an area of 2 m × 1 m, are placed in several locations. States and IDs of the sensors are recorded twice a second, and the time interval between the state changes is computed for each sensor. With the analysis of sensor data, a closed set of movements, such as movement in the room, movement in the bathroom, lying down in the bed, getting up, and leaving the room, are categorized into 24 groups.

Barger et al. (2005) suggests another system to monitor the whole home. A PIR sensor is placed in each room, and the sensor location, activation time, the length of time spent in the room, and the activity level associated with each observation are stored. Activity level is defined as the number of sensor firings in a room per time spent in the room. Mixture models are applied to the training data (collected separately for work and off days) to develop a probabilistic model of all event types. These models are then used to identify the type of test events including sleeping, changing clothes, computer or television use, meal activities, and so on. Other unsupervised learning methods such as *k*-means clustering and self-organizing maps can also be preferred instead of mixture models.

In Yin and Bruckner (2010), a PIR sensor installed in a living room corner is employed to detect the abnormalities in daily activities of an elderly person. The PIR sensor sends the value "1" to the controller if there are activities from the person and the value "0" otherwise. HMMs, a forward algorithm and a Viterbi algorithm, are used to analyze the obtained data sequence. In case of a detection of a certain deviation from the constituted models, an alarm is sent to the caregiver. In Suryadevara and Mukhopadhyay (2014) a wireless sensor network including PIR, chair, bed, toilet, and couch sensors is suggested to determine the wellness of the elderly. Time-stamped sensor activities are recorded and fed to predefined wellness functions. A similar approach can be found in Virone et al. (2008).

Gait analysis can provide valuable information regarding the health status of an individual, for example, it may be a precursor of a disability or cognitive decline. Hagler et al. (2010) aims to measure the gait velocity with the help of PIR sensors that are sequentially installed on the ceiling. The time indices at which the sensors are fired, the locations and the sensor IDs are used as features. The gait velocity, in this manner, is shown to be estimated with a tolerable error.

Hayes et al. (2008) make use of a similar setup and a method to assess the walking speed. In addition, they track the total time of walking and the activity patterns with the help of PIR sensors installed in each room. A binary sequence including the sensors' states is formed first, and then a wavelet analysis is carried out to distinguish between different patterns. The proposed method is tested on two groups of people: healthy and those with mild cognitive impairment (MCI). It is shown that the average walking speed and the 24-hour wavelet variance differ noticeably for the healthy and MCI groups. That is, the proposed system may be utilized for the early diagnosis of the MCI.

15.5.2 Fall Detection

Unexpected falls are one of the major common health problems facing elderly people today. Most of the commercial fall detectors use wearable sensors. However, PIR sensors can also be utilized for this task.

Toreyin et al. (2008) combine sound, PIR, and vibration sensors to detect falls. Each sensor data is processed separately, and the results are fused to reach a common decision at the end. The system is tested for a variety of events such as speech only, normal walking only, and both speech and normal walking for cross validation. Falls are, indeed, detected by the sound and vibration sensors. While analyzing the sound signals to determine the class of the test signal, zero-crossing numbers and HMMs extracted from the wavelet coefficients of the signal are used. Processing the vibration sensor signals is somehow easier because the responses of the vibration sensor to walking and falling events exhibit sharp differences in amplitude and duration. Accuracy rates can be enhanced by interpreting the events occurring after the detection of a fall. Following the fall alarm issued by the combination of sound and vibration sensors, there should be no activity in the surveillance area, or the only activity should be due to a pet. That is, if a human activity is detected following the detection of a fall, then it is probably a false alarm. Human and pet motions are discriminated by a PIR sensor. A set of HMMs is built for the two classes, and a similar decision mechanism as in Erden et al. (2012) is employed. The final decision is made by the logical *and* operator where the inputs are the individual decisions of each sensor. It is experimentally shown that the vibration-only and sound-only systems produce no errors in detecting falls when a PIR sensor is introduced to the system.

Yazar et al. (2013) present a multisensory system to detect falls. The system includes PIR and vibration sensors. Falls are detected by a vibration sensor using single-tree complex wavelet transform, and false alarms are prevented by a PIR sensor by means of eliminating nonhuman vibration signal sources such as falling objects and slamming doors. A HMM-based analysis is carried out again to analyze the PIR sensor signal. In a more recent study, Yazar et al. (2014) describe another way of detecting falls with the same type of sensors but with a different setup and a method. In this study, two PIR sensors are placed on a wall such that one of them views the lower part of the human body, and the other views the upper part. If the lower PIR sensor detects a human motion whereas the upper PIR sensor does not, then a fall is reported. The decisions of all the three sensors are fused by the WTA algorithm.

15.5.3 Vital Sign Monitoring

Vital sign monitoring is a trending application area for PIR sensors. Respiratory rate, which is one of the most important vital signs, is a good predictor of the physical and physiological situation of the individuals and can be detected by PIR sensors using appropriate signal processing techniques.

In Hers et al. (2013), a system called breath motion detection system (BMDS) is described. BMDS is not intended to measure the respiratory rate but it is to detect respiratory abnormalities. The system is validated by comparing the results to those of thoracic movements tracked by piezoelectric belts and of pressure obtained by nasal cannula. The prototype includes an array of six PIR sensors that are located at a distance to a person lying in bed, around the chest of an individual. Sensor signals are transferred to a laptop computer as a means of an USB acquisition device and analyzed each second. Each frame of data signal is characterized due to the root-mean-square (RMS) value of the signal obtained after some filtering operations. If the RMS value of the signal is greater than 625 mV, then the window is labeled as *move*, which means that the sensor has detected a significant movement. If it is greater than 156 mV, then the window is labeled as *good* to indicate that the respiratory movements are clearly identifiable. If the RMS is below this range, then further processing is required to make a distinction between a signal and a noise. That is, a spectral analysis is carried out based on the Fast Fourier Transform (FFT) of the signal, and *good* or *detect* decision is made. The label *detect* refers to a respiratory abnormality, that is, sleep apnea.

Erden et al. (2015b) suggest another method for detection of the respiratory movements using PIR sensors. This method also allows measuring the respiratory rate with a reasonable error. Three PIR sensors are placed onto a stand near the bed such that the thoracic movements are visible in different lying positions, that is, supine, prone, and side lying. An additional vibration sensor is employed to collect the vibrations in bed and help achieving higher accuracy rates.

The idea of the study is that, because breathing is almost a periodic activity, once the period of the acquired signals is detected, then the breathing rate can be determined as well. To do so, the sensor signals are divided into windows of 1 minute, because the breathing rate is desired to be measured on a per minute basis. A four-stage wavelet decomposition follows in order to focus on the frequency content of interest. Breathing is a low-frequency activity with the highest frequency of 1 Hz, so the subsignal obtained after wavelet decomposition should carry 0–1 Hz frequency subband of the original signal. To allow fast numerical implementations, Lagrange wavelets are used. Each decomposition stage consists of two digital filters and two down samplers as depicted in Figure 15.6. The first filter $g[.]$ is the high-pass filter, and the second one $h[.]$ is the low-pass filter with coefficients $\{1/4, -1/2, 1/4\}$ and $\{1/4, 1/2, 1/4\}$, respectively. If the original signal $x[n]$ is sampled at 20 Hz, then the detailed feature D_1 carries 10–20 Hz, whereas the approximate feature A_1 carries 0–10 Hz frequency subband of $x[n]$. Obviously 0–1 Hz subband signal can be obtained by constructing a wavelet tree of four stages. Note that other filter banks may also be used for wavelet analysis, but Lagrange wavelets are much more efficient.

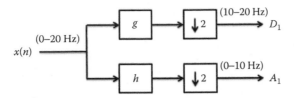

FIGURE 15.6
A single-stage wavelet decomposition.

An alternative way to decompose a signal into different frequency bandwidths is to use empirical mode decomposition (EMD). The main difference between EMD and wavelet decomposition is that EMD decomposes a signal into intrinsic oscillations in an adaptive manner and is fully data driven, whereas wavelet decomposition uses predefined filters. Moreover, EMD requires more complicated data processing operations. The wavelet signal or the intrinsic mode function (IMF), the output of EMD, is applied to an average magnitude difference function (AMDF) for the detection of period, if exists. AMDF is widely used in speech signal processing and defined as follows:

$$\text{AMDF}[k] = \frac{1}{N} \cdot \sum_n |u[n] - u[n-k]| \qquad (15.6)$$

where:
$u[n]$ is the input function, the wavelet signal in this case
N is the number of samples

The corresponding time index of the first local minimum of AMDF gives the period of $u[n]$. An AMDF signal due to an individual taking breath approximately in every 1.94 s is shown in Figure 15.7.

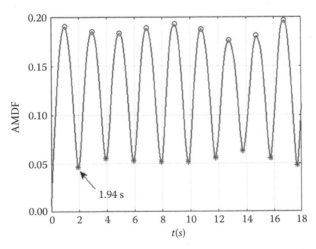

FIGURE 15.7
AMDF signal corresponding to a human taking breath approximately in 1.94 s.

Note that the Equation 15.6 is valid when there is a single input. As the case here is a plurality of sensors, AMDF definition is updated as follows:

$$\text{AMDF}[k] = \frac{1}{N} \cdot \sum_{i=1}^{L} \sum_{n} \propto_i \left| u_i[n] - u_i[n-k] \right| \tag{15.7}$$

where:
$\propto_i$'s are weights
L is the number of inputs to the system

The modified AMDF exhibits a local minimum when at least one of the sensor signals is periodic. Further information can be extracted by investigating the AMDF characteristics. For example, if AMDF does not exhibit a local minimum, then a respiratory disorder can be concluded.

15.6 Conclusion

In this chapter, we introduced IR sensors, in particular the PIR sensors, as an alternative or a complementary sensor for radars. PIR sensors are cheap, small, simple, and very low-power sensors. The passive nature of PIR sensors allows continuous operations for 10 years from a lithium battery. Commercially available PIR detectors detect the changes in the amount of IR radiation in their FOV and convert these changes into a voltage. This voltage is compared to some other predetermined values to trigger an alarm. In this way, PIR sensors have been widely utilized in intruder detection and smart control of the lighting systems. Applications of the PIR sensors are not limited to binary on/off mode operations. It is also possible to achieve more advanced tasks by processing the continuous-time analog signals, rather than the digital "1"s and "0"s, captured from the sensor circuitry. Features extracted from the continuous-time signals are trained first, and then fed to classifiers so that different types of human motion can be recognized. This brings PIR sensor technology to the forefront of many current and potential applications in indoor monitoring for home safety, home automation, and health status monitoring.

PIR sensors and radars have similar operating principles and features such as contactless sensing and privacy protection. However, they have their own advantages and disadvantages, which can make one an alternative to the other. The choice on the type of technology to be utilized depends on the flexibility/functionality and the cost. PIR sensors provide inexpensive, simple, and practical solutions for indoor monitoring. They are capable of recognizing different IR sources and the motion patterns from a single source (with the proper signal processing techniques), but up to a certain extent. It is not easy to distinguish similar types of motions and/or sources with PIR sensors, which in turn may cause intolerable false alarm rates. The sensitivity of PIR sensors to outside environmental changes may also increase the false alarms; however, this is out of the scope of this discussion. Moreover, it is important to note that the smart applications of the PIR sensors are currently in the prototype phase and not commercialized yet. On the other hand, radar technologies, which are decreasing in cost, can be utilized to detect the motion as well as to determine the speed and even the direction of an object regardless of its heat. They require more advanced signal processing techniques but in turn provide much richer information about the surveillance area. Radar systems can be said to have no range limitations in indoor monitoring, whereas the PIR sensors are capable of viewing subjects only within a distance up to 10 m.

There are situations where it is very useful to combine PIR sensors with radar sensors, especially when it comes to waking a radar from a low-power sleep state. That is, a PIR sensor can be exploited to wake a radar only when additional information is needed from the target. By this means, a PIR sensor, along with the signal processing techniques mentioned in this chapter, can offer a compelling power savings in comparison with leaving the radar on all the time.

References

Albers, B. W., and A. K. Agrawal. 1999. Schlieren analysis of an oscillating gas-jet diffusion flame. *Combustion and Flame* 119: 84–94.

Bai, Y. W., and Y. T. Ku. 2008. Automatic room light intensity detection and control using a microprocessor and light sensors. *IEEE Transactions on Consumer Electronics* 54 (3): 1173–1176.

Banerjee, S., F. Steenkeste, P. Couturier, M. Debray, and A. Franco. 2003. Telesurveillance of elderly patients by use of passive infra-red sensors in a "smart" room. *Journal of Telemedicine and Telecare* 9: 23–29.

Barger, T. S., D. E. Brown, and M. Alwan. 2005. Health-status monitoring through analysis of behavioral patterns. *IEEE Transactions on Systems Man and Cybernetics Part A Systems and Humans* 35: 22–27.

Chernbumroong, S., S. Cang, A. Atkins, and H. Yu. 2013. Elderly activities recognition and classification for applications in assisted living. *Expert Systems with Applications* 40 (5): 1662–1674.

Erden, F., A. Z. Alkar, and A. E. Cetin. 2015a. A robust system for counting people using an infrared sensor and a camera. *Infrared Physics and Technology* 72: 127–134.

Erden, F., A. Z. Alkar, and A. E. Cetin. 2015b. Contact-free measurement of respiratory rate using infrared and vibration sensors. *Infrared Physics and Technology* 73: 88–94.

Erden, F., and A. E. Cetin. 2014. Hand gesture based remote control system using infrared sensors and a camera. *IEEE Transactions on Consumer Electronics* 60 (4): 675–680.

Erden, F., E. B. Soyer, B. U. Toreyin, and A. E. Cetin. 2010. VOC gas leak detection using pyro-electric infrared sensors. In *IEEE International Conference on Acoustics, Speech and Signal Processing*, pp. 1682–1685.

Erden, F., B. U. Toreyin, E. B. Soyer, I. Inac, O. Gunay, K. Kose, A. E. Cetin. 2012. Wavelet based flickering flame detector using differential PIR sensors. *Fire Safety Journal* 53: 13–18.

Erden, F., S. Velipasalar, A. Z. Alkar, and A. E. Cetin. 2016. Sensors in assisted living: A survey of signal and image processing methods. *IEEE Signal Processing Magazine* 33 (1): 36–44.

Fang, J.-S., Q. Hao, D. J. Brady, B. D Guenther, and K. Y. Hsu. 2006. Real-time human identification using a pyroelectric infrared detector array and hidden Markov models. *Optics Express* 14 (15): 6643–6658.

Ha, K. N., K. C. Lee, and S. Lee. 2006. Development of PIR sensor based indoor location detection system for smart home. In *SICE-ICASE International Joint Conference*, pp. 2162–2167.

Hagler, S., D. Austin, T. L. Hayes, J. Kaye, and M. Pavel. 2010. Unobtrusive and ubiquitous in-home monitoring: A methodology for continuous assessment of gait velocity in elders. *IEEE Transactions on Biomedical Engineering* 57: 813–820.

Hayes, T. L., F. Abendroth, A. Adami, M. Pavel, T. A. Zitzelberger, and J. A. Kaye. 2008. Unobtrusive assessment of activity patterns associated with mild cognitive impairment. *Alzheimer's & Dementia: The Journal of the Alzheimer's Association* 4: 395–405.

Hers, V., D. Corbugy, I. Joslet, P. Hermant, J. Demarteau, B. Delhougne, G. Vandermoten, and J. P. Hermanne. 2013. New concept using passive infrared (PIR) technology for a contactless detection of breathing movement: A pilot study involving a cohort of 169 adult patients. *Journal of Clinical Monitoring and Computing* 27: 521–529.

Luo, X., B. Shen, X. Guo, G. Luo, and G. Wang. 2009. Human tracking using ceiling pyroelectric infrared sensors. In *2009 IEEE International Conference on Control and Automation*, pp. 1716–1721.

Moghavvemi, M., and L. C. Seng. 2004. Pyroelectric infrared sensor for intruder detection. *IEEE Region 10 Conference TENCON* 4: 656–659.

Mrazovac, B., M. Z. Bjelica, I. Papp, and N. Teslic. 2011. Smart audio/video playback control based on presence detection and user localization in home environment. In *Second Eastern European Regional Conference on the Engineering of Computer Based Systems*, pp. 44–53.

Nam, M.-Y., M. Z. Al-Sabbagh, and C.-G. Lee. 2006. Real-time indoor human/object tracking for inexpensive technology-based assisted living. In *IEEE Real-Time Systems Symposium*, pp. 5–8.

Song, B., H. Choi, and H. S. Lee. 2008. Surveillance tracking system using passive infrared motion sensors in wireless sensor network. In *IEEE International Conference on Information Networking*, pp. 1–5.

Suryadevara, N. K., and S. C. Mukhopadhyay. 2014. Determining wellness through an ambient assisted living environment. *IEEE Intelligent Systems* 29 (3): 30–37.

Toreyin, B. U., E. B. Soyer, I. Onaran, and A. E. Cetin. 2008. Falling person detection using multisensor signal processing. *Eurasip Journal on Advances in Signal Processing* 2008: 1–7.

Tsai, C. H., Y. W. Bai, C. A. Chu, C. Y. Chung, and M. B. Lin. 2011. PIR-sensor-based lighting device with ultra-low standby power consumption. *IEEE Transactions on Consumer Electronics* 57: 1157–1164.

Virone, G., M. Alwan, S. Dalal, S. W. Kell, B. Turner, J. A. Stankovic, and R. Felder. 2008. Behavioral patterns of older-adults in assisted living. *IEEE Transactions on Information Technology in Biomedicine* 12 (3): 387–398.

Wojtczuk, P., A. Armitage, T. D. Binnie, and T. Chamberlain. 2011. PIR sensor array for hand motion recognition. In *SENSORDEVICES 2011: The Second International Conference on Sensor Device Technologies and Applications*, IARIA, Red Hook, NY, pp. 99–102.

Yazar, A., F. Erden, and A. E. Cetin. 2014. Multi-sensor ambient assisted living system for fall detection. In *IEEE International Conference on Acoustics, Speech, and Signal Processing*, pp. 1–3.

Yazar, A., F. Keskin, B. U. Töreyin, and A. E. Cetin. 2013. Fall detection using single-tree complex wavelet transform. *Pattern Recognition Letters* 34 (15): 1945–1952.

Yin, G., and D. Bruckner. 2010. Daily activity learning from motion detector data for ambient assisted living. In *3rd IEEE International Conference on Human System Interaction*, pp. 89–94.

Yun, J., and S.-S. Lee. 2014. Human movement detection and identification using pyroelectric infrared sensors. *Sensors* 14 (5): 8057–8081.

Zappi, P., E. Farella, and L. Benini. 2010. Tracking motion direction and distance with pyroelectric IR sensors. *IEEE Sensors Journal* 10 (9): 1486–1494.

Index

Note: Page numbers followed by f and t refer to figures and tables, respectively.